计算机技术开发与应用丛书

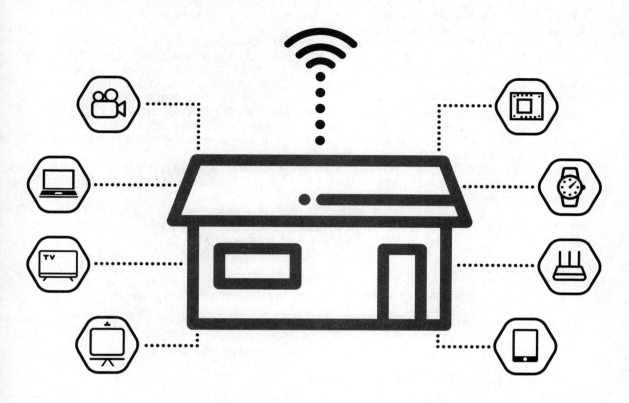

# HarmonyOS应用开发实战
## （JavaScript版）

徐礼文 ◎ 著

清华大学出版社
北京

## 内 容 简 介

本书详细讲解 HarmonyOS ArkUI（方舟开发框架）的两大 UI 框架：ArkUI JS（类 Web 范式框架）和 ArkUI ETS（声明式开发范式框架）。通过大量案例带领开发者深入掌握 HarmonyOS 应用开发和基于 OpenHarmony 3.0 LTS 的智能家居方向应用开发。

本书分六篇共 20 章。第一篇为开发准备篇，共 2 章，介绍 HarmonyOS 的系统特性、架构和应用开发环境搭建；第二篇为 ArkUI JS UI 篇，共 4 章，深入浅出地介绍 ArkUI JS 框架、内置组件、服务接口等，其中第 5 章通过一个分布式游戏案例深入讲解鸿蒙分布式应用开发的技巧，第 6 章深入讲解原子化服务和服务卡片的开发；第三篇为 JavaScript API 篇，共 5 章，深入讲解 ArkUI JavaScript API；第四篇为 ArkUI ETS UI 篇，共 3 章，深入讲解 ArkUI ETS，其中第 12 章系统讲解 ArkUI 声明式框架的开发语言 TypeScript，第 14 章深入介绍 ArkUI ETS 实战：华为商城 App 开发；第五篇为 OpenHarmony 篇，共 3 章，介绍 OpenHarmony 3.0 LTS 的源码下载、编译、烧录、北向和南向应用开发；第六篇为提高篇，共 3 章，介绍轻设备端 JavaScript 框架和富设备端 JavaScript 框架的原理，其中第 20 章详细介绍如何开发一个类 Web 范式的组件，并提交 Gitee OpenHarmony 仓库。

学习本书内容，需要具备一定的 HTML、CSS、JavaScript 基础知识，希望本书能够对读者学习使用鸿蒙开发者框架构建美观、快速、跨终端的移动应用程序有所帮助。本书适合 HarmonyOS 应用开发爱好者，以及嵌入式爱好者阅读。

本书封面贴有清华大学出版社防伪标签，无标签者不得销售。
版权所有，侵权必究。举报：010-62782989，beiqinquan@tup.tsinghua.edu.cn。

**图书在版编目（CIP）数据**

HarmonyOS 应用开发实战：JavaScript 版/徐礼文著. —北京：清华大学出版社，2022.2
（计算机技术开发与应用丛书）
ISBN 978-7-302-60031-2

Ⅰ. ①H… Ⅱ. ①徐… Ⅲ. ①移动终端－操作系统－程序设计 Ⅳ. ①TN929.53

中国版本图书馆 CIP 数据核字（2022）第 021630 号

责任编辑：赵佳霓
封面设计：吴　刚
责任校对：时翠兰
责任印制：丛怀宇

出版发行：清华大学出版社
网　　址：http://www.tup.com.cn，http://www.wqbook.com
地　　址：北京清华大学学研大厦 A 座　　邮　编：100084
社 总 机：010-83470000　　邮　购：010-83470235
投稿与读者服务：010-62776969，c-service@tup.tsinghua.edu.cn
质量反馈：010-62772015，zhiliang@tup.tsinghua.edu.cn
课件下载：http://www.tup.com.cn，010-83470236

印　装　者：天津安泰印刷有限公司
经　　销：全国新华书店
开　　本：186mm×240mm　　印　张：35　　字　数：785 千字
版　　次：2022 年 3 月第 1 版　　印　次：2022 年 3 月第 1 次印刷
印　　数：1～2000
定　　价：129.00 元

产品编号：094257-01

# 序
## FOREWORD

在以手机为中心的移动互联网时代，增长红利已见顶。移动产业正在从手机竞争向全场景多设备竞争转移，IoT设备的增长成为移动互联的新引擎。在目前快速增长的IoT市场中，智能设备普遍存在连接步骤复杂、生态无法共享、数据难以互通、能力难以协同等问题。此外，操作系统的碎片化也制约了多设备场景化体验。

HarmonyOS作为万物互联时代的操作系统，用革命性的分布式技术带来"超级终端"。设备间的硬件互助、资源共享为消费者带来全新的全场景体验。一次开发，多端部署，让应用开发者可以不用担心设备的形态差异，可将主要精力聚焦于业务逻辑，更好地开发应用。统一OS，弹性部署，让设备开发者可根据设备的资源灵活地裁剪OS，满足不同设备对OS的要求。

在2020年9月10日的华为开发者大会上，HarmonyOS开源项目OpenHarmony正式对公众开源，并托管在开放原子开源基金会。自2020年9月开源以来，OpenHarmony开源社区已陆续发布了1.0初始版本、2.0 Canary版本、3.0 LTS版本等，支持轻量带屏设备的开发。OpenHarmony社区已成为Gitee平台热度最高的开源项目之一。2022年OpenHarmony社区将继续完善应用界面和图形渲染能力，完善分布式能力基座，更丰富的多媒体、联网、通话能力，更多的第三方库及HDF驱动，让开发者能基于富设备开发体验更好的分布式应用。

徐礼文老师有非常丰富的移动应用全栈开发经验及教学经验，是国内研发H5与Android、iOS混合开发工具发起人之一，也是最早参与HarmonyOS课程开发及书籍写作的老师之一。徐老师对HarmonyOS的技术理解非常深入，在51CTO已发布9篇鸿蒙应用开发视频课程，受到社区开发者的广泛好评，衷心感谢徐老师对HarmonyOS的支持和奉献。

本书详细介绍了最新的ArkUI（方舟开发框架）的新特性及案例，对开发者开发轻应用、富应用，尤其是智能家居应用会有很好的启发。此外，本书也基于最新的OpenHarmony 3.0 LTS版本，介绍了如何基于3.0 LTS进行端到端的源码下载、编译、烧录和应用开发，以及如何向OpenHarmony的Gitee代码仓提交PR，值得关注OpenHarmony/HarmonyOS的开发者深入学习。

操作系统、编译器、编程语言、数据库等"根"技术是需要长期投入的系统工程,软件生态需要操作系统等基础软件及应用软件均衡发展才能枝繁叶茂,其中最重要的操作系统生态需要合作伙伴及广大开发者一起参与应用开发、设备开发、开源开发才能繁荣昌盛。希望有越来越多的开发者体验、使用、贡献 OpenHarmony,有更多的合作伙伴能把 OpenHarmony 应用到千行百业,让 OpenHarmony 未来能成为"数字中国"的"数字底座"。

刘 果

开放原子开源基金会 OpenHarmony 项目导师

# 前言
PREFACE

随着 IoT 产业的快速发展，消费者手中拥有越来越多的智能设备，但当前智能设备的操作系统仍然是以面向单设备的体验为主，没有发挥出多设备协同的优势，少量设备互联的体验难以在整个生态内系统性地规模复制。

HarmonyOS 作为万物互联时代的操作系统，用革命性的分布式技术带来"超级终端"，实现多终端协同，为用户提供"超级应用"和"超级服务"。对于设备厂家，基于 HarmonyOS，可以更便捷地加入 IoT 生态，实现设备智能化的产业升级。对于应用厂家，基于全新的操作系统 HarmonyOS 可将应用触达更多类终端（包括手机），发掘更多流量入口，吸引更多用户，实现从智能手机市场到全新万物互联市场的变革。

在 2020 年 9 月 10 日的华为开发者大会上，HarmonyOS 开源项目 OpenHarmony 正式对公众开源，并托管在开放原子开源基金会，首个版本支持 128KB～128MB 设备。自 9 月开源以来，得力于广大开发者的关注及贡献，OpenHarmony 开源社区已陆续发布了 1.0 初始版本、3.0 LTS 版本，完善了 SIG 申请及项目孵化机制、第三方库组件组织代码库及 Codelabs 样例代码库，并已成为 Gitee 平台热度最高的开源项目之一。

2021 年，升级到 HarmonyOS 的华为自有品牌手机超过 1 亿部，OpenHarmony 生态也逐步建立起来，华为和美的打造的智能家居操作系统已发布。HarmonyOS 正在逐步改变国内产业生态和国际产业竞争格局。

希望本书能够帮助 HarmonyOS 应用开发爱好者打开国产操作系统应用开发之门，正如华为消费者业务 CEO 余承东先生所说："没有人能够熄灭满天星光，每位开发者都是华为公司要汇聚的星星之火，星星之火可以燎原。"HarmonyOS 正聚集每位开发者的努力和贡献，HarmonyOS 作为底层基石为企业数字化赋能，其芯片、人工智能等技术通过云端输出，为全球数字化创新带来了崭新的活力。基于 HarmonyOS 打造的 HMS 移动生态将是继安卓、iOS 后全球第三大移动应用生态。

## 本书特色

本书详细讲解 HarmonyOS ArkUI 的两大 UI 开发框架：ArkUI JS 和 ArkUI ETS。通过大量案例带领开发者深入掌握 HarmonyOS 轻设备（ArkUI JS）、富设备（ArkUI ETS）和智能家居（OpenHarmony 3.0 LTS）方向应用开发，通过游戏案例教学的方式，让读者快速学习和掌握基于 HarmonyOS 的应用开发。同时本书提供了大量的代码示例，读者可以通过这些例子理解知识点，也可以直接在开发实战中修改并应用这些代码。另外，笔者专门为本书中的游戏实战章节录制了高清配套教学视频，并提供了本书涉及的源代码，以便于读者高效、直观地学习。

## 本书内容

本书分六篇共 20 章，主要内容如下：

第一篇，共 2 章，介绍 HarmonyOS 的系统特性、架构和应用开发环境搭建；

第二篇，共 4 章，深入浅出地介绍 ArkUI JS 框架内置组件、服务接口等，其中第 5 章通过一个分布式游戏案例深入讲解鸿蒙分布式应用开发的技巧，第 6 章深入讲解原子化服务和服务卡片的开发；

第三篇，共 5 章，深入讲解 ArkUI JavaScript API；

第四篇，共 3 章，深入讲解 ArkUI 声明式 UI 框架（ArkUI ETS），其中第 12 章系统讲解 ArkUI 声明式框架的开发语言 TypeScript，第 14 章深入介绍 ArkUI ETS 实战：华为商城 App 开发；

第五篇，共 3 章，介绍 OpenHarmony 3.0 LTS 的源码下载、编译、烧录、北向和南向混合应用开发；

第六篇，共 3 章，介绍轻设备端 JavaScript 框架和富设备端 JavaScript 框架的原理，其中第 20 章详细介绍如何开发一个类 Web 范式的组件，并提交 Gitee OpenHarmony 仓库。

## 本书读者对象

学习本书内容需要具备一定的 HTML、CSS、JavaScript 基础知识，希望本书能够对读者学习使用鸿蒙开发者框架构建美观、快速、跨终端的移动应用程序有所帮助。本书适合 HarmonyOS 应用开发爱好者，以及嵌入式爱好者阅读。

## 致谢

感谢华为杭州研究所 JS 开发框架负责人吴勇辉、HarmonyOS JS 框架架构师马家骏在写作本书过程中提出的宝贵改进意见，以及在写作本书过程中华为上海研究所于小飞提供的支持与帮助。

目前 HarmonyOS ArkUI 框架版本迭代较快，本书所涉及的内容可能在未来鸿蒙的框架版本中有所变化，望大家谅解。

徐礼文<br>2022 年 1 月

本书源代码

# 目录
CONTENTS

## 第一篇 开发准备篇

### 第 1 章 HarmonyOS 系统简介 ... 3
- 1.1 HarmonyOS 的设计目标 ... 3
  - 1.1.1 5G 万物互联时代 ... 4
  - 1.1.2 物联网操作系统碎片化 ... 4
  - 1.1.3 下一代操作系统的发展方向 ... 5
- 1.2 HarmonyOS 技术特性 ... 6
  - 1.2.1 分布式架构 ... 6
  - 1.2.2 操作系统可裁剪 ... 7
  - 1.2.3 一套代码多端运行 ... 7
- 1.3 HarmonyOS 技术架构 ... 8
  - 1.3.1 内核层 ... 8
  - 1.3.2 系统服务层 ... 8
  - 1.3.3 架构层 ... 10
  - 1.3.4 应用层 ... 10
- 1.4 HarmonyOS 与 LiteOS ... 10
  - 1.4.1 LiteOS-A 简介 ... 12
  - 1.4.2 LiteOS-M 简介 ... 12
- 1.5 OpenHarmony 生态 ... 13
  - 1.5.1 Android 与 AOSP ... 13
  - 1.5.2 HarmonyOS 与 OpenHarmony ... 13
- 1.6 HarmonyOS 与 Fuchsia OS ... 16
  - 1.6.1 Fuchsia OS 系统架构 ... 17
  - 1.6.2 Fuchsia OS 与产业 ... 20
- 1.7 本章小结 ... 20

### 第 2 章 开发环境搭建 ... 21
- 2.1 鸿蒙应用开发环境搭建 ... 21
  - 2.1.1 下载和安装 Node.js ... 21
  - 2.1.2 下载和安装 DevEco Studio ... 25

2.1.3 运行 Hello World ······ 27
2.2 鸿蒙应用程序运行调试 ······ 32
　　2.2.1 在远程模拟器中运行应用 ······ 32
　　2.2.2 在 Simulator 中运行应用 ······ 34
2.3 使用真机设备运行应用 ······ 35
　　2.3.1 手动真机签名流程 ······ 36
　　2.3.2 自动化真机签名流程 ······ 42
2.4 本章小结 ······ 46

## 第二篇　ArkUI JS UI 篇

## 第 3 章　ArkUI JS 框架详细讲解 ······ 49
3.1 ArkUI JS 框架介绍 ······ 49
　　3.1.1 ArkUI JS 框架的特征 ······ 50
　　3.1.2 ArkUI JS 架构介绍 ······ 50
　　3.1.3 ArkUI JS 运行流程 ······ 51
3.2 创建一个 ArkUI JS 项目 ······ 51
　　3.2.1 新建 ArkUI JavaScript 项目 ······ 52
　　3.2.2 编写界面布局 ······ 54
　　3.2.3 编写界面逻辑代码 ······ 54
　　3.2.4 通过模拟器预览效果 ······ 55
3.3 项目目录结构 ······ 59
　　3.3.1 项目整体结构 ······ 59
　　3.3.2 项目的配置文件 ······ 61
　　3.3.3 资源文件的使用方式 ······ 66
3.4 页面布局 ······ 69
　　3.4.1 Flexbox 布局 ······ 69
　　3.4.2 Grid 网格布局 ······ 85
3.5 语法详细讲解 ······ 101
　　3.5.1 HML 语法 ······ 101
　　3.5.2 CSS 语法 ······ 107
　　3.5.3 JS 逻辑 ······ 112
　　3.5.4 多语言支持 ······ 114
3.6 内置组件 ······ 116
　　3.6.1 容器组件 ······ 116
　　3.6.2 基础组件 ······ 143
　　3.6.3 媒体组件 ······ 166
　　3.6.4 画布组件 ······ 172
3.7 自定义组件 ······ 174
　　3.7.1 自定义组件定义 ······ 175

  3.7.2　自定义组件事件与交互 176
 3.8　本章小结 179

# 第 4 章　ArkUI JS 与 Java 混合开发 180
 4.1　JavaScript 调用 Service Ability 180
  4.1.1　JS 端调用远端 Service Ability 180
  4.1.2　JS 端订阅远端 Service Ability 184
 4.2　JS 端调用音乐播放 Service Ability 188
  4.2.1　申请分布式使用权限 190
  4.2.2　创建 Java 端 Service Ability 191
  4.2.3　音乐播放器前端的 UI 195
  4.2.4　封装 JS 前端调用 Service Ability 的方法 198
  4.2.5　JS 端调用 Service Ability 的方法 200
  4.2.6　音乐播放器遥控 UI 201
  4.2.7　音乐播放器遥控逻辑实现 203
  4.2.8　通过实体音量键控制远程设备音量 211
  4.2.9　JS 端订阅 Service Ability 中的播放状态 212
  4.2.10　本节小结 217
 4.3　JavaScript 项目混合 Java UI 开发 217
  4.3.1　JS Ability 和 Java Ability 跳转 218
  4.3.2　JS 端调用相机拍照功能 222

# 第 5 章　ArkUI JS 游戏开发案例 227
 5.1　飞机大战游戏介绍 227
 5.2　飞机大战游戏分析 229
  5.2.1　游戏性能问题分析 229
  5.2.2　游戏角色分析 229
 5.3　飞机大战核心算法 230
  5.3.1　碰撞检测算法 231
  5.3.2　子弹飞行算法 234
 5.4　飞机大战游戏界面实现 235
  5.4.1　游戏主界面 236
  5.4.2　游戏控制手柄界面 239
 5.5　飞机大战核心代码实现——单机篇 244
  5.5.1　加载游戏资源 244
  5.5.2　太空背景动画 246
  5.5.3　游戏动画入口 247
  5.5.4　绘制游戏主角 247
  5.5.5　绘制游戏敌机 249
  5.5.6　绘制子弹对象 252
  5.5.7　绘制爆炸效果 255

```
        5.5.8   操作主角飞机 ································································ 256
   5.6  飞机大战核心代码实现——鸿蒙篇 ········································ 256
        5.6.1   多设备间游戏流转 ·············································· 256
        5.6.2   实现游戏远程控制 ·············································· 259
   5.7  本章小结 ·············································································· 266

第6章  原子化服务和服务卡片开发 ········································ 267
   6.1  什么是原子化服务 ································································ 267
        6.1.1   原子化服务特征 ·················································· 267
        6.1.2   原子化服务与传统应用的区别 ·························· 268
        6.1.3   原子化服务上架流程 ·········································· 269
        6.1.4   原子化服务开发要求 ·········································· 269
        6.1.5   原子化服务开发流程 ·········································· 270
   6.2  什么是服务卡片(Service Widget) ········································ 273
        6.2.1   服务卡片定义 ······················································ 274
        6.2.2   服务卡片的三大特征 ·········································· 274
        6.2.3   服务卡片的设计规范 ·········································· 275
        6.2.4   服务卡片的整体架构 ·········································· 277
   6.3  服务卡片开发详细讲解 ······················································· 278
        6.3.1   创建JavaScript服务卡片 ····································· 278
        6.3.2   服务卡片界面实现 ·············································· 287
        6.3.3   服务卡片数据绑定 ·············································· 297
        6.3.4   服务卡片数据更新 ·············································· 299
        6.3.5   服务卡片跳转事件和消息事件 ·························· 303

                    第三篇  JavaScript API 篇

第7章  基本功能接口 ····································································· 309
   7.1  页面路由 ·············································································· 309
        7.1.1   页面路由用法 ······················································ 310
        7.1.2   页面路由动画 ······················································ 311
   7.2  应用上下文 ·········································································· 314
   7.3  日志打印 ·············································································· 314
   7.4  应用配置 ·············································································· 315
   7.5  窗口 ······················································································ 315
   7.6  弹框 ······················································································ 317
   7.7  动画 ······················································································ 318
   7.8  剪贴板 ·················································································· 320
第8章  网络与媒体接口 ································································ 322
   8.1  网络访问 ·············································································· 322
   8.2  WebSocket ············································································ 324
```

8.3 上传和下载 ......326
8.4 媒体 ......327

## 第9章 分布式能力接口 ......329

9.1 分布式迁移 ......329
  9.1.1 申请分布式迁移权限 ......329
  9.1.2 通过 FeatureAbility 发起迁移 ......329

9.2 分布式拉起 ......332
  9.2.1 申请分布式迁移权限 ......332
  9.2.2 允许以显式的方式拉起远程或本地的 FA ......332
  9.2.3 拉起远程带返回值的 FA ......333
  9.2.4 分布式 API 在 FA 中的生命周期 ......334

9.3 文件数据管理 ......335
  9.3.1 轻量级存储 ......335
  9.3.2 文件管理 ......336

## 第10章 系统设备接口 ......340

10.1 消息通知 ......340
10.2 地理位置 ......341
10.3 设备信息 ......342
10.4 应用管理 ......343
10.5 媒体查询 ......343
10.6 振动 ......344

## 第11章 多实例管理 ......346

11.1 多实例接口 ......346
  11.1.1 多 Ability 实例管理 ......346
  11.1.2 多 Ability 之间跳转 ......347

11.2 使用 NPM 安装 JavaScript 模块 ......351

## 第四篇 ArkUI ETS UI 篇

## 第12章 ArkUI ETS 开发语言入门 ......355

12.1 ArkUI TypeScript 介绍 ......356
12.2 ets-loader 编译 ETS ......356
12.3 TypeScript 基础数据类型 ......357
  12.3.1 布尔值 ......357
  12.3.2 数字 ......357
  12.3.3 字符串 ......357
  12.3.4 数组 ......358
  12.3.5 元组 ......358
  12.3.6 枚举 ......359
  12.3.7 any ......359

12.3.8 void ......360
12.3.9 null 和 undefined ......360
12.3.10 never ......361
12.4 TypeScript 高级数据类型 ......361
12.4.1 泛型 ......362
12.4.2 交叉类型 ......364
12.4.3 联合类型 ......365
12.5 TypeScript 面向对象特性 ......365
12.5.1 类 ......366
12.5.2 接口 ......369
12.6 TypeScript 装饰器 ......373
12.6.1 属性装饰器 ......373
12.6.2 方法装饰器 ......374
12.6.3 参数装饰器 ......374
12.6.4 类装饰器 ......375
12.7 TypeScript 模块与命名空间 ......376
12.7.1 模块 ......376
12.7.2 命名空间 ......378

# 第 13 章 ArkUI ETS 框架详细讲解 ......380

13.1 框架特点 ......380
13.2 组件化设计 ......381
13.2.1 组件装饰器 @Component ......381
13.2.2 组件的内部私有状态 @State ......383
13.2.3 组件的输入和输出属性 ......385
13.2.4 单向同步父组件状态 @Prop ......387
13.2.5 双向同步状态 @Link ......389
13.2.6 自定义组件的生命周期函数 ......390
13.2.7 跨组件数据传递 @Consume 和 @Provide ......393
13.2.8 监听变量状态变更 @Watch ......394
13.2.9 自定义组件方法 @Builder ......396
13.2.10 统一组件样式 @Extend ......396
13.3 状态管理仓库 ......397
13.3.1 持久化数据管理 ......399
13.3.2 环境变量 Environment ......399
13.3.3 AppStorage 与组件同步 ......400
13.4 渲染控制语法 ......401
13.4.1 条件渲染 if...else... ......401
13.4.2 循环渲染 ForEach ......401
13.5 动画效果 ......402

13.5.1　属性动画 402
　　　13.5.2　显式动画 403
　　　13.5.3　转场动画 403
　　　13.5.4　手势处理 408
　13.6　框架结构详细讲解 411
　　　13.6.1　文件组织 411
　　　13.6.2　JS标签配置 412
　　　13.6.3　app.ets 413
　　　13.6.4　资源访问 414
　　　13.6.5　像素单位 414
　13.7　界面布局 415
　　　13.7.1　Flex布局 415
　　　13.7.2　Grid布局 417
　　　13.7.3　堆叠布局 419
　　　13.7.4　栅格布局 421
　13.8　基础组件 422
　　　13.8.1　Text组件 423
　　　13.8.2　Button组件 424
　　　13.8.3　Image组件 425
　　　13.8.4　List组件 427
　　　13.8.5　Swiper组件 430
　　　13.8.6　Tabs组件 432
　　　13.8.7　Scroll组件 433
　　　13.8.8　AlertDialog组件 435
　　　13.8.9　自定义弹框 436

# 第14章　ArkUI ETS UI开发案例 439
　14.1　华为商城框架封装 439
　　　14.1.1　公共组件封装 440
　　　14.1.2　公共数据接口封装 441
　14.2　商城首页实现 442
　　　14.2.1　头部组件 444
　　　14.2.2　头部滚动 446
　　　14.2.3　轮播广告 446
　　　14.2.4　导航菜单 448
　　　14.2.5　限时购 450
　14.3　商城商品分类页实现 451
　　　14.3.1　中间左侧分类区 452
　　　14.3.2　中间右侧商品区 456
　14.4　商品详情页实现 458

14.4.1 头部商品图片轮播区 ········································································ 460
14.4.2 商品价格展示栏 ············································································· 460
14.4.3 商品底部购买栏 ············································································· 461

## 第五篇　OpenHarmony 篇

### 第 15 章　OpenHarmony 基础 ·································································· 465
15.1　OpenHarmony 介绍 ········································································ 465
15.2　OpenHarmony 3.0 LTS 编译与烧录 ···················································· 466
　　15.2.1　编译环境搭建 ·········································································· 466
　　15.2.2　标准系统编译和烧录 ································································· 469

### 第 16 章　OpenHarmony 应用开发详细讲解 ················································ 474
16.1　配置 OpenHarmony SDK ································································· 474
16.2　创建 OpenHarmony 工程 ································································· 475
　　16.2.1　选择项目模板 ·········································································· 475
　　16.2.2　创建 ArkUI JS 项目 ································································· 476
16.3　配置 OpenHarmony 应用签名信息 ····················································· 477
　　16.3.1　生成密钥和证书请求文件 ··························································· 477
　　16.3.2　生成应用证书文件 ···································································· 480
　　16.3.3　生成应用 Profile 文件 ······························································· 481
　　16.3.4　配置应用签名信息 ···································································· 482
16.4　推送并将 HAP 安装到开发板/设备 ···················································· 483
　　16.4.1　OpenHarmony 命令行启动 hdcd ················································· 484
　　16.4.2　下载 hdc_std 工具 ···································································· 484
　　16.4.3　配置环境变量（Windows） ······················································· 485
　　16.4.4　使用 hdc_std 安装 HAP ···························································· 485
　　16.4.5　Hi3516DV300 的运行 ······························································· 485
　　16.4.6　hdc_std 连接不到设备 ······························································· 485

### 第 17 章　OpenHarmony "HiSpark 智能赛车" ············································· 487
17.1　鸿蒙 HiSpark 智能赛车游戏介绍 ······················································ 487
17.2　HiSpark 智能赛车端实现 ································································ 489
　　17.2.1　HiSpark 赛车配置 WiFi 网络 ····················································· 489
　　17.2.2　HiSpark 赛车电机驱动 ······························································ 490
　　17.2.3　HiSpark 赛车操作控制 ······························································ 493
17.3　将赛车控制模块添加到鸿蒙源码并编译 ············································· 496
　　17.3.1　添加赛车控制模块代码 ······························································ 497
　　17.3.2　编译 OpenHarmony 源码 ···························································· 497
　　17.3.3　烧录 OpenHarmony ··································································· 498
17.4　鸿蒙 HAP 端控制赛车实现 ······························································ 500
　　17.4.1　赛车控制手柄界面实现 ······························································ 501

17.4.2 将赛车控制手柄设置为横屏模式 505
17.4.3 Java 端通过 Service Ability 发送指令 506
17.4.4 赛车控制手柄界面逻辑实现 508
17.5 本章小结 511

## 第六篇 提 高 篇

### 第 18 章 轻鸿蒙端 JavaScript 框架 515
18.1 JerryScript 轻量级引擎 515
18.1.1 编译 JerryScript 515
18.1.2 运行 JerryScript 516
18.2 轻量级 JS 核心开发框架 517
18.2.1 JS Framework 518
18.2.2 组件绑定实现 520
18.2.3 路由实现 520
18.2.4 图形绘制层 521
18.2.5 渲染流程 521

### 第 19 章 富鸿蒙端 JavaScript 框架 522
19.1 QuickJS 引擎 522
19.1.1 安装基础编译环境 522
19.1.2 通过 Git 下载 QuickJS 源码 523
19.1.3 编译 QuickJS 523
19.1.4 编译验证 JS 523
19.2 Google V8 引擎 523
19.3 ArkUI JS Engine 框架 524
19.4 新方舟编译器（ArkCompiler 3.0） 527

### 第 20 章 类 Web 范式组件设计与开发 529
20.1 JavaScript 端组件设计 529
20.1.1 前端组件效果 529
20.1.2 组件的详细设计 530
20.2 JS 的界面解析 530
20.2.1 在 dom_type 中增加新组件的属性定义 531
20.2.2 新增 DOMMyCircle 类 531
20.3 后端的布局和绘制 535
20.3.1 新增 MyCircleComponent 类 535
20.3.2 新增 MyCircleElement 类 537
20.3.3 新增 RenderMyCircle 类 537
20.3.4 新增 FlutterRenderMyCircle 类 539

# 第一篇 开发准备篇

# 第1章 HarmonyOS 系统简介

鸿蒙操作系统(HarmonyOS)是一款面向物联网全场景的分布式操作系统,如图1.1所示。鸿蒙操作系统不同于现有的Android、iOS、Windows、Linux等操作系统,它设计的初衷是解决在5G万物互联时代,各个系统间的连接问题。鸿蒙操作系统面向的是$1+8+N$的全场景设备,能够根据不同内存级别的设备进行弹性组装和适配,可实现跨硬件设备间的信息交互。

图1.1 HarmonyOS 操作系统

"鸿蒙"名字源于华为公司内部一个研究操作系统内核的项目代号。

"鸿蒙操作系统"的英文名字 HarmonyOS,Harmony 之意为和谐,引申为世界大同、和合共生,是中华文明一直秉持的理念。"鸿蒙"有盘古开天辟地之意,"鸿蒙初辟原无性,打破顽空须悟空",鸿蒙生态刚刚起步,需要华为、国内外企业的共同努力,需要众多"悟空"共同推动构建更加绚丽多彩的世界。华为的鸿蒙,中国的鸿蒙,必将成为世界的鸿蒙。

## 1.1 HarmonyOS 的设计目标

尽管 HarmonyOS 是在美国对华为公司实施制裁时从"备胎"提前转正,但是实际上华为公司在2012年就开始规划自有操作系统"鸿蒙",这是华为公司面对以5G技术推动的产业革命和国外技术风险做出的提前布局,HarmonyOS是华为公司专门为5G万物互联时代打造的战略性产品,创造性地通过分布式技术打造一个万物互联互通的物联网操作系统。

HarmonyOS的设计目标是为解决5G智能物联网时代操作系统严重碎片化问题,同时也是应对国外技术封锁和制裁下的自力更生,确保了华为在未来国际竞争中的商业安全和信息安全,同时为国家操作系统自主可控和信息安全提供了有力保障。

### 1.1.1  5G万物互联时代

以"超高网速、低延时高可靠、低功率海量连接"为特征的5G(第五代移动通信系统的简称)万物互联时代的到来,传统的面向单一设备的开源操作系统Android和闭源操作系统iOS都很难满足人们在不同场景下的需求,如图1.2所示,在产业层面亟须一款专门为5G时代定制的操作系统来推动产业的长远发展。

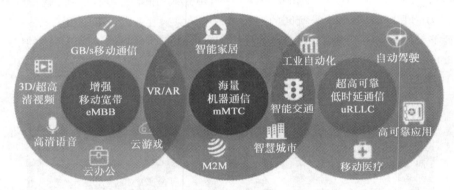

图1.2  5G三大应用场景(eMBB、mMTC、uRLLC)

在5G技术的大背景下,物联网、移动计算、智能家居、智能手机、可穿戴设备、智慧城市、无人驾驶汽车、智慧医疗、VR(虚拟现实技术,英文名称:Virtual Reality,缩写为VR)等被认为是受益最大的领域。目前,公认的5G技术适用的三大应用场景为mMTC(超高带宽引领下的智能物联网产业)、eMBB(超高清流媒体引领下的视频流产业)、uRLLC(需要5G高效低时延特点的产业,如车联网、自动化产业等),如图1.2所示。

针对未来的5G技术发展,华为制定了"1+8+N的5G全场景战略",1代表智能手机,8代表大屏、音箱、眼镜、手表、车机、耳机、平板等,围绕着关键的八类设备,周边还有合作伙伴开发的N类领域,围绕着智能家居、穿戴、办公、影音、娱乐等,华为将致力搭建一套更加完善的5G服务生态体系。

### 1.1.2  物联网操作系统碎片化

随着近几年智能物联网产业的高速发展,物联网领域的深层次问题亟待解决,物联网目前落地的痛点是下游应用场景与需求的高度碎片化,物联网终端异构、网络通信方式与操作系统平台多样化,对设备之间互联互通的实现造成较大挑战,操作系统的碎片化阻碍了万物互联时代的业务创新。

HarmonyOS 的定位就是万物互联时代的操作系统，创造性地通过分布式技术，以及高性能的软总线技术，将多个物理上相互分离的设备融合成一个"超级终端"。按需调用、组合不同设备的软硬件能力，为用户带来最适合其所在场景的智慧体验。即使用户切换场景，智慧体验也能跨终端迁移，无缝流转。

HarmonyOS 通过软总线和分布式技术打通了不同设备之间的壁垒，让内容无缝流转。例如在出行领域，HarmonyOS 可以通过手机、手表、车机的协同，优化出行体验。在等待网约车时，用户不需要频繁掏出手机查看车辆动态，车牌号、车辆位置等信息会在手表上实时同步，抬手可见。

家庭智能化产品中带 IoT（物联网）功能的设备越来越多，如电冰箱、豆浆机、摄像头等，访问不同的 IoT 设备需要安装不同的 App，基本上每个 IoT 设备都有一个 App，导致手机上 App 众多，操作和查找起来都非常不方便，由于这个原因，实际中，App 的安装率不到 10%，而安装的 App 的使用率不到 5%，HarmonyOS 可以大大简化 IoT 设备的访问。

HarmonyOS 碰一碰能力（OneHop Engine）通过 NFC 解决 App 跨设备接续难、设备配网难、传输难的问题，并能够和后台智能系统结合起来，进行相关操作推荐，如结合个人运动健康数据推荐合适的豆浆配方，智能冰箱推荐菜品的保存温度等。

### 1.1.3　下一代操作系统的发展方向

操作系统经历了 60 多年的发展，历经多代，如表 1.1 所示，从最早期的多任务操作系统，如 MULTICS 和 UNIX，到适用于个人计算机的多处理器操作系统，如 Linux 和 Windows，再到最近十多年广泛流行的移动操作系统，如 iOS 和 Android，其核心技术已经非常成熟，软件复杂度也达到了上亿行代码的规模。操作系统的每一次大发展必定跟计算机硬件的发展密切相关。随着物联网时代的到来，操作系统必将迎来新的发展。

表 1.1　每一代操作系统的特征

| 所属年代 | 第几代 | 产业环境 | OS 驱动力、需求 | 典型操作系统 |
| --- | --- | --- | --- | --- |
| 20 世纪 60 年代 | 第一代 | 大型机 | 多用户、多任务 | MULTICS、UNIX |
| 20 世纪 80 年代末—20 世纪 90 年代初 | 第二代 | SMP 硬件架构、虚拟内存 | 硬件架构 | Windows NT、Linux、386BSD |
| 2007 年 | 第三代 | 设备、PDA、智能手机 | 设备通信 | MbedOS、RT-Thread、LiteOS、FreeRTOS |
| 2017 年至今 | 第四代 | 各种物联网设备使用场景 | 大量 IoT 设备需要管理，分散的设备，AI 算法 | HarmonyOS、Fuchsia OS |

在第二代（PC 时代）和第三代（手机时代），人们依赖一个单一设备实现网络连接和智能计算，但是在 5G 时代，连接网络和具备计算能力的终端数量呈几何基数增长，尽管 PC 和

手机依然是工作和生活的主力装备,但是在更多场景会有越来越多的连接和计算在更多其他设备(包括边缘设备)上完成。操作系统所管理的设备的概念外延就扩展了。以往的操作系统通常对单一设备进行管理,但是未来的操作系统需要对处于连接状态的分布式多终端进行统一管理。

因此,HarmonyOS 并非移动智能操作系统,而是面向未来全场景的分布式操作系统。

从技术角度尤其是设计理念来看,HarmonyOS 和 Android 有本质区别,虽然都是基于 Linux 内核,但是 HarmonyOS 采用多内核设计,同时基于分布式架构和组件化设计,能够实现弹性部署(不同设备选取原生操作系统的不同组件进行拼装)、同时支持实时(无人驾驶车机)和分时(生活娱乐)、虚拟化快速连接(不同终端从底层 OS 已被联通)。

HarmonyOS 是一款"面向未来"、面向全场景(移动办公、运动健康、社交通信、媒体娱乐等)的分布式操作系统。在传统的单设备系统能力的基础上,HarmonyOS 提出了基于同一套系统能力、适配多种终端形态的分布式理念,能够支持手机、平板、智能穿戴、智慧屏、车机等多种终端设备。

对消费者而言,HarmonyOS 能够将生活场景中的各类终端进行能力整合,可以实现不同的终端设备之间的快速连接、能力互助、资源共享,匹配合适的设备、提供流畅的全场景体验等。

对应用开发者而言,HarmonyOS 采用了多种分布式技术,使应用程序的开发实现与不同终端设备的形态差异无关。这能够让开发者聚焦于上层业务逻辑,以便更加便捷、高效地开发应用。

对设备开发者而言,HarmonyOS 采用了组件化的设计方案,可以根据设备的资源能力和业务特征进行灵活裁剪,满足不同形态的终端设备对于操作系统的要求。

## 1.2 HarmonyOS 技术特性

HarmonyOS 具有三大技术特性:硬件互助,资源共享(分布式架构)、一次开发,多端部署(一套代码适配各种终端)、统一 OS,弹性部署(系统可裁剪)。

### 1.2.1 分布式架构

硬件互助,资源共享:基于分布式软总线技术,如图 1.3 所示,结合分布式设备虚拟化平台实现不同设备的资源融合、设备管理、数据处理,使多种设备共同形成一个超级虚拟终端。

任务自动匹配后执行于不同硬件,从而让任务能够连续地在不同设备间流转,充分发挥不同设备的资源优势。分布式数据管理基于分布式软总线的能力,实现应用程序数据和用户数据的分布式管理。用户数据不再与单一物理设备绑定,业务逻辑与数据存储分离,应用跨设备运行时数据无缝衔接,为打造一致、流畅的用户体验创造了基础条件。

图 1.3 HarmonyOS 分布式架构

## 1.2.2 操作系统可裁剪

统一 OS,弹性部署,如图 1.4 所示,HarmonyOS 通过组件化和小型化等设计方法,支持多种终端设备按需弹性部署,能够适配不同类别的硬件资源和功能需求。支撑通过编译链关系去自动生成组件化的依赖关系,形成组件树依赖图,支撑产品系统的便捷开发,从而降低硬件设备的开发门槛。

HarmonyOS 支持多种组件配置方案,实现了组件可选、组件内功能集可选、组件间依赖关系可关联。

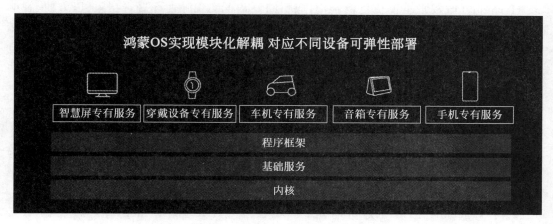

图 1.4 HarmonyOS 系统可裁剪

## 1.2.3 一套代码多端运行

一次开发、多端部署,如图 1.5 所示,HarmonyOS 提供了用户程序框架、Ability 框架及 UI 框架,支持在应用开发过程中对多终端的业务逻辑和界面逻辑进行复用,能够实现应用的一次开发、多端部署，提升了跨设备应用的开发效率。

图 1.5　HarmonyOS 一套代码多端运行

## 1.3　HarmonyOS 技术架构

　　HarmonyOS 整体的分层结构自下而上依次为内核层、系统服务层、应用框架层、应用层。HarmonyOS 基于多内核设计，系统功能按照"系统→子系统→功能/模块"逐级展开，在多设备部署场景下，各功能模块组织符合"抽屉式"设计，即功能模块采用 AOP(面向切面编程)的设计思想，可根据实际需求裁剪某些非必要的子系统或功能/模块，如图 1.6 所示。

　　HarmonyOS 实现了模块化耦合，对应不同设备可实现弹性部署，使其可以方便、智能地适配 GB、MB、KB 等由低到高的不同内存规模设备，可以便捷地在诸如手机、智慧屏、车机、穿戴设备等 IoT 设备间实现数据的流转与迁移，同时兼具了小程序的按需使用，过期自动清理的突出优点。

### 1.3.1　内核层

　　内核层基于 Linux 系统设计，主要包括内核子系统和驱动子系统。

　　内核子系统：HarmonyOS 采用多内核设计，支持针对不同资源受限设备选用适合的 OS 内核。KAL(Kernel Abstract Layer，内核抽象层)通过屏蔽多内核差异，对上层提供基础的内核能力，包括进程/线程管理、内存管理、文件系统、网络管理和外设管理等。

　　驱动子系统：包括 HarmonyOS 驱动框架(HDF)，HarmonyOS 驱动框架是 HarmonyOS 硬件生态开放的基础，提供了统一的外设访问能力和驱动开发、管理框架。

### 1.3.2　系统服务层

　　系统服务层是 HarmonyOS 的核心能力集合，通过框架层对应用程序提供服务。该层包含以下几个部分。

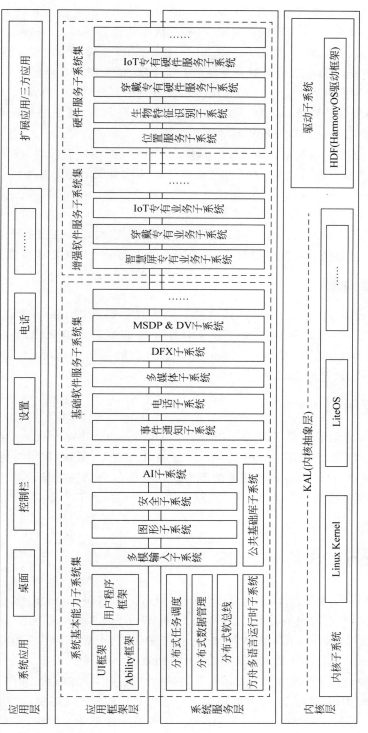

图 1.6　华为 HarmonyOS 发展历程

（1）系统基本能力子系统集：为分布式应用在 HarmonyOS 多设备上的运行、调度、迁移等操作提供了基础能力，由分布式软总线、分布式数据管理、分布式任务调度、方舟多语言运行时、公共基础库、多模输入、图形、安全、AI 等子系统组成。其中，方舟多语言运行时提供了 C/C++/JavaScript 多语言运行时和基础的系统类库，也为使用自研的方舟编译器静态化的 Java 程序（应用程序或框架层中使用 Java 语言开发的部分）提供运行时。

（2）基础软件服务子系统集：为 HarmonyOS 提供了公共的、通用的软件服务，由事件通知、电话、多媒体、DFX、MSDP&DV 等子系统组成。

（3）增强软件服务子系统集：为 HarmonyOS 提供了针对不同设备的、差异化的能力增强型软件服务，由智慧屏专有业务、穿戴专有业务、IoT 专有业务等子系统组成。

（4）硬件服务子系统集：为 HarmonyOS 提供了硬件服务，由位置服务、生物特征识别、穿戴专有硬件服务、IoT 专有硬件服务等子系统组成。

根据不同设备形态的部署环境，基础软件服务子系统集、增强软件服务子系统集、硬件服务子系统集内部可以按子系统粒度裁剪，每个子系统内部又可以按功能粒度裁剪。

### 1.3.3 架构层

框架层为 HarmonyOS 的应用程序提供了 Java/C/C++/JavaScript 等多语言的用户程序框架和 Ability 框架，以及各种软硬件服务对外开放的多语言框架 API；同时为采用 HarmonyOS 的设备提供了 C/C++/JavaScript 等多语言的框架 API，但不同设备支持的 API 与系统的组件化裁剪程度相关。

### 1.3.4 应用层

应用层包括系统应用和第三方非系统应用。HarmonyOS 的应用由一个或多个 FA（Feature Ability）或 PA（Particle Ability）组成。其中，FA 有 UI 界面，提供与用户交互的能力，而 PA 无 UI 界面，提供后台运行任务的能力及统一的数据访问抽象。基于 FA/PA 开发的应用，能够实现特定的业务功能，支持跨设备调度与分发，为用户提供一致、高效的应用体验。

## 1.4 HarmonyOS 与 LiteOS

Huawei LiteOS 目前作为 HarmonyOS 内核的一部分，HarmonyOS 可以根据不同的 ROM 大小选择使用不同的内核版本。Huawei LiteOS 始于 2012 年，是为支持华为终端产品而开发的嵌入式操作系统，后来在华为 Mate 系列、P 系列、荣耀系列手机和可穿戴产品上批量应用。2016 年 9 月华为正式发布 LiteOS 开源版本。

Huawei LiteOS 是华为面向 IoT 领域的开源的物联网实时操作系统，如图 1.7 所示，支持 ARM（Advanced RISC Machine，进阶精简指令集机器）、RISC-V（RISC-V 是加州大学伯克利分校开发的一种特定指令集架构，严格地说，并不是一种全新的架构，它与 ARM 同属

RISC(Reduced Instruction Set Computer,精简指令集)等主流的 CPU 架构,遵循 BSD-3 开源许可协议,具备轻量级(最小内核大小仅为 6KB)、低功耗、互联互通、组件丰富、快速开发等能力,可广泛应用于智能家居、个人穿戴、车联网、城市公共服务、制造业等领域,为开发者提供"一站式"完整的软件平台,有效降低开发门槛、缩短开发周期。

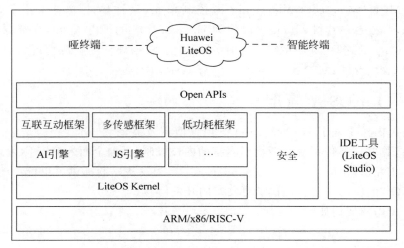

图 1.7 华为 LiteOS 系统架构

Huawei LiteOS 支持多种芯片架构,如表 1.2 所示,如 Cortex-M series、Cortex-A series 等,可以快速移植到多种硬件平台。Huawei LiteOS 也支持 UP(单核)与 SMP(多核)模式,即支持在单核或者多核的环境上运行。

表 1.2 Huawei LiteOS 支持的架构

| 架 构 | 系 列 |
| --- | --- |
| ARM | Cortex-M0 |
|  | Cortex-M0+ |
|  | Cortex-M3 |
|  | Cortex-M4 |
|  | Cortex-M7 |
|  | Cortex-A7 |
|  | Cortex-A9 |
| ARM64 | Cortex-A53 |
| RISC-V | RV32 |
| C-SKY | CK802 |

说明:ARM 公司在经典处理器 ARM 11 以后的产品改用 Cortex 命名,并分成 A、R 和 M 三类,旨在为各种不同的市场提供服务。

Cortex 系列属于 ARM v7 架构，这是到 2010 年为止 ARM 公司最新的指令集架构。（2011 年，ARM v8 架构在 TechCon 上推出）ARM v7 架构定义了三大分工明确的系列：A 系列面向尖端的基于虚拟内存的操作系统和用户应用；R 系列针对实时系统；M 系列对微控制器。

LiteOS 既可以作为一款 RTOS（Real Time Operating System，实时操作系统）运行在资源受限的 MCU（Microcontroller Unit，微控制单元，又称单片微型计算机 Single Chip Microcomputer 或者单片机）上，也可以作为 HarmonyOS 的子内核运行在资源丰富的 SoC（System on Chip 的缩写，称为系统级芯片）平台上。根据硬件的资源情况，LiteOS 又可以分为 LiteOS-A（内存≥1MB）和 LiteOS-M（内存≥128KB）。

### 1.4.1　LiteOS-A 简介

LiteOS-A 内核是基于 Huawei LiteOS 内核演进发展的新一代内核，是面向 IoT 领域构建的轻量级物联网操作系统。新增了丰富的内核机制、更加全面的 POSIX 标准接口及统一驱动框架 HDF（Harmony Driver Foundation）等，为设备厂商提供了更统一的接入方式，为 HarmonyOS 的应用开发者提供了更友好的开发体验。

LiteOS-A 内核架构如图 1.8 所示。

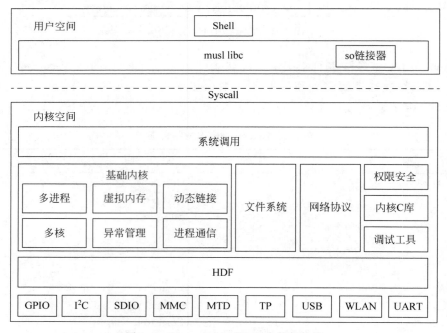

图 1.8　Huawei LiteOS-A 内核架构图

### 1.4.2　LiteOS-M 简介

LiteOS-M 内核是面向 IoT 领域构建的轻量级物联网操作系统内核，具有小体积、低功耗、高性能的特点，其代码结构简单，主要包括内核最小功能集、内核抽象层、可选组件及工程目录等，分为硬件相关层及硬件无关层，硬件相关层提供统一的 HAL（Hardware Abstraction Layer）接口，提升硬件易适配性，以及不同编译工具链和芯片架构的组合分类，以此满足 AIoT 类型丰富的硬件和编译工具链的拓展。

LiteOS-M 内核架构如图 1.9 所示。

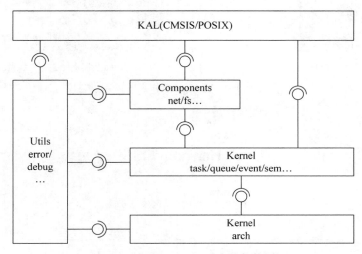

图 1.9　华为 LiteOS-M 内核架构图

## 1.5　OpenHarmony 生态

OpenHarmony 是 HarmonyOS 的社区开源版本，与谷歌公司的 AOSP（Android 开源项目）的作用是一致的。

OpenHarmony 项目由开放原子开源基金会负责社区化开源运营，HarmonyOS 是华为公司基于 OpenHarmony 开源版定制开发的商用发行版。开放原子开源基金会成立的目的是支持更多企业基于 OpenHarmony 定制开发自己的商用发行版本。

### 1.5.1　Android 与 AOSP

目前手机中运行的"安卓系统"通常是指 AOSP（Android Open Source Project，Android 开源项目）+ GMS（谷歌服务框架，商用收费）。这两部分构成了安卓开发者使用的基础 SDK，也是所有安卓 App 的基础，如图 1.10 所示。

### 1.5.2　HarmonyOS 与 OpenHarmony

HarmonyOS 实际上由 3 个部分组成：OpenHarmony（开源鸿蒙操作系统）、HMS（华为移动服务）在内的闭源应用与服务，以及其他开放源代码，如图 1.11 所示。

图 1.10　Android 操作系统体系

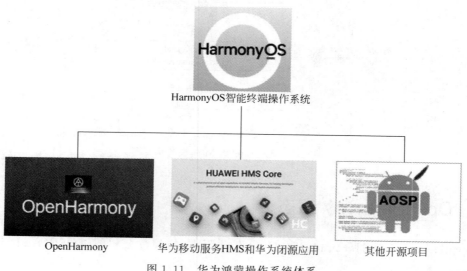

图 1.11　华为鸿蒙操作系统体系

华为手机操作系统包括 OpenHarmony(开源免费)＋HMS(华为移动服务,商用收费),HMS 是对标谷歌 GMS 的商业产品,用于支持开发者为华为手机开发 App。为了兼容现有的 Android 生态,HMS 的许多接口设计尽量兼容了 GMS。

开放原子开源基金会是在民政部注册的致力于开源产业公益事业的非营利性独立法人机构。开放原子开源基金会的服务范围包括开源软件、开源硬件、开源芯片及开源内容等,为各类开源项目提供中立的知识产权托管,保证项目的持续发展不受第三方影响,通过开放治理寻求更丰富的社区资源的支持与帮助,包括募集并管理资金,提供法律、财务等专业支持。

2020 年 9 月 10 日华为将 HarmonyOS 2.0 源码捐赠给开放原子开源基金会孵化,并为其命名为 OpenHarmony 1.0,通过 Gitee 托管代码并对外开放下载。OpenHarmony 开源项目主要遵循 Apache 2.0 等商业友好的开源协议,所有企业、机构与个人均可基于 OpenHarmony 开源代码开发自己的商业发行版,HarmonyOS 与 OpenHarmony 的发展路标如图 1.12 所示。

# 第1章 HarmonyOS系统简介

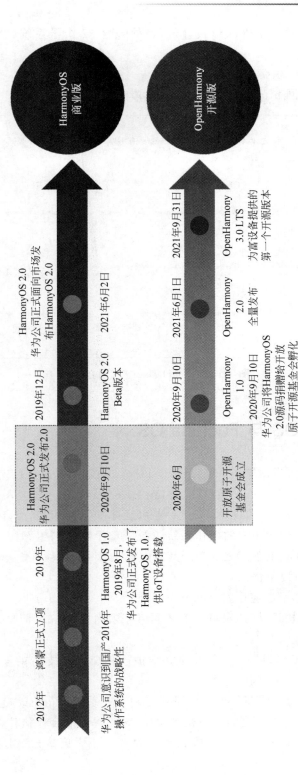

图 1.12 HarmonyOS 与 OpenHarmony 发展历程

OpenHarmony 代码以组件的形式开放，目前包含 32 个子系统、281 个代码仓库，如图 1.13 所示。

Gitee 网站 https://gitee.com/openharmony。

OpenHarmony 官网 https://openharmony.io/。

图 1.13　OpenHarmony 代码仓库

## 1.6　HarmonyOS 与 Fuchsia OS

Fuchsia OS 是谷歌开发的操作系统，与基于 Linux 内核的 Chrome OS 和 Android 等不同，Fuchsia OS 基于新的名为 Zircon 的微内核，受 Little Kernel 启发，用于嵌入式系统，主要使用 C 语言和 C++ 编写。Fuchsia OS 的设计目标之一是可运行在众多的设备上，包括手机和计算机。

Fuchsia OS 是谷歌使用单一的操作系统去统一整个物联网生态圈的一种新的操作系统架构模式。Fuchsia OS 的独特之处在于其微内核、模块化设计。

Fuchsia OS 主要有以下优点：

（1）原生进程沙箱，解决应用安全和分发问题。

（2）全新微内核架构设计，Fuchsia OS 不再基于宏内核 Linux，而是采用了谷歌自己研发的全新微内核 Zircon。

（3）稳定的驱动接口，硬件厂商可独立维护硬件驱动。支持的架构包括 x86-64 和 ARM 64，支持的设备从 IoT 到服务器。

（4）系统模块化、分层，设备厂商可以灵活定制专有系统，Fuchsia OS 的另一大特点是其模块化系统体系结构与模块化的应用程序设计，这使谷歌的合作伙伴与硬件制造商可以用自己的模块替代 Fuchsia OS 的单个系统级别，从而在不影响其他级别功能的同时改进或扩展 Fuchsia OS,增强了系统可扩展性。

（5）Vulkan 图形接口、3D 桌面渲染 Scenic、全局光照。

（6）Flutter 应用开发框架,Flutter SDK 能在 Android 运行,兼容 Android 应用,是使用 Dart 和 Flutter 作为界面开发的语言和框架。此外,与安卓相比,Fuchsia OS 无论是存储器还是内存等硬件要求都大幅降低。

### 1.6.1 Fuchsia OS 系统架构

从 Fuchsia OS 技术架构来看,内核层 Zircon 的基础是 Little Kernel(简称 LK),LK 是专为嵌入式应用中小型系统设计的内核,代码简洁,适合嵌入式设备和高性能设备,例如 IoT、移动可穿戴设备等,如图 1.14 所示。

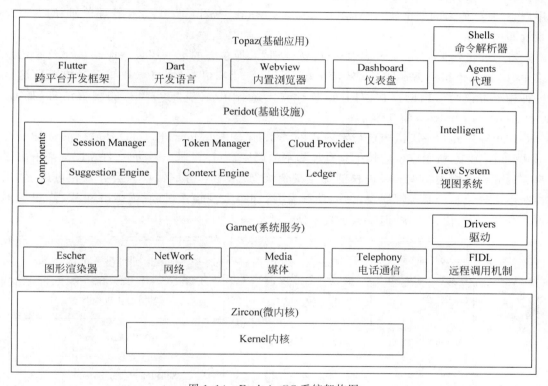

图 1.14 Fuchsia OS 系统架构图

Fuchsia OS 的系统架构也基于模块化。操作系统由 4 个独立级别模块组成,每个级别都有其自己的任务：Zircon、Garnet、Peridot 和 Topaz。

下面逐一认识一下 Fuchsia OS 的系统架构模块。

**1. 微内核（Zircon）**

Fuchsia OS 并非基于 Linux 内核,而是基于一个新的名为 Zircon /ˈzɜːk(ə)n/ 的微内核架构开发,Zircon Kernel 最早由 LK(Little Kernel)的一个分支发展而来。

LittleKernel 是一个适用于嵌入式设备和 BootLoader（BootLoader 是嵌入式系统在加电后执行的第一段代码，在它完成 CPU 和相关硬件的初始化之后，再将操作系统映像或固化的嵌入式应用程序装载到内存中，然后跳转到操作系统所在的空间，启动操作系统运行）等场景的微型操作系统（微内核），提供了线程调度、互斥量和定时器等支持。在嵌入式 ARM 平台，Little Kernel 的核心大约 15～20KB。部分芯片厂商（如高通、联发科等）的 Android 操作系统使用 Little Kernel 作为其 BootLoader 和 TEE（Trusted Execution Environment）的安全区运行环境。

不同于为微控制器设计的 Little Kernel（微控制通常只有非常有限的 RAM，少量的外设及运行任务），Zircon 的设计目标是运行在具备更强处理能力的智能手机及个人计算机上。Zircon 仅支持 64 位处理器系统，在 Little Kernel 的基础上增加了进程的概念、添加了用户态、基于能力的安全模型、MMU（Memory Management Unit，内存管理单元）的支持及系统调用等。

Zircon 包含 Fuchsia OS 的内核、设备管理器、最核心的第一层设备驱动程序及底层系统库（如 libc 和 launchpad）。此外，Zircon 还提供 FIDL（Fuchsia OS 接口定义语言），这是一种用于进程间通信的协议。FIDL 是独立于编程语言的，能与流行的编程语言（如 C、C++、Dart、Go 和 Rust）进行连接。

这里简单介绍一下微内核与宏内核的关系，微内核（Micro Kernel）是内核的一种精简形式，如图 1.15 所示。微内核中只有最基本的调度、内存管理。驱动、文件系统等都采用用户态的守护进程去实现。微内核的优点是超级稳定，驱动等的错误只会导致相应进程"死掉"，不会导致整个系统都崩溃，做驱动开发时，如果发现错误，则只需 Kill 进程，修正后重启进程就行了，比较方便，但缺点是效率低。

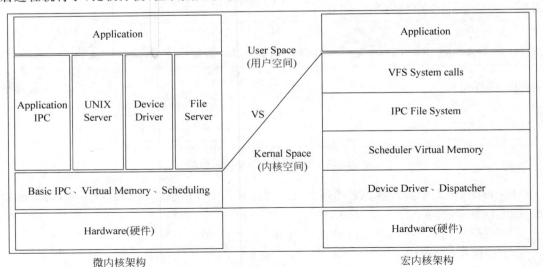

图 1.15 微内核与宏内核比较图

宏内核(Monolithic Kernel)简单来讲,就是把很多东西都集成进内核,例如 Linux 内核,除了最基本的进程、线程管理、内存管理外,文件系统、驱动、网络协议等都在内核里面。宏内核的优点是效率高。缺点是稳定性差,开发过程中的 Bug 经常会导致整个系统崩溃,具体区别如表 1.3 所示。

表 1.3 微内核与宏内核比较

| 比较项 | 微内核 | 宏内核 |
| --- | --- | --- |
| 基本概念 | 用户服务和内核服务运行在不同的地址空间中 | 用户服务和内核服务运行在相同的地址空间中 |
| 尺寸 | 比较小 | 比微内核大 |
| 执行速度 | 容易扩展 | 不容易扩展 |
| 可扩展性 | 单个服务崩溃不影响全局 | 单个服务崩溃往往意味着整个系统崩溃 |
| 代码开发 | 需要开发的代码量大 | 平台提供的代码多,相对需要开发的代码量少 |
| 系统举例 | QNX、Symbian、L4Linux、Singularity、K42、Mac OS X、PikeOS、Minix、Coyotos | Linux、BSDS(FreeBSD、penBSD、NetBSD)、Microsoft Windows(95、98、Me)、Solaris、OS-9、AIX、DOS、HP-UX、OpenvMS、XTS-400 等 |

**2. 系统服务(Garnet)**

Garnet 是基于 Zircon 内核之上的系统服务层。提供了设备级别的各种系统服务和网络,以及媒体和图形服务,例如:软件安装、系统管理及与其他系统的通信。Garnet 包含图形渲染器 Escher、程序包管理、更新系统 Amber 及文本和代码编辑器 Xi。

**3. 基础设施(Peridot)**

Peridot 用于处理模块化应用程序设计,Peridot 的另外两个重要组件可直接用于模块 Ledger 和 Maxwell。

Ledger:Ledger 是基于云的存储系统(分布式存储系统),它为每个 Fuchsia 组件(模块或代理)提供单独的数据存储,可在不同设备之间同步,这使用户可以在当前 Fuchsia OS 的设备上继续停留在其他 Fuchsia OS 设备上的相应位置。

Maxwell:通过 Maxwell,谷歌在 Fuchsia OS 中集成了一个组件,该组件向用户提供了人工智能。就像 Fuchsia OS 一样,Maxwell 具有模块化设计。AI 系统由一系列代理组成,这些代理分析用户的行为及其所使用的内容,在后台确定合适的信息,并将建议转发给操作系统。例如,应加载哪些模块或故事以适合用户在特定时间的行为。谷歌语言助手也是 AI 组件的一部分,该组件将在 Fuchsia OS 项目的框架内以代码 Kronk 的形式进一步开发。

**4. 基础应用(Topaz)**

Topaz 作为基础应用层,Topaz 提供 Flutter 支持,而有了 Flutter 的支持,各种华丽的应用程序可以提供日常使用的功能齐全的应用程序。例如,现在最令人印象深刻的当然是 Armadillo UI,它是 Fuchsia OS 主要用户界面和主屏幕,Armadillo 现在被 Ermine 取代,这是一个面向开发人员的 shell,专为测试的明确目的而设计 Fuchsia OS 的应用。

### 1.6.2　Fuchsia OS 与产业

HarmonyOS 与 Fuchsia OS 均为搭建物联网生态而推出的系统,而搭建完善的生态体系需要接入不同领域的合作伙伴,因此,HarmonyOS 与 Fuchsia OS 未来竞争方向主要是在合作伙伴上。

谷歌目前计划将 Fuchsia OS 应用于智能手机、智能家居市场与笔记本电脑市场,由于 Fuchsia OS 的进展较慢,华为 HarmonyOS 目前已经在将近 2 亿台华为设备上安装并运行,同时华为通过 HarmonyOS Connect 计划已经接入了大量第三方产品,两者差异如表 1.4 所示。

表 1.4　HarmonyOS 与 Fuchsia OS 比较

| 比较项 | HarmonyOS | Fuchsia OS |
| --- | --- | --- |
| 应用场景 | 手机、计算机、智能手表、手环、智慧屏、智能音箱、路由器等智能设备 | 智能家居、移动终端等智能嵌入式设备,未来可能被应用于智能手机与笔记本电脑 |
| 特征 | 实现跨终端无缝协同体验。统一的系统 IDE 支撑着开发人员只需一次开发,便可以实现将应用部署到不同的设备上,可大幅提高开发效率 | 与目前 Android 相比,无论是存储器还是内存之类的硬件要求都大幅降低,可以看出这是一款面向物联网的家用电器的系统 |
| 内核机制 | 目前基于多内核设计 | 基于微内核 Magenta(后期改名为 Zircon)的新内核 |

## 1.7　本章小结

鸿蒙操作系统作为国产物联网开源操作系统,将会不断推进中国物联网行业的发展,它不仅给行业带来了新的动力和发展机会,同时不断地影响和改变着人们的生活。学习和掌握鸿蒙操作系统开发给软件、硬件开发者提供了新的机会。

# 第 2 章 开发环境搭建

本章介绍如何配置鸿蒙应用开发环境、下载并安装集成开发工具 DevEco Studio、鸿蒙模拟器的使用、真机调试证书申请,以及如何使用 Scrcpy Android 投屏软件进行真机测试。通过本章的学习,读者可以逐步搭建好鸿蒙应用开发所需的相关环境,为后续章节的学习做好准备工作。

## 2.1 鸿蒙应用开发环境搭建

鸿蒙应用开发环境的搭建分两步,分别是安装 Node.js 和下载并安装 DevEco Studio。

### 2.1.1 下载和安装 Node.js

Node.js 发布于 2009 年 5 月,由 Ryan Dahl 开发,提供了基于 Chrome V8 引擎的 JavaScript 运行环境,使用了一个事件驱动、非阻塞式 I/O 模型,是一个让 JavaScript 运行在服务器端的开发平台。

Node.js 应用于开发鸿蒙 JS 应用程序和运行鸿蒙预览器功能,是开发 HarmonyOS 应用过程中必备的软件。

下载并安装 Node.js,需选择 LTS 版本 12.0.0 及以上,Windows 64 位对应的软件包如图 2.1 所示。Node.js 安装包及源码下载网址为 https://nodejs.org/en/download/。

双击下载后的软件包进行安装,根据安装向导完成 Node.js 的安装。Mac 系统在安装软件过程中需要输入用户系统密码来授权系统运行安装新软件。

Windows 系统具体的安装步骤如下。

步骤 1:双击下载后的安装包 node-v14.16.0-x64.msi 进行安装,如图 2.2 所示。

步骤 2:勾选接受协议选项,单击 Next(下一步)按钮,如图 2.3 所示。

步骤 3:Node.js 默认安装目录为 C:\Program Files\nodejs\,可以根据需要修改目录,并单击 Next(下一步)按钮,如图 2.4 所示。

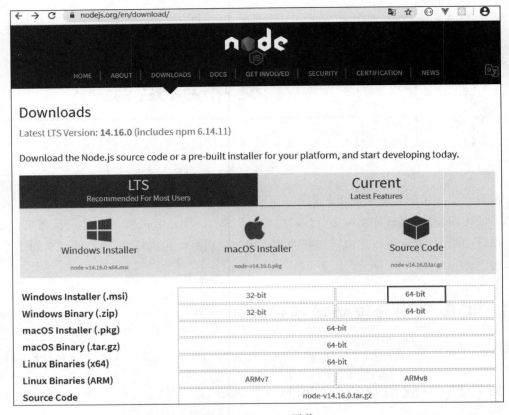

图 2.1　Node.js 下载

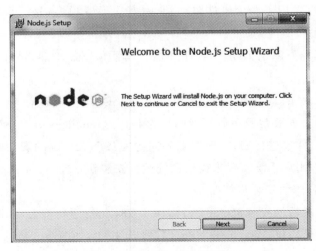

图 2.2　运行 Node.js 安装包

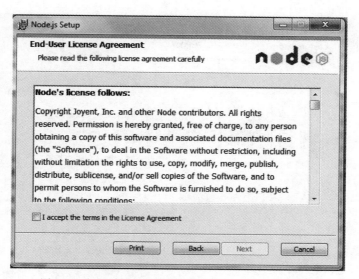

图 2.3　勾选接受协议选项,单击 Next(下一步)按钮

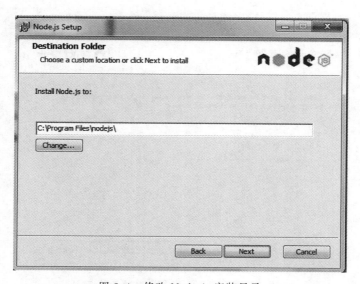

图 2.4　修改 Node.js 安装目录

步骤 4:单击树形图标来选择需要的安装模式,一般保持默认模式即可,然后单击 Next(下一步)按钮,如图 2.5 所示。

步骤 5:单击 Install(安装)按钮开始安装 Node.js。也可以单击 Back(返回)按钮来修改先前的配置,然后单击 Next(下一步)按钮,如图 2.6 所示。

步骤 6:单击 Finish(完成)按钮退出安装向导,如图 2.7 所示。

步骤 7:打开 cmd 命令窗口并输入 node -v 命令,便可查看 Node.js 的版本号,如

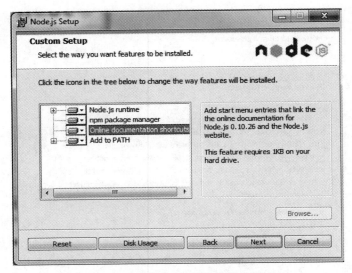

图 2.5 安装模式(选择默认模式即可)

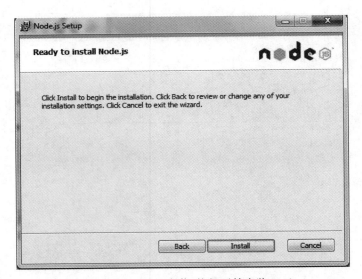

图 2.6 单击 Install(安装)按钮开始安装 Node.js

图 2.8 所示,表示 Node.js 安装成功了。

Mac OS 上安装 Node.js 可以通过以下两种方式安装:

(1) 在官方下载网站下载 pkg 安装包,直接安装即可。

(2) 使用 brew 命令安装,命令如下:

```
brew install node
```

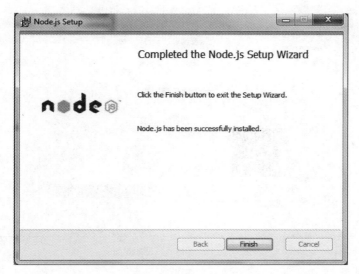

图 2.7　单击 Finish(完成)按钮退出安装向导

图 2.8　查看 Node.js 版本号

## 2.1.2　下载和安装 DevEco Studio

　　HUAWEI DevEco Studio(以下简称 DevEco Studio)是基于 IntelliJ IDEA Community 开源版本打造的,面向华为终端全场景多设备的一站式集成开发环境(IDE),为开发者提供工程模板创建、开发、编译、调试、发布等 E2E 的 HarmonyOS 应用开发服务。通过 DevEco Studio,开发者可以更高效地开发具备 HarmonyOS 分布式能力的应用,进而提升创新效率。

　　下面介绍如何下载并安装 DevEco Studio 开发工具,具体的步骤如下:

　　步骤 1:登录 HarmonyOS 应用开发门户,单击右上角注册按钮,注册开发者账号。可以访问如下地址 https://id1.cloud.huawei.com/CAS/portal/login.html 登录成功后。再访问 HUAWEI DevEco Studio 产品页,下载 DevEco Studio 安装包,如图 2.9 所示。

　　步骤 2:进入 HUAWEI DevEco Studio 产品页,下载 DevEco Studio 安装包,如图 2.10 所示。

　　步骤 3:Windows 用户双击下载的 deveco-studio-xxxx.exe 文件,进入 DevEco Studio 安装向导,在如下安装选项界面勾选 DevEco Studio launcher 后,单击 Next 按钮,如图 2.11 所示,直至安装完成。

　　步骤 4:Mac 用户双击下载的 deveco-studio-xxxx.dmg 软件包。

图 2.9　华为账号登录页面

图 2.10　DevEco Studio 2.1 下载

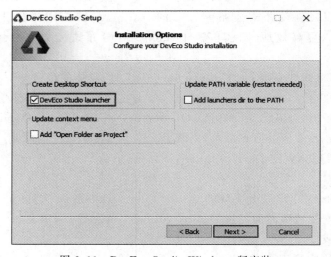

图 2.11　DevEco Studio Windows 版安装

步骤5：在安装界面中，将 DevEco-Studio.app 拖曳到 Applications 中，如图 2.12 所示，等待安装完成。

图 2.12　DevEco Studio Mac 安装

步骤6：安装完成后，先不要勾选 Run DevEco Studio 选项，接下来需要根据开发环境，检测和配置开发环境，如图 2.13 所示。

图 2.13　DevEco Studio 安装检测

步骤7：DevEco Studio 的编译构建依赖 JDK，DevEco Studio 预置了 Open JDK，版本为 1.8，安装过程中会自动安装 JDK。

## 2.1.3　运行 Hello World

DevEco Studio 开发环境配置完成后，可以通过运行 Hello World 工程来验证环境配置是否正确。以 Wearable 工程为例，在 Wearable 远程模拟器中运行该工程。

我们按步骤运行一个 Hello World 程序，步骤如下。

步骤1：打开 DevEco Studio，在欢迎页单击 Create HarmonyOS Project，如图 2.14 所示，创建一个新工程。

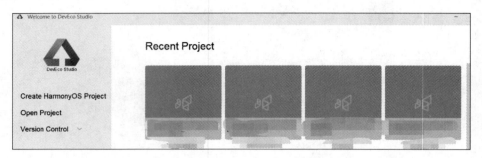

图 2.14  DevEco Studio 欢迎页

步骤2：选择设备类型和模板，以 Wearable 为例，选择 Empty Feature Ability(Java)，单击 Next 按钮，如图 2.15 所示。

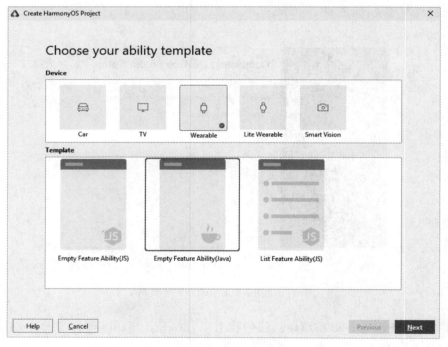

图 2.15  选择模板

步骤3：填写项目相关信息，保持默认值即可，单击 Finish 按钮。

步骤4：工程创建完成后，DevEco Studio 会自动对工程进行同步，同步成功后如

图2.16所示。首次创建工程时会自动下载Gradle工具，时间较长，请耐心等待。

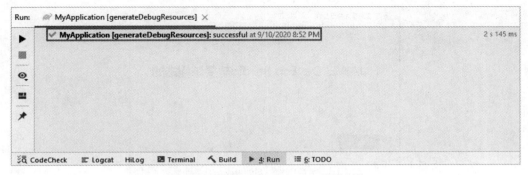

图2.16　安装DevEco Gradle编译环境

步骤5：在DevEco Studio菜单栏，选择Tools→HVD Manager。首次使用模拟器时需下载模拟器相关资源，此时需单击OK按钮，等待资源下载完成后，单击模拟器界面左下角的Refresh按钮，如图2.17所示。

图2.17　安装下载模拟器

步骤6：在浏览器中会弹出华为开发者联盟账号登录界面，输入已实名认证的华为开发者联盟账号的用户名和密码进行登录，如图2.18所示。

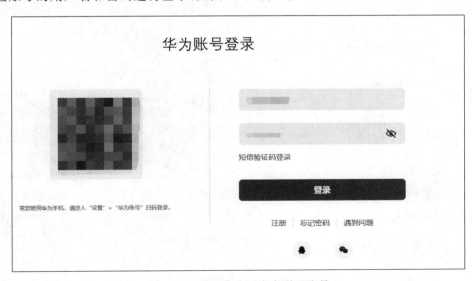

图2.18　登录华为开发者联盟账号

步骤7:登录后,单击界面的"允许"按钮进行授权,如图2.19所示。

图2.19 登录华为账号授权

步骤8:在设备列表中,选择Wearable设备,并单击放大按钮,运行模拟器,如图2.20所示。

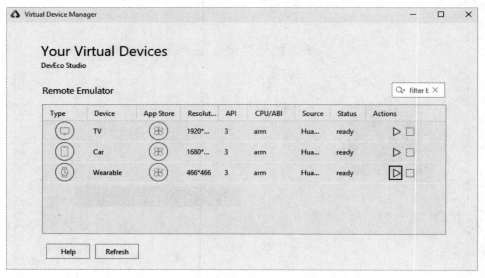

图2.20 选择模拟器

步骤9：单击 DevEco Studio 工具栏中的放大按钮运行工程，或使用默认快捷键 Shift+F10（Mac 为 Control+R）运行工程。

步骤10：在弹出的 Select Deployment Target 界面选择已启动的模拟器，单击 OK 按钮，如图 2.21 所示。

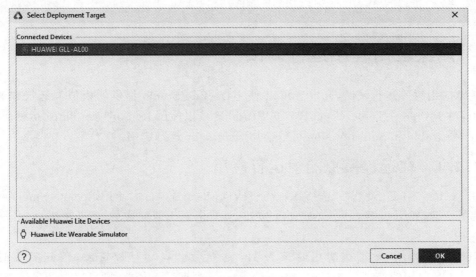

图 2.21　启动模拟器

步骤11：DevEco Studio 会启动应用的编译构建，完成后应用即可运行在模拟器上，如图 2.22 所示。

图 2.22　模拟器运行

注意：下载 JS SDK 时，通常 JS 下载比较缓慢。对于国内用户，可以将 npm 仓库设置为华为云仓库，代码如下，在命令行工具中执行如下命令，重新设置 npm 仓库地址后，再执行 JS SDK 的下载。

```
npm config set registry https://mirrors.huaweicloud.com/repository/npm/
```

## 2.2 鸿蒙应用程序运行调试

鸿蒙应用程序运行调试，开发者可以通过 DevEco Studio 提供的模拟器运行和调试鸿蒙应用，对于 Phone、Tablet、Car、TV 和 Wearable 可以使用 Remote Emulator 运行应用，对于 Lite Wearable 和 Smart Vision 可以使用 Simulator 运行应用。

### 2.2.1 在远程模拟器中运行应用

在 DevEco Studio 菜单栏，选择 Tools→HVD Manager。首次使用 Remote Emulator 时需下载相关资源，如图 2.23 所示，单击 OK 按钮，等待资源下载完成后，重新单击 Tools→HVD Manager。

Remote Emulator 每次使用时长为 1h，到期后会自动释放资源，因此需及时完成 HarmonyOS 应用的调试。如果 Remote Emulator 到期后被释放，则可以重新申请资源。

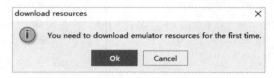

图 2.23　等待资源下载

在浏览器中会弹出华为开发者联盟账号登录界面，输入已实名认证的华为开发者联盟账号的用户名和密码进行登录。

注意：使用 DevEco Studio 远程模拟器需要华为开发者联盟账号进行实名认证，建议在注册华为开发者联盟账号后，立即提交实名认证审核，认证方式包括"个人实名认证"和"企业实名认证"。

登录后，需单击界面的"允许"按钮进行授权，如图 2.19 所示。

单击已经连接的 Remote Emulator 设备的运行按钮 ▷，如图 2.24 所示，启动远程模拟设备（同一时间只能启动一个设备）。

单击 DevEco Studio 的 Run→Run'模块名称'或 ▶，或使用默认快捷键 Shift+F10（Mac 为 Control+R）。

在弹出的 Select Deployment Target 界面选择已启动的 Remote Emulator 设备，如图 2.25 所示，单击 OK 按钮。

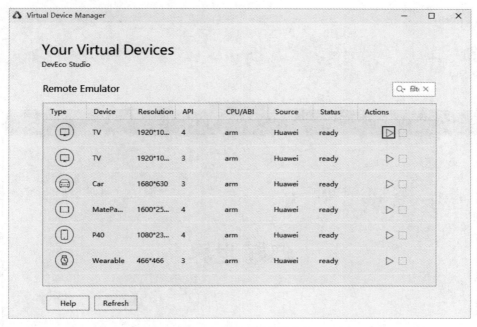

图 2.24　选择远程虚拟设备

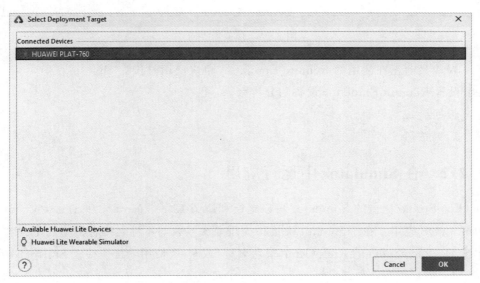

图 2.25　选择识别的设备名称

DevEco Studio 会启动应用的编译构建,完成后应用即可运行在 Remote Emulator 上,如图 2.26 所示。

Remote Emulator 侧边栏按钮的作用如下。

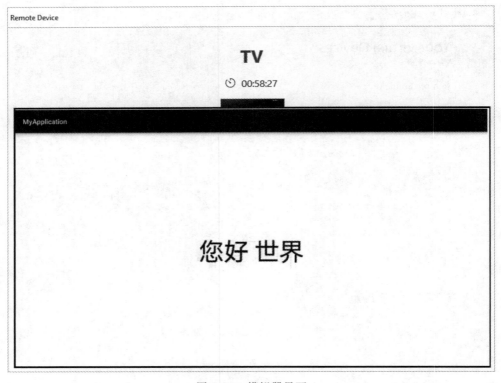

图 2.26　模拟器界面

×：释放当前正在使用的 Remote Emulator，单次使用时长为 1h。

▫：设置 Remote Emulator 设备的分辨率。

○：返回设备主界面。

◁：后退按钮。

## 2.2.2　在 Simulator 中运行应用

DevEco Studio 提供的 Simulator 可以运行和调试 Lite Wearable 和 Smart Vision 设备的 HarmonyOS 应用。在 Simulator 上运行应用兼容签名与不签名两种类型的 HAP。

单击 DevEco Studio 的 Run→Run'模块名称'或 ▶，或使用默认快捷键 Shift＋F10（Mac 为 Control＋R）。

在弹出的 Select Deployment Target 界面的 Available Huawei Lite Devices 设备列表中，选择需要运行的设备，如图 2.27 所示，单击 OK 按钮。

DevEco Studio 会启动应用的编译构建，完成后应用即可运行在 Simulator 上，如图 2.28 所示。

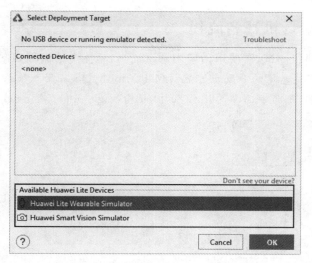

图 2.27　Huawei Lite Devices 设备列表

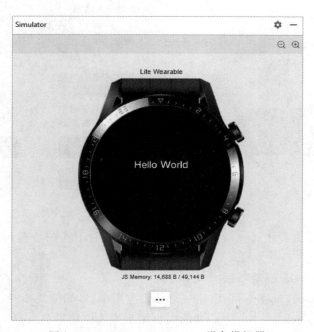

图 2.28　Huawei Lite Devices 设备模拟器

## 2.3　使用真机设备运行应用

使用模拟器测试后,还需要在真机进行测试。真机测试首先需要申请应用调试证书,2.3.1 节将详细讲解申请真机测试证书的详细流程。

### 2.3.1 手动真机签名流程

使用真机进行项目测试,需要申请应用调试证书,具体申请流程如图 2.29 所示。

**注意**：目前只有受邀请的开发者才能访问 HarmonyOS 应用相关菜单,如果 AGC 页面未展示文档中的菜单,则可联系华为运营人员(邮箱：agconnect@huawei.com)。

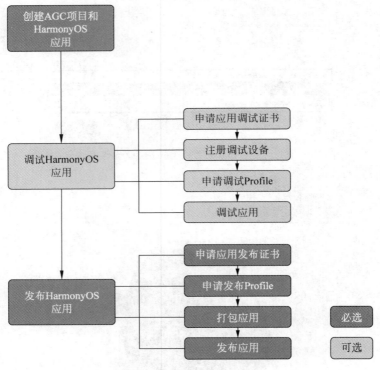

图 2.29　真机设备测试流程

申请真机测试的流程相对比较复杂,开发人员可按照下面的步骤一步一步地完成申请流程。

**1. 创建 HarmonyOS 应用项目**

首先需要通过 DevEco Studio 创建一个鸿蒙应用项目,如图 2.30 所示。

这里需要注意,Package Name 后面的内容用于生成应用签名信息。

**2. 使用 DevEco Studio 生成证书请求文件**

在主菜单栏单击 Build→Create Key Store,如图 2.31 所示。

在 Generate Key 界面中,继续填写密钥信息,然后单击 Generate Key and CSR,如图 2.32 所示。

在弹出的窗口中,单击 CSR File Path 对应的 图标,选择 CSR 文件存储路径,如图 2.33 所示。

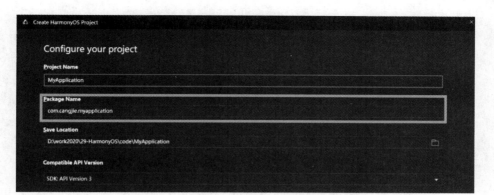

图 2.30 创建项目

图 2.31 生成 p12 文件

图 2.32 生成 CSR 文件

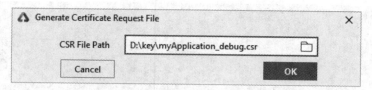

图 2.33 保存 CSR 文件

单击 OK 按钮，创建 CSR 文件，成功后工具会同时生成密钥文件(.p12)和证书请求文件(.csr)，如图 2.34 所示。

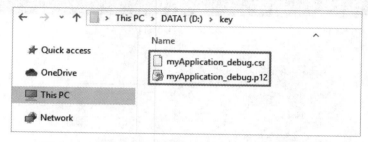

图 2.34 生成的密钥文件(.p12)和证书请求文件(.csr)

### 3. 申请应用调试证书

登录 AppGallery Connect 网站，选择模块"用户与访问"，如图 2.35 所示。

图 2.35 选择"用户与访问"

在左侧导航栏选择"证书管理"，进入证书管理页面，单击"新增证书"按钮，如图 2.36 所示。

在弹出的"新增证书"窗口，填写要申请的证书信息，单击"提交"按钮，如图 2.37 所示。

在左侧导航栏选择"设备管理"，进入设备管理页面，单击右上角的"添加设备"按钮，如图 2.38 所示。

在弹出窗口中填写设备信息，如图 2.39 所示。

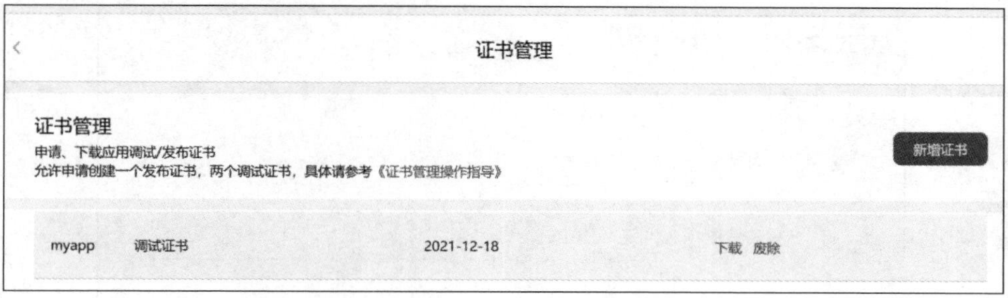

图 2.36 选择"新增证书"

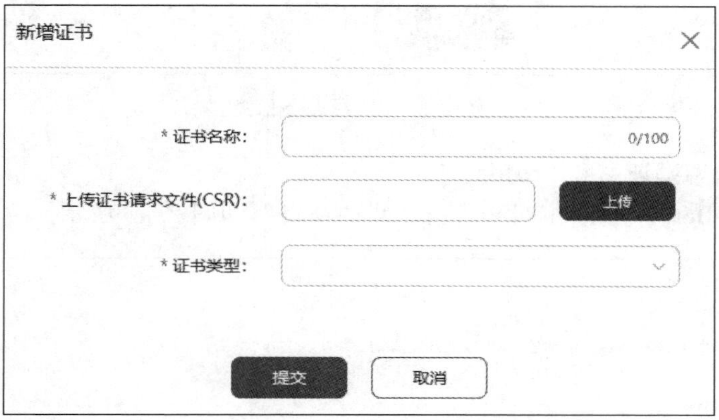

图 2.37 填写证书信息

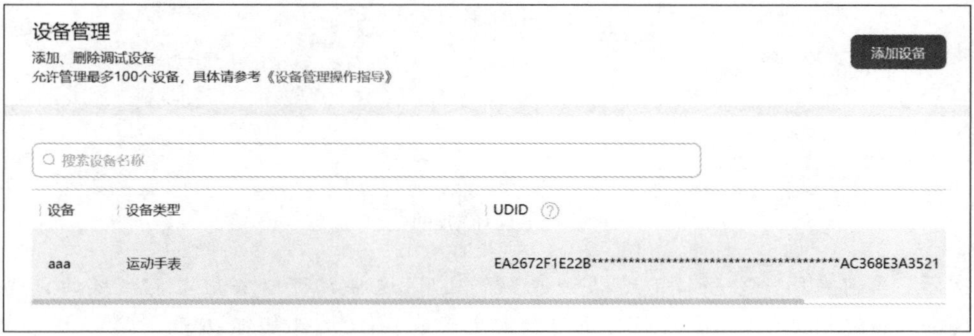

图 2.38 选择"添加设备"

通过 adb 命令可查看 UDID 信息,命令如下:

```
adb shell dumpsys DdmpDeviceMonitorService
#或者
adb shell bm get - udid
```

图 2.39 填写设备信息

### 4．申请应用调试证书和 Profile

登录 AppGallery Connect 网站，选择"我的项目"，如图 2.40 所示。

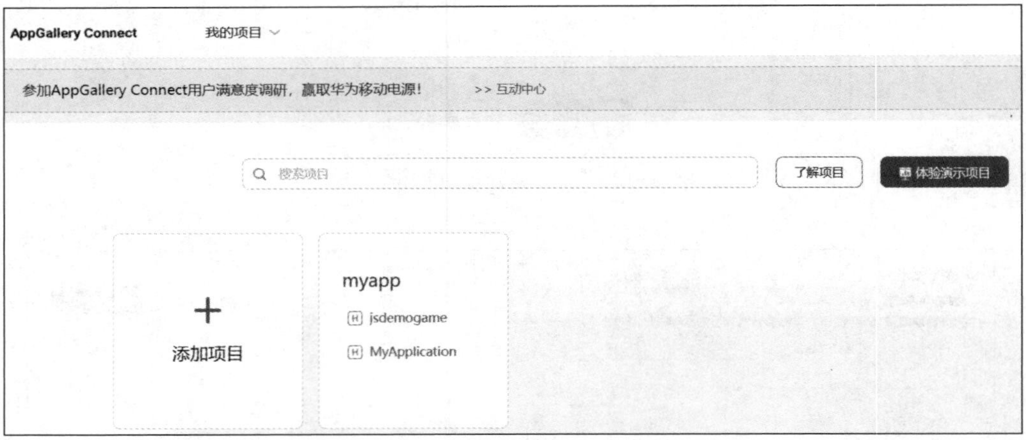

图 2.40 选择"我的项目"

**提示**：当前在同一个项目下可以创建多个应用，这样就可以共用之前生成的证书请求文件(.csr)和密钥文件(.p12)，新的应用只需生成 Profile 文件就可以了。

找到你的项目，单击创建的 HarmonyOS 应用。

选择 HarmonyOS 应用→HAP Provision Profile 管理，进入"管理 HAP Provision Profile"页面，单击右上角的"添加"按钮，如图 2.41 所示。

在弹出的 HarmonyAppProvision 信息窗口添加调试 Profile，如图 2.42 所示。

调试 Profile 申请成功后，"管理 HAP Provision Profile"页面会展示 Profile 名称、Profile 类型、添加的证书和失效日期。下载生成的 Profile 文件如图 2.43 所示。

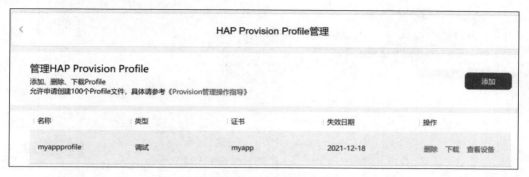

图 2.41　选择添加 HAP Provision Profile

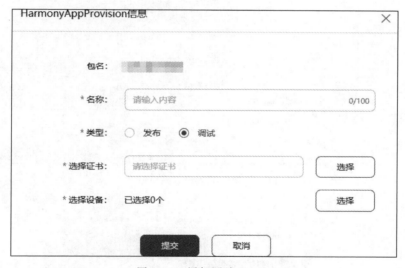

图 2.42　添加调试 Profile

**5. 构建类型为 Debug 的 HAP**

打开 File→Project Structure,可在 Modules→entry(模块名称)→Signing Configs→debug 窗口中配置指定模块的调试签名信息,如图 2.44 所示。

在主菜单栏,单击 Build→Build App(s)/HAP(s)→Build Debug HAP(s),生成已签名的 Debug HAP,如图 2.45 所示。

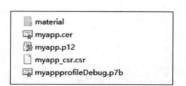

图 2.43　所有的证书文件列表

**6. 运行程序并在真机查看**

单击 Build Debug HAP(s),编译运行后可将签名好的 HAP 安装到连接的真机上,如图 2.46 所示。

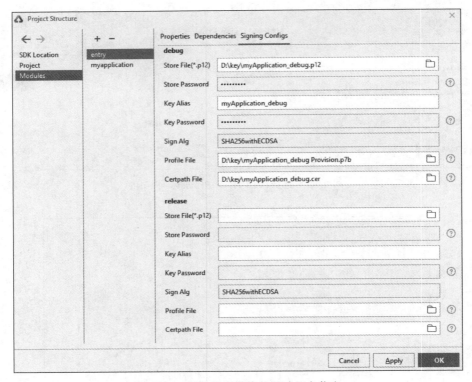

图 2.44　配置指定模块的调试签名信息

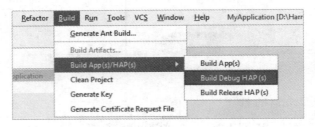

图 2.45　生成已签名的 Debug HAP

## 2.3.2　自动化真机签名流程

下面介绍如何通过 DevEco Studio 完成自动签名，具体步骤如下。

### 1. 连接真机设备

确保 DevEco Studio 与真机设备已连接，真机连接成功后如图 2.47 所示。

如果同时连接多个设备，则在使用自动化签名时会同时将这多个设备的信息写到证书文件中。

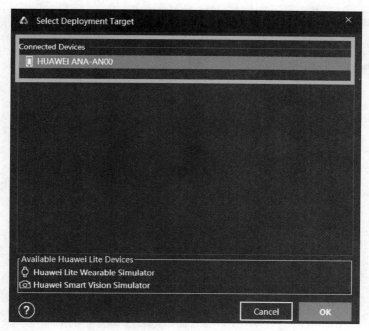

图 2.46　通过选择真机运行

图 2.47　真机连接成功效果图

### 2. 通过开发者账号登录

进入 File→Project Structure→Project→Signing Configs 界面，单击 Sign In 按钮进行登录，登录成功后获取认证和签名信息，如图 2.48 所示。

默认情况下 Automatically generate signing 是选中的，登录成功后，可以再次选择此按钮，IDE 会自动根据连接的真机获取签名文件信息。

### 3. 在 AppGallery Connect 中创建项目和应用

创建项目和应用的具体步骤如下。

(1) 登录 AppGallery Connect，创建一个项目。

(2) 在项目中，创建一个应用。如果是非实名认证的用户，则可单击左侧导航下方的"HAP Provision Profile 管理"界面的 HarmonyOS 应用按钮。

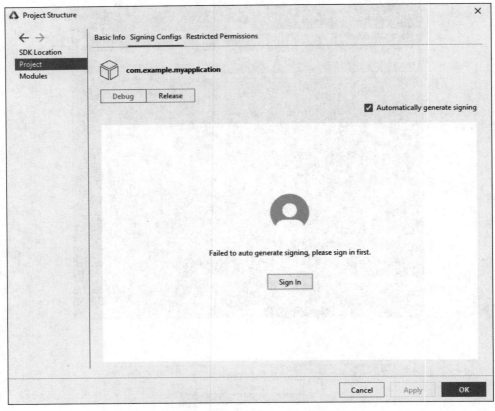

图 2.48　登录后获取认证信息

如果项目中没有应用,则可单击"添加应用"按钮进行创建,如图 2.49 所示。

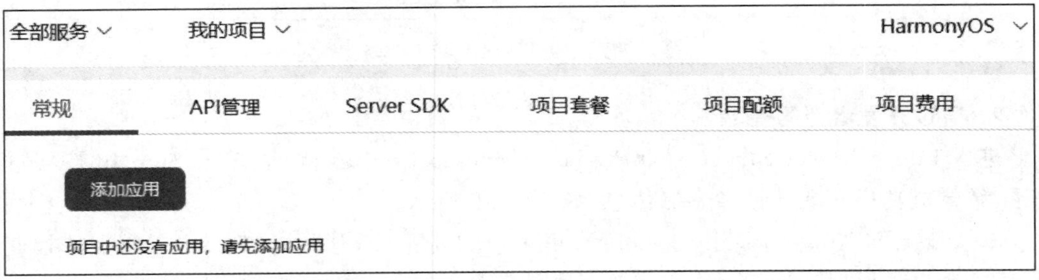

图 2.49　添加应用

如果项目中已有应用,则可展开顶部应用列表框,单击"添加应用"按钮,如图 2.50 所示。

(3) 需要填写的应用信息如下。
- 选择平台:选择 App(HarmonyOS 应用)。

图 2.50　单击"添加应用"按钮

- 支持设备：选择调试的设备类型。
- 应用包名：必须与 config.json 文件中的 bundleName 取值保持一致。
- 应用名称、应用分类、默认语言需要根据实际需要进行设置。

应用包名在 AppGallery Connect 上必须保持唯一，不能与其他应用包名（包含所有用户的包名）冲突，如图 2.51 所示。如果在创建应用时修改了该字段，则应在创建完成后同步修改工程中各模块对应的 bundleName 字段。

图 2.51　添加应用

### 4. 单击 Try Again 按钮即可自动进行签名

返回 DevEco Studio 的自动签名界面，单击 Try Again 按钮即可自动进行签名。自动生成签名所需的密钥(.p12)、数字证书(.cer)和 Profile 文件(.p7b)会存放到用户 user 目录下的 .ohos\config 目录下。

如果是非实名认证用户，则需要先接受 HUAWEI Developer Basic Service Agreement 协议，自动签名成功界面如图 2.52 所示。

设置完签名信息后，单击 OK 按钮进行保存，然后可以在工程下的 build.gradle 中查看签名的配置信息，如图 2.53 所示。

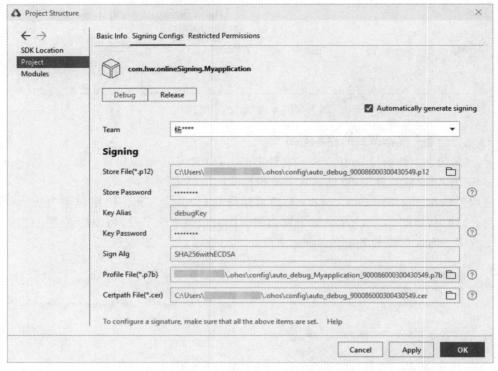

图 2.52　自动签名成功界面

```
ohos {
    signingConfigs { NamedDomainObjectContainer<SigningConfigOptions> it ->
        debug {
            storeFile file('C:\\Users\\          \\.ohos\\config\\auto_debug_900086000300430549.p12')
            storePassword '00000018DB216D38F4A178DB72DE2E7A3D5D4895587C6596DF858A22C72AD02AF255843D7878F826'
            keyAlias = 'debugKey'
            keyPassword '000000183B3E9BDC1B5D3CA0B1403B41FF876071F4C7E74C60F1F24ADA36A382EFE1309E3590E6DE'
            signAlg = 'SHA256withECDSA'
            profile file('C:\\Users\\          \\.ohos\\config\\auto_debug_Myapplication_900086000300430549.p7b')
            certpath file('C:\\Users\\          \\.ohos\\config\\auto_debug_900086000300430549.cer')
        }
    }
}
```

图 2.53　build.gradle 中保存的配置信息

## 2.4　本章小结

本章介绍了鸿蒙应用开发环境搭建、华为开发者账号申请、模拟器使用、真机投屏软件安装及华为真机设备测试证书申请。读者可根据本章的步骤完成自己的鸿蒙应用开发环境的搭建，为后面鸿蒙应用开发做好准备。

# 第二篇　ArkUI JS UI篇

针对不同业务场景，华为开发了两套 JS UI 框架，我们可以按框架推出的先后顺序把该 JS 框架编号为 ArkUI1 和 ArkUI2，也可以按框架开发风格的不同把 ArkUI1 称为 ArkUI JS 类 Web 范式 JS 框架，把 ArkUI2 称为 ArkUI 声明式 TypeScript(简写 TS)UI 框架。

由 ArkUI JS(1.0)框架开发的应用程序可运行在鸿蒙富设备和鸿蒙轻设备上，ArkUI JS 是目前唯一能够运行在 IoT 设备上的 UI 框架。

ArkUI ETS(2.0)框架更加适用于开发复杂业务逻辑，并且对性能要求高的场景下使用，在一些低端 IoT 设备上是无法运行的。

# 第 3 章 ArkUI JS 框架详细讲解

ArkUI JS UI 框架是专门为 JavaScript 开发工程师设计的一套跨设备、高性能的类 Web 应用开发框架，它支持声明式编程和跨设备多态 UI，采用组件化、数据驱动模式进行界面开发，极大地提高了 UI 的开发效率和界面逻辑的复用性。

对于前端开发人员来讲，该框架与前端目前流行的 Vue 2.0 框架的使用方式基本相同，熟悉 Vue.js 的前端开发人员基本可以零成本转到开发鸿蒙 JS 应用程序。

## 3.1　ArkUI JS 框架介绍

ArkUI JS 应用开发框架提供了一套跨平台的类 Web 应用开发框架，通过 Toolkit 将开发者编写的 HML、CSS 和 JS 文件编译并打包成 JS Bundle，解析运行 JS Bundle，生成 native UI View 组件树并进行渲染显示。通过支持第三方开发者使用声明式的 API 进行应用开发，以数据驱动视图变化，避免大量的视图操作，大大降低应用开发难度，提升开发者开发体验，如图 3.1 所示。

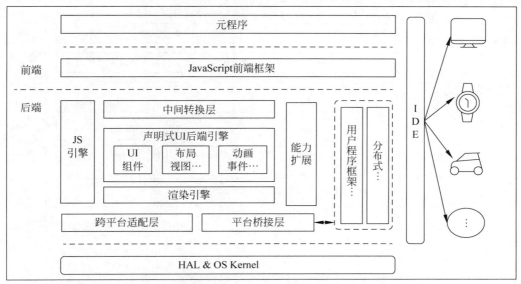

图 3.1　鸿蒙 ArkUI JS 框架

## 3.1.1 ArkUI JS 框架的特征

鸿蒙 ArkUI JS 框架的特征如下。

1. 声明式编程

鸿蒙 JS UI 框架采用类 HTML 和 CSS 声明式编程语言作为页面布局和页面样式的开发语言,页面业务逻辑则支持 ECMAScript 规范的 JavaScript 语言。JS UI 框架提供的声明式编程,可以让开发者避免编写 UI 状态切换的代码,视图配置信息更加直观。

2. 跨设备

开发框架架构上支持 UI 跨设备显示能力,运行时可自动映射到不同的设备类型,开发者无感知,从而降低了开发者多设备适配的成本。

3. 高性能

开发框架包含了许多核心的控件,如列表、图片和各类容器组件等,针对声明式语法进行了渲染流程的优化。

## 3.1.2 ArkUI JS 架构介绍

ArkUI JS UI 应用框架使用 JavaScript 语言实现了一套简单的数据劫持框架,通过数据劫持实现了界面上的组件与数据的分离,实现数据驱动式界面开发,如图 3.2 所示。

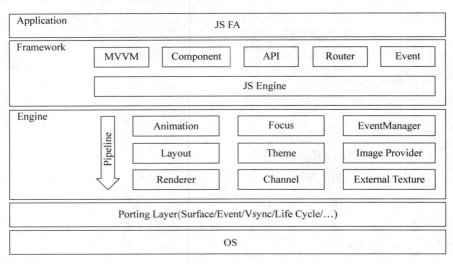

图 3.2 ArkUI JS 架构图

ArkUI JS 前端组件采用类似 Vue.js 2.0 框架设计模式,如图 3.3 所示。

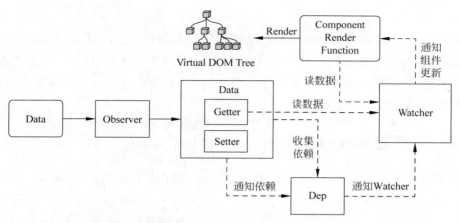

图 3.3　MVVM 模式图

## 3.1.3　ArkUI JS 运行流程

ArkUI JS 框架运行流程如图 3.4 所示。

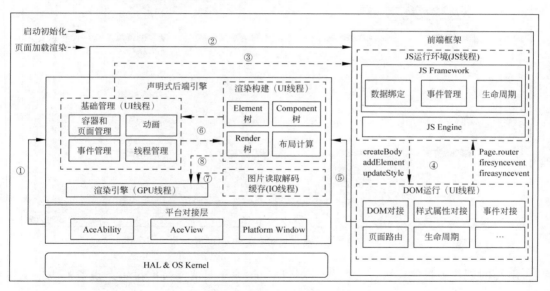

图 3.4　ArkUI JS 运行原理图

## 3.2　创建一个 ArkUI JS 项目

在对 HarmonyOS 应用程序有了一个初步认知之后，我们使用 DevEco Studio 来创建一个项目，把项目运行起来，先从整体上来了解一下 HarmonyOS 项目的整体结构及开发工具的基本使用。

环境和工具配置好后，就可以创建一个项目了。先创建一个项目，从整体上了解一下 HarmonyOS 应用的整体框架。我们在一个布局里放置一个文本框，用于显示一个数字，再添加一个按钮，每次单击按钮让文本框中的数字加 1。通过这样一个小程序简单地演示工具的使用和项目的基本框架。

### 3.2.1　新建 ArkUI JavaScript 项目

选择 File→New→New Project 进行项目创建，会弹出如图 3.5 所示的窗口。

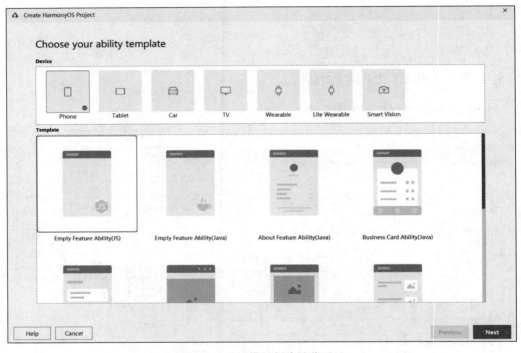

图 3.5　通过模板创建鸿蒙项目

上面的创建项目窗口可分为两块，其中 Device 表示目前支持的设备。设备列表中从左到右依次为手机、平板、车机、智慧屏、穿戴设备和轻型穿戴设备。

因为笔者使用 JavaScript 来开发，所以选择第 1 个，即 Empty Feature Ability(JS)，单击 Next 按钮进入下个页面，如图 3.6 所示。

配置项目名、包名、使用的 SDK 版本及项目的保存路径后，单击 Finish 按钮即可，创建完成后项目会自动构建，构建成功后项目的整体结构如图 3.7 所示。

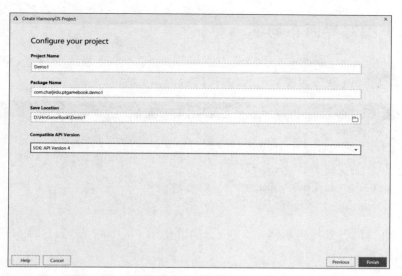

图 3.6　配置项目信息

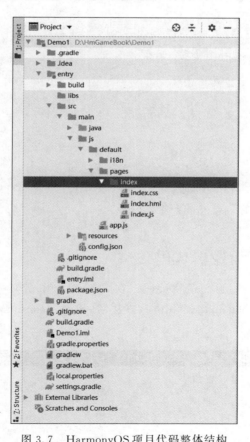

图 3.7　HarmonyOS 项目代码整体结构

## 3.2.2 编写界面布局

界面使用 div、text、button 3 个鸿蒙内置组件。div 表示垂直居中对齐,将 text 的字体设置为 100px,button 默认为胶囊形状,如代码示例 3.1 所示。

**代码示例 3.1 编写界面 HML　Demo1/index.hml**

```html
<div class = "container">
<text class = "title">
        {{num}}
</text>
<button @click = "updateNum">单击 + 1</button>
</div>
```

设置页面的样式,如代码示例 3.2 所示,这里使用 FlexBox 布局,这种布局也是默认的布局方式。

**代码示例 3.2 编写界面样式　Demo1/index.css**

```css
.container {
    flex-direction: column;
    justify-content: center;
    align-items: center;
}

.title {
    font-size: 100px;
}

button {
    width:200px;
    height:60px;
    background-color: cadetblue;
}
```

## 3.2.3 编写界面逻辑代码

在 index.js 文件中编写页面逻辑代码。在 data 对象中定义一个 num 属性,定义一个 updateNum()方法。当调用 updateNum()方法后,设置 this.num++,实现 num 自增加 1,如代码示例 3.3 所示。

**代码示例 3.3 编写界面逻辑代码　Demo1/index.js**

```javascript
export default {
    data: {
        num:0
    },
    onInit() {
```

```
    },
    updateNum(){
        this.num++
    }
}
```

现在可以单击 DevEco Studio 右边的 Preview 预览代码的效果,如图 3.8 所示。

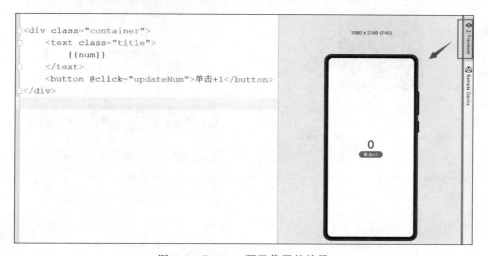

图 3.8　Preview 预览代码的效果

### 3.2.4　通过模拟器预览效果

如果需要使用不同的模拟器进行项目测试,则首先需要登录华为开发者账号。单击 Tools→DevEco Login→Sign in,如图 3.9 所示。

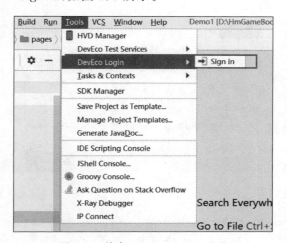

图 3.9　单击 DevEco Login 登录

单击 Sign in 按钮后,会自动弹出华为账号登录的页面,如图 3.10 所示。

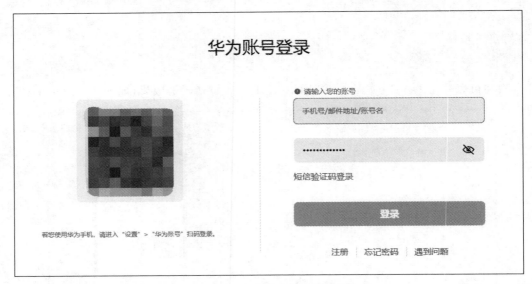

图 3.10 登录华为开发者账号

登录成功后,需要允许 HUAWEI DevEco Studio 访问你的华为账号,如图 3.11 所示。

图 3.11 允许 DevEco Studio 访问华为账号

单击"允许"按钮后，会弹出选择虚拟设备的列表，选择需要的模拟器，单击启动按钮▶，如图3.12所示。

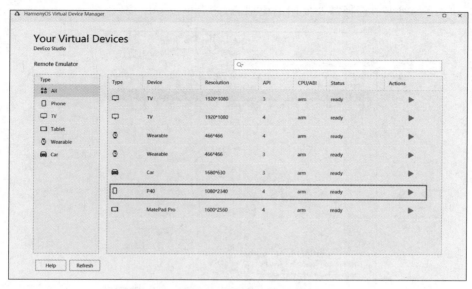

图3.12 选择模拟器

单击启动选择的模拟器后，模拟器的启动效果如图3.13所示。远程模拟器的使用时间是一小时，如果超过一小时，则会自动释放模拟器，如果需要继续使用模拟器，则需要在设备列表中重新启动模拟器。

图3.13 P40模拟器运行效果

启动模拟器后，单击菜单栏下面的启动按钮，如图 3.14 所示，单击启动按钮后，ide 便开始编译，编译成功后会自动打包项目并安装到模拟器上，这里如果有多个模拟器，则可以在已经连接的设备列表中选择需要使用的模拟器，然后单击 OK 按钮。

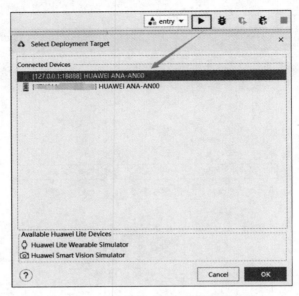

图 3.14　启动编译后打包安装鸿蒙应用

等待编译成功并打包上传后，就可以在模拟器中预览上面页面的显示效果了，如图 3.15 所示。

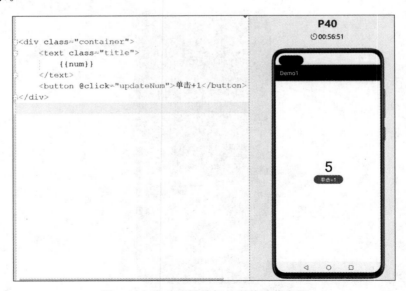

图 3.15　启动编译并打包安装鸿蒙应用

## 3.3 项目目录结构

本节介绍鸿蒙 ArkUI JS 应用开发的项目目录结构及其作用,以及配置文件的基础配置信息。

### 3.3.1 项目整体结构

我们通过 DevEco Studio 创建一个 ArkUI JS 项目后,可在 Project 左边栏中预览项目结构,如图 3.16 所示。

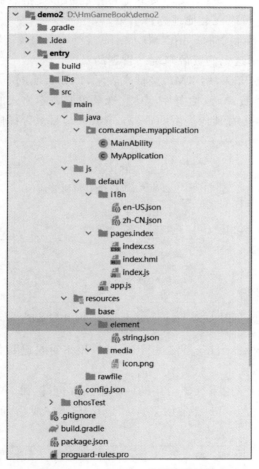

图 3.16　ArkUI JS 项目目录结构

首先有一个 entry 目录,一个应用是由一个或多个 HAP 包所组成的,HAP 包又可以分为 entry 类型和 feature 类型,每个 HAP 包由代码、资源、第三方库及应用配置文件组成,所以代码中的 entry 目录其实是一个应用的 HAP 包,它是 entry 类型的 HAP 包。接着来看这些资源和代码等分布在 entry 包的位置：

（1）在 src/main/java 下以包名命名的文件夹内分布着 Java 代码。这里的代码可以用来创建布局，动态调整布局及为交互提供支撑服务。

（2）js 目录下默认有一个 default 文件夹，每个 default 的同级文件夹都是一个 JS Component(JS 组件)，js 目录下面可以创建多个 JS Component，每个 JS Component 中可以创建多个 JS Page(JS 页面)。

（3）和 java 文件夹同级的 resources 目录下分布着应用资源，在该目录的 base 目录下按资源用途又分为多个文件夹资源。

① element：表示元素资源，该文件夹下主要存放 JSON 格式的文件，主要用来表示字符串、颜色值、布尔值等，可以在其他地方被引用。

② graphic：表示可绘制资源。用 XML 文件来表示，例如我们项目中设置的圆角按钮、按钮颜色等都是通过引用这里的资源来统一管理的。

③ layout：表示布局资源，用 XML 文件来表示，例如页面的布局资源都放在这里。

④ media：表示媒体资源，包括图片、声频、视频等非文本格式的文件。

除了上述的这四类，还有其他类型的资源，resources 目录存储的内容，如图 3.17 所示。

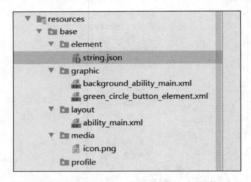

图 3.17　resources 目录存储的内容截图

和 main 目录平级的 test 目录是测试目录，可以用于对自己写的功能添加单元测试，确保代码的正确性。

和 src 平级的 libs 目录用来存储引用的一些第三方包，例如 JAR 包和 SO 包等。

和 entry 目录平级的 build 目录用来存放最终编译完成后的包，也就是 HAP 包，如图 3.18 所示。

图 3.18　HAP 包

最终在该目录下会生成一个 HAP 包。这个 HAP 包中包含了项目中用到的图片、布局、代码和各种资源。

## 3.3.2 项目的配置文件

每个 HAP 包下都包含了该 HAP 包的配置信息,这个配置文件位于 entry/src/main/ 目录下,由工具帮我们生成,命名为 config.json,HarmonyOS 应用配置采用 JSON 格式的形式。下面来看一下这个配置文件中的内容,并简要介绍一下配置的作用。在该配置文件中,主要有 3 个模块,如图 3.19 所示。

```
{
    "app": {"bundleName": "com.example.demo"...},
    "deviceConfig": {},
    "module": {"name": ".MyApplication"...}
}
```

图 3.19 项目的配置文件

(1) app:表示应用的全局配置信息。同一个应用的不同 HAP 包的 app 配置必须保持一致。

(2) deviceConfig:表示应用在具体设备上的配置信息。

(3) module:表示 HAP 包的配置信息。该标签下的配置只对当前 HAP 包生效。

配置文件采用 JSON 格式,其中的属性不分先后顺序,每个属性只允许出现一次。

下面具体看一下项目中出现的配置项都有哪些,以及它们的作用。

图 3.20 app 属性配置

### 1. app 的属性

app 的属性配置如图 3.20 所示,包含应用的包名、开发应用的厂商、版本等信息。

app 的属性的详细说明如表 3.1 所示。

表 3.1 app 的属性的详细说明

| 属 性 | 作 用 |
| --- | --- |
| bundleName | 表示应用的包名,用于标识应用的唯一性。通常采用反转的域名 |
| vendor | 表示开发应用的厂商 |
| version | code 表示内部版本号,用于系统管理版本,对用户不可见,name 表示应用的版本号,用于向用户呈现 |

续表

| 属 性 | 作 用 |
| --- | --- |
| apiVersion | apiVersion：包含 3 个选项。<br>（1）compatible：表示应用运行需要的 API 的最低版本。<br>（2）target：表示应用运行需要的 API 的目标版本。<br>（3）releaseType：表示应用运行需要的 API 目标版本的类型，取值为 CanaryN、BetaN 或者 Release，其中，N 代表大于零的整数。<br>Canary：受限发布的版本<br>Beta：公开发布的 Beta 版本<br>Release：公开发布的正式版本 |
| deviceConfig | 表示应用在具体设备上的配置信息 |

### 2. module 的属性

module 配置项属性表示 HAP 包的配置信息，该标签下的配置只对当前 HAP 包生效，如图 3.21 所示。

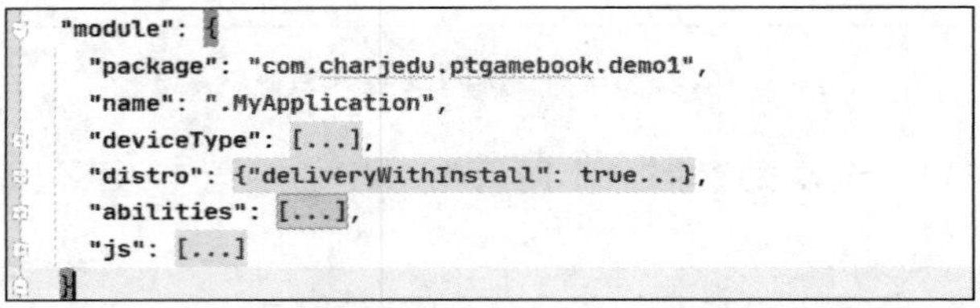

```
"module": {
    "package": "com.charjedu.ptgamebook.demo1",
    "name": ".MyApplication",
    "deviceType": [...],
    "distro": {"deliveryWithInstall": true...},
    "abilities": [...],
    "js": [...]
}
```

图 3.21　module 的属性配置

module 的属性的详细说明如表 3.2 所示。

表 3.2　module 的属性的详细说明

| 属 性 | 作 用 |
| --- | --- |
| package | "com.example.demo"，<br>表示 HAP 包的结构名称，在应用内应保证唯一性。采用反向域名格式 |
| name | "com.example.demo"，<br>表示 HAP 包的结构名称，在应用内应保证唯一性。采用反向域名格式 |
| deviceType | ["phone"]<br>表示允许 Ability 运行的设备类型。phone 表示手机 |
| distro | {<br>"deliveryWithInstall": true,<br>　　//表示当前 HAP 包是否支持随应用安装。true 为支持随应用安装<br>"moduleName": "entry",<br>　　//表示当前 HAP 包的名称<br>"moduleType": "entry"<br>　　//表示当前 HAP 包的类型，包括两种类型 entry 和 feature<br>}<br>表示 HAP 包发布的具体描述 |

续表

| 属 性 | 作 用 |
|---|---|
| abilities | ```<br>//表示当前模块内的所有 Ability<br>[<br>    {<br>      "skills": [<br>            {<br>"entities": ["entity.system.home"],<br>"actions": ["action.system.home"]<br>            }<br>        ],<br>"orientation": "unspecified",<br>"name": "com.example.demo.MainAbility",<br>"icon": "$media:icon",<br>"description": "$string:mainability_description",<br>"label": "first_demo",<br>"type": "page",<br>"launchType": "standard"<br>    }<br>]<br>``` |
| js | ```<br>[<br>    {<br>        //注册的所有页面,第1个为首页<br>"pages": [<br>"pages/index/index"<br>        ],<br>        //js component 的名称<br>"name": "default",<br>        //设计稿的参考设置<br>"window": {<br>"designWidth": 720,<br>"autoDesignWidth": false<br>        }<br>    }<br>]<br>``` |

### 3. module 下 abilities 的属性

module 下 abilities 的属性值是一个数组,表示当前模块内的所有 Ability 的配置,如图 3.22 所示。

```json
"abilities": [
  {
    "skills": [
      {
        "entities": [
          "entity.system.home"
        ],
        "actions": [
          "action.system.home"
        ]
      }
    ],
    "name": "com.example.myapplication.MainAbility",
    "icon": "$media:icon",
    "description": JS_Phone_Empty Feature Ability,
    "label": MyApplication,
    "type": "page",
    "launchType": "standard"
  }
],
```

图 3.22  abilities 的属性配置

每个 ability 对应一个配置对象，属性的详细说明如表 3.3 所示。

表 3.3  module 下 abilities 的属性

| 属性 | 作用 |
| --- | --- |
| skills | [<br>{<br>"entities": [<br>//表示能够接收的 Intent 的 Ability 的类别(如视频、桌面应用等)，可以包含一<br>//个或多个 entity<br>"entity.system.home"<br>],<br>"actions": [<br>"action.system.home"<br>//表示能够接收的 Intent 的 action 值，可以包含一个或多个 action<br>]<br>}<br>],<br>表示 Ability 能够接收的 Intent 的特征 |
| orientation | 表示该 Ability 的显示模式，这里表示由系统自动判断方向 |
| name | 表示 Ability 名称。取值可采用反向域名方式表示，由包名和类名组成，也可以用"."开头的形式表示 |
| icon | 表示 Ability 图标资源文件的索引，$ media 表示引用 media 目录下的 icon 资源 |
| description | 表示对 Ability 的描述 |
| label | 表示 Ability 对用户显示的名称，也就是你的应用安装用户设备后显示的名称 |

续表

| 属性 | 作用 |
|---|---|
| type | 表示 Ability 的 Type 类型,可以为 page、service 或 data |
| launchType | 表示 Ability 的启动模式,支持 standard 和 singleton 两种模式,standard 表示可以有多个实例,singleton 则表示只有一个实例 |

**4. 设置 Ability 配置不同的主题**

可以为每个 Ability 配置不同的主题,在 abilities 数组中的每个 Ability 的配置后面可以添加一个 metaData 项,在 value 中可以设置不同的主题。例如添加图 3.23 中的主题后,页面就不会显示头部横条了,如图 3.23 所示。

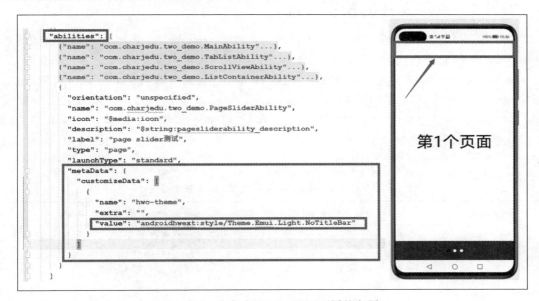

图 3.23 为每个 Ability 配置不同的主题

代码如下:

```
"metaData" :{
"customizeData":[
    {
"extra": "",
"value":"androidhwext:style/Theme.Emui.Light.NoTitleBar"
    }
  ]
}
```

也可以使用不同的主题,可以选择的部分主题如图 3.24 所示。

```
可以选择的部分主题风格：
androidhwext:style/Theme.Emui.Dialog                        // 对话框模式
androidhwext:style/Theme.Emui.NoTitleBar                    // 不显示应用程序标题栏
androidhwext:style/Theme.Emui.Light                         // 背景为白色
androidhwext:style/Theme.Emui.Light.NoTitleBar              // 白色背景并无标题栏
androidhwext:style/Theme.Emui.Light.NoTitleBar.Fullscreen   // 白色背景，无标题栏，全屏
androidhwext:style/Theme.Emui.Translucent                   // 透明效果
androidhwext:style/Theme.Emui.Translucent.NoTitleBar        // 半透明并无标题栏
androidhwext:style/Theme.Emui.Panel                         // 面板风格显示
androidhwext:style/Theme.Emui.Light.Panel                   // 平板风格显示
```

图 3.24 可以选择的部分主题

### 3.3.3 资源文件的使用方式

应用的资源文件（字符串、图片、声频等）统一存放于 resources 目录下，便于开发者使用和维护。resources 目录包括两大类目录，一类为 base 目录与限定词目录，另一类为 rawfile 目录。

base 目录下面可以创建资源组目录，包括 element（元素资源）、media（媒体资源，包括图片、声频、视频等非文本格式的文件）、animation（动画资源）、layout、graphic（图形）、profile（表示其他类型文件，以原始文件形式保存）。目录中的资源文件除 profile 目录中的文件外会被编译成二进制文件，并赋予资源文件 ID，可以用 ResourceTable 引用，如表 3.4 所示。

表 3.4 element（元素资源）详细表

| 资源组目录 | 目录说明 | 资源文件 |
| --- | --- | --- |
| element | 表示元素资源，以下每一类数据都采用相应的 JSON 文件来表征<br>boolean：布尔型<br>color：颜色<br>float：浮点型<br>intarray：整型数组<br>integer：整型<br>pattern：样式<br>plural：复数形式<br>strarray：字符串数组<br>string：字符串 | element 目录中的文件名称建议与下面的文件名保持一致。每个文件中只能包含同一类型的数据<br>boolean.json<br>color.json<br>float.json<br>intarray.json<br>integer.json<br>pattern.json<br>plural.json<br>strarray.json<br>string.json |

限定词目录需要开发者自行创建。目录名称由一个或多个表征应用场景或设备特征的限定词组合而成，包括语言、文字、国家或地区、横竖屏、设备类型和屏幕密度等 6 个维度，限定词之间通过下画线（_）或者半字线（-）连接。开发者在创建限定词目录时，需要掌握限定词目录的命名要求及与限定词目录、设备状态的匹配规则。

## 1. 限定词目录的命名要求

限定词的组合顺序：语言\_文字\_国家或地区-横竖屏-设备类型-屏幕密度。开发者可以根据应用的使用场景和设备特征，选择其中的一类或几类限定词组成目录名称。

限定词的连接方式：语言、文字、国家或地区之间采用下画线(\_)连接，除此之外的其他限定词之间均采用半字线(-)连接。例如：zh\_Hant\_CN、zh\_CN-car-ldpi。

限定词的取值范围：每类限定词的取值必须符合表3.5中的条件，否则将无法匹配目录中的资源文件。

表 3.5 每类限定词的取值范围

| 限定词类型 | 含义与取值说明 |
| --- | --- |
| 语言 | 表示设备使用的语言类型，由两个小写字母组成。例如：zh 表示中文，en 表示英语。<br>详细取值范围，参见 ISO 639-1(ISO 制定的语言编码标准) |
| 文字 | 表示设备使用的文字类型，由 1 个大写字母(首字母)和 3 个小写字母组成。例如：Hans 表示简体中文，Hant 表示繁体中文。详细取值范围，参见 ISO 15924(ISO 制定的文字编码标准) |
| 国家或地区 | 表示用户所在的国家或地区，由 2～3 个大写字母或者 3 个数字组成。例如：CN 表示中国，GB 表示英国。<br>详细取值范围，参见 ISO 3166-1(ISO 制定的国家和地区编码标准) |
| 横竖屏 | 表示设备的屏幕方向，取值如下。<br>    vertical：竖屏<br>    horizontal：横屏 |
| 设备类型 | 表示设备的类型，取值如下。<br>    car：车机<br>    tv：智慧屏<br>    wearable：智能穿戴 |
| 屏幕密度 | 表示设备的屏幕密度(单位为 dpi)，取值如下。<br>    sdpi：表示小规模的屏幕密度(Small-scale Dots Per Inch)，适用于 dpi 取值为(0, 120]的设备。<br>    mdpi：表示中规模的屏幕密度(Medium-scale Dots Per Inch)，适用于 dpi 取值为(120, 160]的设备。<br>    ldpi：表示大规模的屏幕密度(Large-scale Dots Per Inch)，适用于 dpi 取值为(160, 240]的设备。<br>    xldpi：表示特大规模的屏幕密度(Extra Large-scale Dots Per Inch)，适用于 dpi 取值为(240, 320]的设备。<br>    xxldpi：表示超大规模的屏幕密度(Extra Extra Large-scale Dots Per Inch)，适用于 dpi 取值为(320, 480]的设备。<br>    xxxldpi：表示超特大规模的屏幕密度(Extra Extra Extra Large-scale Dots Per Inch)，适用于 dpi 取值为(480, 640]的设备 |

rawfile 目录支持创建多层子目录，目录名称可以自定义，文件夹内可以自由放置各类资源文件。rawfile 目录的文件不会根据设备状态去匹配不同的资源。目录中的资源文件会被直接打包进应用，而不经过编译，并且不可以用 ResourceTable 引用。

**2. 资源文件的使用方法**

（1）base 目录与限定词目录中的资源文件可通过指定资源类型（type）和资源名称（name）来引用。

通过 Java 引用资源文件的不同方式如下。

- 普通资源引用：ResourceTable.type_name。
- 系统资源引用：ohos.global.systemres.ResourceTable.type_name。
- 引用 string.json 文件中类型为 String、名称为 app_name 的资源。
- 引用 color.json 文件中类型为 Color、名称为 red 的资源。

代码如下：

```
ResourceManager resourceManager = getResourceManager();
resourceManager.getElement(ResourceTable.String_app_name).getString();
resourceManager.getElement(ResourceTable.Color_red).getColor();
```

XML 文件引用资源文件的不同方式如下。

- 资源文件：$type:name。
- 系统资源：$ohos:type:name。
- 引用 string.json 文件中类型为 String、名称为 app_name 的资源。

代码如下：

```
<?xml version = "1.0" encoding = "utf-8"?>
<DirectionalLayout xmlns:ohos = "http://schemas.huawei.com/res/ohos"
    ohos:width = "match_parent"
    ohos:height = "match_parent"
    ohos:orientation = "vertical">
<Text ohos:text = " $ string:app_name"/>
</DirectionalLayout>
```

（2）rawfile 目录中的资源文件可通过指定文件路径和文件名称来引用。

在 Java 文件中，引用一个路径为 resources/rawfile/、名称为 example.js 的资源文件，代码如下：

```
Resource resource = null;
String rawfileUrl = "resources/rawfile/example.js"
try {
    resource = getResourceManager().getRawFileEntry(rawfileUrl).openRawFile();
    InputStreamReader inputStreamReader = new InputStreamReader(resource,"utf-8");
    BufferedReader bufferedReader = new BufferedReader(inputStreamReader);
    String lineTxt = "";
    while ((lineTxt = bufferedReader.readLine()) != null){
```

```
            System.out.println(lineTxt);
        }
    }catch (Exception e){}
```

## 3.4 页面布局

ArkUI JS UI 目前支持 Flexbox 与 Grid 两种布局方式,Flexbox 是默认容器组件的布局方式。

### 3.4.1 Flexbox 布局

Flex 容器是一种当页面需要适应不同的屏幕大小及设备类型时确保元素拥有恰当的行为的布局方式。引入弹性盒布局模型的目的是提供一种更加有效的方式来对一个容器中的子元素进行排列、对齐和分配空白空间。

**1. Flex 容器介绍**

Flex 容器布局中有两个重要的概念:Flex 容器和 Flex 项目,如图 3.25 所示。在 Flex 容器中默认存在两条轴,水平主轴(Main Axis)和垂直的交叉轴(Cross Axis),这是默认的设置,可以通过修改使垂直方向变为主轴,使水平方向变为交叉轴。

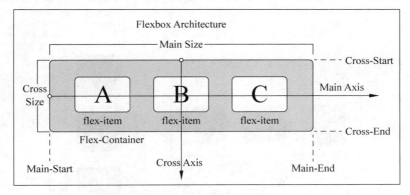

图 3.25 Flex 容器效果图

在容器中的每个单元块被称为 Flex-item,每个项目占据的主轴空间为(Main Size),占据的交叉轴的空间为(Cross Size)。

**2. Flex 容器属性**

鸿蒙并没有完全支持所有 Flex 容器的所有属性,Flex 容器的属性分为两类:父容器的属性和子项的属性,所有的属性值几乎和排列、对齐、占地面积 3 类特性相关,如表 3.6 和表 3.7 所示。

表 3.6　鸿蒙支持的父容器属性

| 属性 | 作用 | 特性分类 |
| --- | --- | --- |
| flex-direction | 定义子项在容器内的排列方向 | 排列 |
| flex-wrap | 定义子项在容器内的换行效果 | 排列 |
| justify-content | 定义子项在容器内以水平方式对齐 | 对齐 |
| align-items | 定义子项在容器内以垂直方式对齐 | 对齐 |
| align-content | 定义多行子项在容器内整体以垂直方式对齐 | 对齐 |

表 3.7　鸿蒙支持的子容器属性

| 属性 | 作用 | 特性分类 |
| --- | --- | --- |
| flex-grow | 当子项宽度之和不足父元素宽度时,定义子项拉伸的比例 | 占用面积 |
| flex-shrink | 当子项宽度之和超过父元素宽度时,定义子项缩放的比例 | 占用面积 |
| flex-basis | 定义子项的初始宽度,若子项宽度之和超过父元素宽度,则子项按照 flex-basis 的比例缩放 | 占用面积 |

首先,介绍父容器属性的特点和使用方法:

**1) flex-direction 属性**

flex-direction 属性决定主轴的方向(项目的排列方向),目前鸿蒙只支持 row 和 column 两个值,不支持 row-reverse 和 column-reverse,如图 3.26 所示。

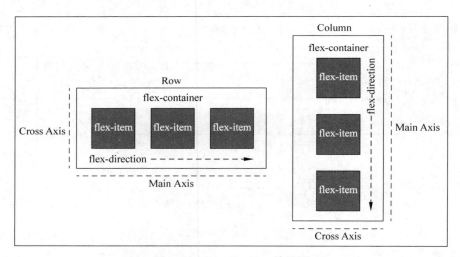

图 3.26　flex-direction 效果图

代码如下:

```
.flex-container { display: flex; flex-direction: column; }
.flex-item { width: 100px; margin: 10px; }
```

## 2) flex-wrap 属性

默认情况下,项目都排在一条线(又称轴线)上。flex-wrap 属性用于控制容器内元素如何换行。目前鸿蒙 flex-wrap 的值只支持 wrap(换行)和 nowrap(不换行),如图 3.27 所示。

代码如下:

```
.flex - container { display: flex; flex - wrap: wrap; }
.flex - item { width: 100px; margin: 10px; }
```

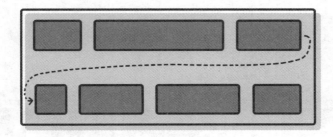

图 3.27　将 flex-wrap 设置为 wrap 效果

## 3) justify-content 属性

justify-content 定义元素在主轴上的对齐方式,有 5 个值如下面代码所示,每个值的设置效果如图 3.28 所示。

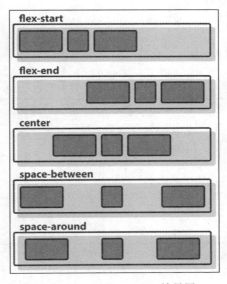

图 3.28　justify-content 效果图

代码如下：

```
.container {
    justify-content: flex-start | flex-end | center | space-between | space-around;
}
```

**4）align-items 属性**

align-items 定义元素在交叉轴上的对齐方式，有 5 个值如下面代码所示，每个值的设置效果如图 3.29 所示。

代码如下：

```
.container {
    align-items: stretch | flex-start | flex-end | center | baseline;
}
```

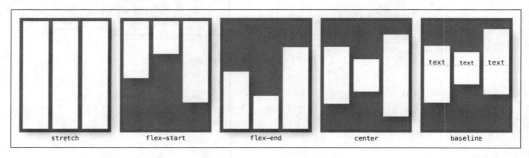

图 3.29 align-items 效果图

align-items 属性的作用如表 3.8 所示。

表 3.8 align-items 属性的作用

| 属性 | 作用 |
| --- | --- |
| stretch | 默认值，如果项目未设置高度或设置为 auto，将占满整个容器的高度（最高的元素定义了容器的高度） |
| flex-start | 对齐交叉轴的起始端（cross-start） |
| flex-end | 对齐交叉轴的结束端（cross-end） |
| center | 沿交叉轴居中对齐 |
| baseline | 以项目的第一行文字的基线对齐 |

**5）align-content 属性**

align-content 定义多行元素在垂直方向上的对齐方式，有 6 个值如下面代码所示，每个值的对齐效果如图 3.30 所示。

代码如下:

```
.container {
  align-content: stretch | flex-start | flex-end | center |
                 space-between | space-around;
}
```

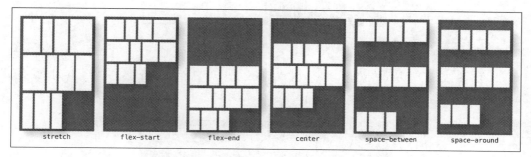

图 3.30　align-content 属性效果

align-content 属性的作用如表 3.9 所示。

表 3.9　align-content 属性的作用

| 属　　性 | 作　　用 |
| --- | --- |
| stretch | 默认值,每行拉伸占满整个交叉轴 |
| flex-start | 对齐交叉轴的起始端(cross-start) |
| flex-end | 对齐交叉轴的结束端(cross-end) |
| center | 沿交叉轴居中对齐 |
| space-between | 与交叉轴两端对齐,轴线之间的间隔平均分布 |
| space-around | 每一行上下两端的间隔都相等 |

接下来,介绍子容器属性的特点和使用方法:

**1) flex-grow 属性**

子元素宽度的和小于、大于父元素的宽度,分别对应了 flex-grow 和 flex-shrink 属性生效的情况,即当子元素的宽度的和小于父元素的宽度值时 flex-grow 生效,反之 flex-shrink 生效。

此属性用于定义子容器的伸缩比例,将剩余空间按照该比例分配给子容器,代码如下:

```
.item{
    flex-grow: <number>; /* default 0 */
}
/默认 0(值为 0 时,如果有剩余空间也不放大。值为 1 时放大,值为 2 时是 1 的双倍大小,以此类推)*/
```

将父元素宽度设置为200px,将红色、蓝色div宽度设置为50px。

(1) 如图3.31所示,此图为原始状态,flex-grow的默认值为0,都不分配剩余空间,红色、蓝色div的宽度都为50px。

图3.31 flex-grow属性效果(1)

(2) 如图3.32所示,将蓝色div设置为flex-grow:1后,其分配了剩余空间,红色div宽度还是50px,蓝色div为剩余的100px宽度,宽度变为了150px。

图3.32 flex-grow属性效果(2)

(3) 如图3.33所示,将红色div设置为flex-grow:1,将蓝色div设置为flex-grow:2,两个div按比例1:2分配了剩余空间,红色div分得的宽度为100*1/(1+2)=33.33px,蓝色div分得100*2/(1+2)=66.67px,原始宽度加上分配剩余空间得到的宽度后,红色、蓝色div的宽度变为了83.33px和116.67px。

图3.33 flex-grow属性效果(3)

2) flex-shrink

此属性定义了子容器弹性收缩的比例。把各自的空间按比例分配出去。flex-shrink的默认值为1,当flex-shrink的值为0时,不缩放,代码如下:

```
.item{
    flex - shrink: <number>; /* default 1 */
}
/*默认值为1(如果空间不足,则会缩小,值为0时不缩小)*/
```

将父元素宽度设置为200px,将红色、蓝色div宽度设置为150px(子元素宽度的和为300px,大于父元素宽度200px)。

（1）如图3.34所示，此图为原始状态，flex-shrink默认值为1，红色和蓝色div按1∶1的比例让出自己的宽度(50px)后，两个div的宽度都为100px。

图3.34　flex-shrink属性效果(1)

（2）如图3.35所示，将蓝色div设置为flex-shrink：0后，其不出让自己的空间，红色div让出空间100px，宽度变为50px。

图3.35　flex-shrink属性效果(2)

（3）如图3.36所示，将红色div设置为flex-shrink：1，将蓝色div设置为flex-shrink：2，红色和蓝色div按1∶2的比例让出自己的宽度，红色div让出100/3＝33.33px，蓝色div让出100×2/3＝66.67px，原始宽度减去让出宽度后红色、蓝色div的宽度变为了66.67px、33.33px。

图3.36　flex-shrink属性效果(3)

### 3）flex-basis属性

此属性定义了项目占据的主轴空间，规定的是子元素的基准值。浏览器会根据这个属性计算主轴多余空间或不足空间的大小。它的默认值为auto，即项目的本来大小。取值可以为绝对单位或百分比(可用于栅格效果)，代码如下：

```
.item {
    flex-basis: <length> | auto; /* default auto */
}
/*默认为auto,可以设置px值,也可以设置百分比大小*/
```

项目的原始长度由flex-basis和width属性控制。flex-basis的优先级高于width属性，如果只设置了width属性，则flex-basis为auto，此时项目的原始长度等于width，而如果同时设置了width和flex-basis，则项目的原始长度等于flex-basis，因此用到width的地方可以使用 flex-basis 来代替。flex-basis/width 影响的元素是 flex-grow 和 flex-shrink。

### 4）flex 属性

规定当前组件如何适应父组件中的可用空间。第 1 个值必须是无单位数，用来设置组件的 flex-grow；第 2 个值必须是无单位数，用来设置组件的 flex-shrink；第 3 个值必须是一个有效的宽度值，用来设置组件的 flex-basis，如图 3.37 所示。

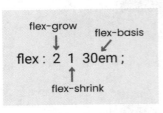

图 3.37　flex 属性

### 3. Flex 容器布局案例

通过 Flex 容器实现以下效果，如图 3.38 所示。下面的效果图从上到下可以分为 3 个模块，都可以使用 Flex 容器布局的方式实现。

这里首先把页面从上到下分为 3 个模块：头部、中间部分、底部。通过设置垂直方向弹性 Flex 布局实现，如图 3.39 所示。

图 3.38　效果图

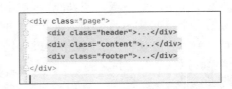

图 3.39　页面结构

首先对页面结构的样式进行设置，代码如下：

```
.page {
    width: 100%;
```

```
    height: 100%;
    background-color: red;
    flex-direction: column;
}
```

有了页面结构后,接下来可以实现头部搜索栏的效果,如图3.40所示。

图 3.40 头部搜索栏

这里使用 Flex 横向布局的方式实现,中间的文字、下拉箭头和文本搜索框分别按水平、垂直轴方向水平和垂直居中显示,如代码示例3.4所示。

代码示例3.4 头部搜索栏

```
<div class = "header">
<div class = "city">
<text>
<span>北京</span>
</text>
<image src = "/common/images/xjt.png"></image>
</div>
<div class = "search">
<image src = "/common/images/search.png"></image>
<text>全城热搜:袁家村</text>
</div>
</div>
```

接下来对头部搜索栏样式进行设置如代码示例3.5所示。

代码示例3.5 头部搜索栏样式设置

```
.header {
    width: 100%;
    height: 60px;
    background-color: #ff6634;
    justify-content: center;
    align-items: center;
    flex-direction: row;
}

.header > .city {
    flex:1;
    justify-content: center;
    align-items: center;
    flex-direction: row;
    text-align: center;
    margin-left: 3px;
```

```css
}
.city image {
    width: 20px;
    margin-top: 5px;
    margin-right: 10px;
    object-fit: contain;
}

.city text {
    font-size: 20px;
    color: #fff;
}

.header>.search {
    flex:2;
    height: 35px;
    border-radius: 20px;
    background-color: #fff;
    justify-content: flex-start;
    align-items: center;
    flex-direction: row;
    text-align: center;
    margin-right: 30px;
}

.search image {
    margin-left: 10px;
    margin-right: 10px;
    width: 20px;
    object-fit: contain;
}

.search text {
    font-size: 16px;
}
```

接下来,实现中间部分,中间部分有两部分内容,2 行 5 列的菜单导航和 2 行 3 列的图文列表,这两部分可通过 flex-wrap 的设置自动实现换行效果,如图 3.41 所示。

图 3.41  2 行 5 列的菜单导航

中间部分的页面结构如代码示例 3.6 所示。

**代码示例 3.6　中间部分实现**

```
< div class = "content">
< div class = "container">
< div class = "item">
< image src = "/common/images/nav01.png"></image>
< text >
                    美食
</text >
</div >
< div class = "item">
< image src = "/common/images/nav02.png"></image>
< text >
                    电影/演出
</text >
</div >
< div class = "item">
< image src = "/common/images/nav03.png"></image>
< text >
                    酒店
</text >
</div >
< div class = "item">
< image src = "/common/images/nav04.png"></image>
< text >
                    休闲娱乐
</text >
</div >
< div class = "item">
< image src = "/common/images/nav05.png"></image>
< text >
                    外卖
</text >
</div >
< div class = "item">
< image src = "/common/images/nav06.png"></image>
< text >
                    婚纱摄影
</text >
</div >
< div class = "item">
< image src = "/common/images/nav07.png"></image>
< text >
                    丽人/美发
</text >
</div >
< div class = "item">
< image src = "/common/images/nav08.png"></image>
```

```html
            <text>
                            周边游
            </text>
        </div>
        <div class="item">
            <image src="/common/images/nav09.png"></image>
            <text>
                            KTV
            </text>
        </div>
        <div class="item">
            <image src="/common/images/nav10.png"></image>
            <text>
                            火锅
            </text>
        </div>
    </div>
    <div class="best">
        <div class="best-item">
            <image src="/common/best/best01.png"></image>
            <div class="title">
                <text>
                            元气人气美食
                </text>
            </div>
            <div class="sub-title">
                <text>
                            吃完活力一天
                </text>
            </div>
        </div>
        <div class="best-item">
            <image src="/common/best/best02.png"></image>
            <div class="title">
                <text>
                            外卖券大放送
                </text>
            </div>
            <div class="sub-title">
                <text>
                            测一测领红包
                </text>
            </div>
        </div>
        <div class="best-item">
            <image src="/common/best/best03.png"></image>
            <div class="title">
                <text>
```

```html
                9.9元起
            </text>
        </div>
        <div class="sub-title">
            <text>
                热门喜剧大片
            </text>
        </div>
    </div>
    <div class="best-item">
        <image src="/common/best/best04.png"></image>
        <div class="title">
            <text>
                小黄人再归来
            </text>
        </div>
        <div class="sub-title">
            <text>
                神偷奶爸3
            </text>
        </div>
    </div>
    <div class="best-item">
        <image src="/common/best/best05.png"></image>
        <div class="title">
            <text>
                躺着吃小龙虾
            </text>
        </div>
        <div class="sub-title">
            <text>
                简直爽爆了
            </text>
        </div>
    </div>
    <div class="best-item">
        <image src="/common/best/best06.png"></image>
        <div class="title">
            <text>
                萌娃喜欢的
            </text>
        </div>
        <div class="sub-title">
            <text>
                轻奢亲子酒店
            </text>
        </div>
    </div>
</div>
</div>
```

对中间部分样式进行设置，如代码示例3.7所示。

代码示例3.7 中间样式实现

```css
.content {
    flex: 1;
    width: 100%;
    background-color: #ccc;
    flex-direction: column;
}

.content > .container {
    display: flex;
    flex-direction: row;
    justify-content: space-around;
    align-items: center;
    flex-wrap: wrap;
    width: 100%;
    background-color: #fff;
    padding: 5px;
}

.content > .best {
    margin-top: 6px;
    background-color: #fff;
    width: 100%;
    flex-direction: row;
    flex-wrap: wrap;
    justify-content: space-between;
    align-items: center;
}

.best .best-item {
    width: 120px;
    height: 140px;
    flex-direction: column;
    justify-content: space-between;
    align-items: center;
    padding: 6px;
}

.best-item image {
    object-fit: cover;
}

.best-item .title {
    display: flex;
    height: 60px;
    text-align: center;
    justify-content: center;
```

```css
    align - items: center;
}

.best - item .title text {
    font - size: 16px;
}

.best - item .sub - title {
    display: flex;
    height: 40px;
    text - align: center;
    justify - content: center;
    align - items: center;
}

.best - item .sub - title text {
    font - size: 14px;
}

.container .item {
    flex: 1;
    height: 90px;
    display: flex;
    justify - content: center;
    align - items: center;
    flex - direction: column;
}

.item image {
    width: 60px;
    height: 60px;
    border - radius: 30px;
}

.item text {
    margin - top: 5px;
    font - size: 14px;
}
```

最后，实现底部的菜单导航，如图 3.42 所示。

图 3.42　底部的菜单导航

实现底部菜单导航页面结构如代码示例3.8所示。

代码示例3.8 底部菜单实现

```html
<div class = "footer">
<div class = "tab-item">
<image src = "/common/navs/f01.png"></image>
<text>
            首页
</text>
</div>
<div class = "tab-item">
<image src = "/common/navs/f02.png"></image>
<text>
            品质优惠
</text>
</div>
<div class = "tab-item">
<image src = "/common/navs/f03.png"></image>
<text>
            发现
</text>
</div> 123
<div class = "tab-item">
<image src = "/common/navs/f04.png"></image>
<text>
            我的
</text>
</div>
</div>
```

底部导航的样式实现如代码示例3.9所示。

代码示例3.9 底部导航样式实现

```css
.footer {
    width: 100%;
    height: 60px;
    background-color: #efefef;
    flex-direction: row;
    align-items: center;
    justify-content: space-between;
}
.footer > .tab-item {
    flex: 1;
    height: 50px;
    flex-direction: column;
    align-items: center;
    justify-content: center;
}
```

```
.tab-item > image {
    width: 30px;
    height: 30px;
    border-radius: 15px;
}
.tab-item > text {
    margin-top: 3px;
    font-size: 14px;
}
```

### 3.4.2 Grid 网格布局

网格布局(Grid)是最强大的 CSS 布局方案。它将网页划分成一个个网格,可以任意组合不同的网格,做出各种各样的布局。以前,通过复杂的 CSS 框架才能达到此效果,鸿蒙 JS UI 支持 Grid 布局的部分属性,如图 3.43 所示。

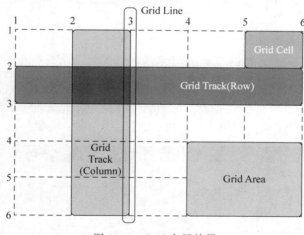

图 3.43 Grid 布局效果

**1. Grid 布局介绍**

CSS 网格布局(Grid)是一套二维的页面布局系统,它的出现将完全颠覆页面布局的传统方式。传统的 CSS 页面布局一直不够理想。包括 table 布局、浮动、定位及内联块等方式,从本质上都是 Hack 的方式,并且遗漏了一些重要的功能(例如:垂直居中)。Flexbox 的出现部分地解决了上述问题,但 Flexbox 布局是为了解决简单的一维布局,适用于页面局部布局,而 Grid 天然就是为了解决复杂的二维布局而出现的,适用于页面的整体布局。在实际工作中,Grid 和 Flexbox 不但不矛盾,而且还能很好地结合使用。

**2. Grid 布局与 Flexbox 布局的差异**

Grid 与 Flexbox 布局的共同点是元素均存放在一个父级容器内,尺寸与位置受容器影响。最核心的区别是 Flexbox 布局使用单坐标轴的布局系统,而 Grid 布局则使用二维布局,使元素可以在两个维度上进行排列,如图 3.44 所示。

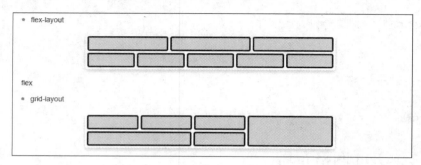

图 3.44　Grid 布局 VS Flexbox 布局

Flexbox 布局很明显是一维布局,元素在容器中都采用横向或者纵向的方式进行排列,并不能跨越维度进行排列,而 Grid 布局相比于 Flexbox 布局,很明显是二维布局,Grid 布局不仅可以在横向上像 Flexbox 那样进行排列,某些子元素还可以跨越维度,同时可以在横向和纵向上进行布局。

**3. Grid 网格的基本概念**

前面对 Grid 有了一个大概的了解后,我们来介绍一下 Grid 中比较重要的几个术语。

**1) 网格容器**

网格容器(grid-container),类似于 Flex 的容器,我们可以通过添加 display:grid 将一个元素设置成一个网格容器。例如代码示例 3.10 中的 container 就是一个网格容器。

**代码示例 3.10**
```
<div class = "container">
    <div class = "item item1">
    <text> 1 </text>
    </div>
    <div class = "item item2">
    <text> 2 </text>
    </div>
    <div class = "item item3">
        <text> 3 </text>
    </div>
</div>
```

**2) 行和列**

容器里面的水平区域称为"行"(row),垂直区域称为"列"(column),如图 3.45 所示。
图 3.45 中,水平的深色区域就是行,垂直的深色区域就是列。

**3) 网格项目**

网格项目(grid-item)就是网格容器中的一个子元素。例如代码示例 3.11 中的 item 就

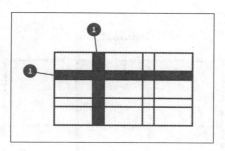

图 3.45　行和列

是一个网格项目,但要注意,sub-item 不是一个网格项目。

代码示例 3.11
```
< div class = "container">
< div class = "item"></div>
< div class = "item">
< div class = "sub - item"></div>
</div>
    < div class = "item"></div>
</div>
```

4）网格线

划分网格的线,称为网格线(grid-line)。应该注意的是,当我们定义网格时,定义的是网格轨道,而不是网格线。Grid 会为我们创建带编号的网格线来让我们来定位每个网格元素。$m$ 列有 $m+1$ 根垂直的网格线,$n$ 行有 $n+1$ 根水平网格线。例如图 3.46 中就有 4 根垂直网格线。一般而言按从左到右和从上到下以 1、2、3……的顺序进行编号排序。当然也可以按从右到左和从下到上,按照 $-1$、$-2$、$-3$……的顺序进行编号排序。

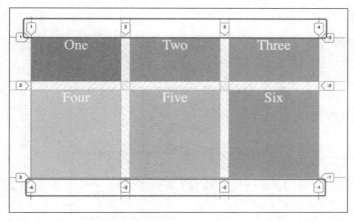

图 3.46　网格线

### 5）网格单元格

网格单元格（grid-cell）就是网格容器中划分出来的最小单元，如图 3.47 所示。

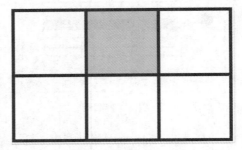

图 3.47　网格单元格

### 6）网格轨道

网格轨道（grid-track）就是由若干个网格单元格组成的横向或者纵向区域，如图 3.48 所示。

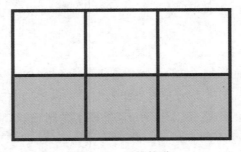

图 3.48　网格轨道

### 7）网格区域

网格区域（grid-area）也是由若干个网格单元格组成的区域，但是不同于网格轨道，它的规格不局限于单个维度，如图 3.49 所示。

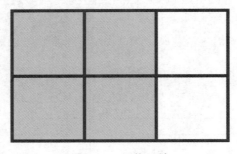

图 3.49　网格区域

### 4. Grid 布局属性

前面对 Grid 布局的一些重要术语进行了介绍，接下来介绍与 Grid 布局相关的基本属性，如表 3.10 和表 3.11 所示。

表 3.10　鸿蒙支持的 Grid Container 的属性

| 属　　性 | 作　　用 |
| --- | --- |
| display | grid：生成一个块级（block-level）网格<br>inline-grid：生成一个行级（inline-level）网格（鸿蒙暂不可用） |
| grid-template-columns | 使用以空格分隔的多个值来定义网格的列和行。这些值表示轨道大小（Track Size），它们之间的空格代表表格线（Grid Line） |
| grid-template-rows | 使用以空格分隔的多个值来定义网格的列和行。这些值表示轨道大小（Track Size），它们之间的空格代表表格线（Grid Line） |
| grid-row-gap | 设置行之间的间距的高度 |
| grid-column-gap | 设置列之间的间距的宽度 |
| grid-auto-flow | 使用框架自动布局算法进行网格的布局，可选值如下。<br>row：逐行填充元素，如果行空间不够，则新增行。<br>column：逐列填充元素，如果列空间不够，则新增列 |

表 3.11　鸿蒙支持的 Grid Items 的属性

| 属　　性 | 作　　用 |
| --- | --- |
| grid-column-start/end | 用于设置当前元素在网格布局中的起止行号，仅当父组件 display 样式为 grid 时生效（仅 div 支持将 display 样式设置为 grid） |
| grid-row-start/end | 用于设置当前元素在网格布局中的起止列号，仅当父组件 display 样式为 grid 时生效（仅 div 支持将 display 样式设置为 grid） |

#### 1）容器属性

容器属性，顾名思义，就是可以在网格容器中添加此属性，是对网格整体进行控制的一系列属性。

我们通过添加 display：grid 可设置一个网格容器。定义二维的网格容器，我们需要定义列和行。创建 3 列和 2 行，我们将使用 grid-template-row 和 grid-template-column 属性。

首先定义页面标签，这里让网格布局容器垂直居中显示，如代码示例 3.12 所示。

代码示例 3.12

```
< div class = "container">
< div class = "control">
< div class = "wrapper">
< div class = "item item1">
< text class = "title"> 1 </text>
</div>
< div class = "item item2">
< text class = "title"> 2 </text>
```

```
</div>
<div class = "item item3">
<text class = "title"> 3 </text>
</div>
<div class = "item item4">
<text class = "title"> 4 </text>
</div>
<div class = "item item5">
<text class = "title"> 5 </text>
</div>
<div class = "item item6">
<text class = "title"> 6 </text>
</div>
</div>
</div>
```

通过样式创建一个 2 行 3 列的网格容器,如代码示例 3.13 所示。

**代码示例 3.13**

```
.wrapper{
    display: grid;
    grid-template-columns: 100px 100px 100px;
    grid-template-rows: 50px 50px;
}
```

grid-template-columns 的 3 个值表示三列,相应的数值表示列宽,即都为 100px。
grid-template-rows 的两个值表示两行,相应的数值表示行高,即都为 50px。
完整样式如代码示例 3.14 所示。

**代码示例 3.14**

```
.container {
    width: 100%;
    height: 100%;
    background-color: #fff;
    justify-content: center;
    align-items: center;
    flex-direction: column;
}

.control {
    width: 600px;
    height: 500px;
    margin: 15px;
}
```

```css
.control .wrapper {
    display: grid;
    grid-template-columns: 200px 50px 100px;
    grid-template-rows: 100px 30px;
    grid-rows-gap: 5px;
    grid-columns-gap: 5px;
}

.item {
    width: 100%;
    height: 100%;
    background-color: grey;
    justify-content: center;
    align-items: center;
}

.title {
    font-size: 20px;
    color: #fff;
}

.item1 {
    background-color: goldenrod;
}

.item2 {
    background-color: blueviolet;
}

.item3 {
    background-color: indianred;
}

.item4 {
    background-color: darkcyan;
}

.item5 {
    background-color: lightcoral;
}

.item6 {
    background-color: deepskyblue;
}
```

效果如图 3.50 所示。

变化一下行高和列宽的值看一下效果，如代码示例 3.15 所示。

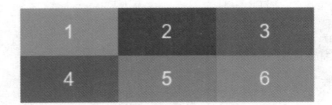

图 3.50 定义行和列

代码示例 3.15

```
.container {
    display: grid;
    grid-template-columns: 200px 50px 100px;
    grid-template-rows: 100px 30px;
}
```

修改行高和列宽后，效果如图 3.51 所示。

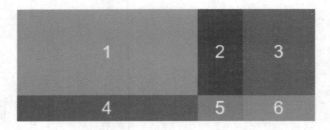

图 3.51 设置行和列的值

从图 3.51 的效果中可以看出每个网格之间是没有间隙的，但是可以通过 grid-rows-gap 设置行间距，如代码示例 3.16 所示。

代码示例 3.16

```
.container {
    display: grid;
    grid-template-columns: 200px 50px 100px;
    grid-template-rows: 100px 30px;
    grid-rows-gap: 5px;
}
```

这里将行间距设置为 5px，效果如图 3.52 所示。

通过 grid-columns-gap 可设置列间距，如代码示例 3.17 所示。

代码示例 3.17

```
.container {
```

```
    display: grid;
    grid-template-columns: 200px 50px 100px;
    grid-template-rows: 100px 30px;
    grid-rows-gap: 5px;
    grid-columns-gap: 5px;
}
```

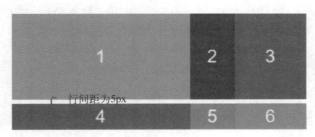

图 3.52　设置行间距

这里将行间距设置为 5px,将列间距设置为 5px,效果如图 3.53 所示。

图 3.53　设置列间距

#### 2) 子元素属性

接下来,创建一个 3×3 的 grid(网格),如代码示例 3.18 所示。

代码示例 3.18

```
.container{
    display: grid;
    grid-template-columns: 100px 100px 100px;
    grid-template-rows: 100px 100px 100px;
}
```

为了更加方便地表示几行几列,上面的代码可以用下面的代码代替,这里 repeat(3,1fr) 的意思是把列分为三份,每份占 1fr。fr 是网格容器剩余空间的等分单位,fr 单位代表网格容器中可用空间的一等份,如代码示例 3.19 所示。

代码示例 3.19

```
.container {
```

```
    display: grid;
    grid-template-columns: repeat(3,1fr);
    grid-template-rows: repeat(3,100px);
}
```

鸿蒙中同样使用CSS3引入了一个新的单位fr(fraction)，中文意思为"分数"，用于替代百分比，因为百分比(小数)存在除不尽的情况，所以用分数表示可以避免多位小数的写法。例如三列等宽的grid可以表示为

```
grid-template-columns: 1fr 1fr 1fr;
```

也可以用repeat方法来简化column或者row的写法，repeat方法接收两个参数，第1个参数表示重复的次数，第2个参数表示重复的内容，所以三列等宽的grid还可以表示为

```
grid-template-columns: repeat(3, 1fr);
```

当要定义的列数很多时，repeat就会变得非常有用，例如要定义一个10列等宽的grid，可以写成repeat(10,1fr)，而不用将1fr重复书写10遍。

效果如图3.54所示。

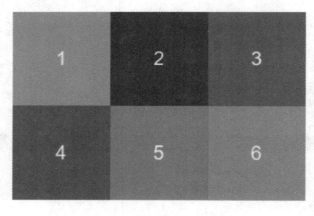

图3.54　3×3的grid(网格)

我们只看到3×2的grid(网格)，而定义的是3×3的grid(网格)。这是因为只有6个items(子元素)来填满这个网格。如果再加3个items(子元素)，则最后一行也会被填满。

要定位和调整items(子元素)的大小，将使用grid-column-start/end和grid-row-start/end属性进行设置如代码示例3.20所示。

**代码示例 3.20**
```
.item1 {
    grid-column-start: 0;
    grid-column-end: 3;
}
```

上面代码的意思是 item1 占据从第一条网格线开始到第四条网格线结束,显示效果如图 3.55 所示。

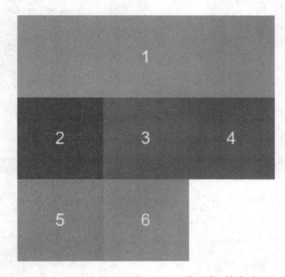

图 3.55 定位和调整 items(子元素)的大小

如果看懂了上面的代码,再来个复杂点的代码巩固一下,如代码示例 3.21 所示。

**代码示例 3.21**
```
.item1 {
    grid-column-start: 0;
    grid-column-end: 1;
}
.item3 {
    grid-row-start: 1;
    grid-row-end: 2;
}
.item4 {
    grid-column-start: 1;
    grid-column-end: 2;
}
```

效果如图 3.56 所示。

图 3.57 所示的十字结构如何排列?这里首先需要确认每个盒子所在的网格。

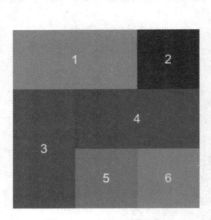

图 3.56 复杂列网格线

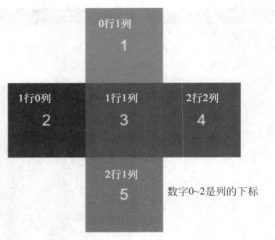

图 3.57 十字排列

样式如代码示例 3.22 所示。

代码示例 3.22

```
.item1 {
    background-color: goldenrod;
    grid-row-start: 0;
    grid-column-start: 1;
}

.item2 {
    background-color: blueviolet;
    grid-row-start: 1;
    grid-column-start: 0;
}

.item3 {
    background-color: indianred;
    grid-row-start: 1;
    grid-column-start: 1;
}

.item4 {
    background-color: darkcyan;
    grid-row-start: 1;
    grid-column-start: 2;
}

.item5 {
    background-color: lightcoral;
    grid-row-start:2;
    grid-column-start: 1;
}
```

item1 所在位置是 1 行 2 列的位置,但是鸿蒙的索引值是从 0 开始的,所示是 grid-row-start：0/grid-column-start：1。

item2 所在位置是 2 行 1 列的位置,但是鸿蒙的索引值是从 0 开始的,所示是 grid-row-start：1/grid-column-start：0。

item3 所在位置是 2 行 2 列的位置,但是鸿蒙的索引值是从 0 开始的,所示是 grid-row-start：1/grid-column-start：1。

item4 所在位置是 2 行 3 列的位置,但是鸿蒙的索引值是从 0 开始的,所示是 grid-row-start：1/grid-column-start：2。

item5 所在位置是 3 行 2 列的位置,但是鸿蒙的索引值是从 0 开始的,所示是 grid-row-start：2/grid-column-start：1。

鸿蒙的 Grid 网格布局看起来和 W3C 的标准类似,但是还是有很多不同的地方需要注意。

**注意**：Grid-column-start/end 和 grid-row-start/end 后面的值和 W3C 定义的值是不一样的,ArkUI JS 中 Grid 的这两个值的索引是从 0 开始计算的。

### 5. Grid 布局案例

这里使用 Grid 布局实现 3.4.1 节中菜单导航的例子,如图 3.58 所示。

图 3.58 Grid 布局实现菜单导航

通过 Grid 布局,首先创建一个 2 行 5 列的网格容器,页面结构如代码示例 3.23 所示。

```
代码示例 3.23
<div class = "container">
<div class = "control">
<div class = "wrapper">
<div class = "item item1">
<text class = "title">1</text>
</div>
<div class = "item item2">
<text class = "title">2</text>
</div>
<div class = "item item3">
<text class = "title">3</text>
</div>
<div class = "item item4">
```

```
< text class = "title"> 4 </text >
</div >
< div class = "item item5">
< text class = "title"> 5 </text >
</div >
< div class = "item item6">
< text class = "title"> 6 </text >
</div >
< div class = "item item1">
< text class = "title"> 7 </text >
</div >
< div class = "item item2">
< text class = "title"> 8 </text >
</div >
< div class = "item item3">
< text class = "title"> 9 </text >
</div >
< div class = "item item4">
< text class = "title"> 10 </text >
</div >
</div >
</div >
</div >
```

样式设置如代码示例 3.24 所示。

**代码示例 3.24**

```
.control .wrapper {
    display: grid;
    grid-template-columns: repeat(5,1fr);
    grid-template-rows: repeat(2,1fr);
}
```

效果如图 3.59 所示。

图 3.59　定义 2 行 5 列网格容器

接下来，给每个网格项添加图片，如代码示例3.25所示。

代码示例3.25

```html
<div class="container">
<div class="control">
<div class="wrapper">
<div class="item item1">
<image src="/common/images/nav01.png"></image>
<text class="title">美食</text>
</div>
<div class="item item2">
<image src="/common/images/nav02.png"></image>
<text class="title">电影/演出</text>
</div>
<div class="item item3">
<image src="/common/images/nav03.png"></image>
<text class="title">酒店</text>
</div>
<div class="item item4">
<image src="/common/images/nav04.png"></image>
<text class="title">休闲娱乐</text>
</div>
<div class="item item5">
<image src="/common/images/nav05.png"></image>
<text class="title">外卖</text>
</div>
<div class="item item6">
<image src="/common/images/nav06.png"></image>
<text class="title">婚纱摄影</text>
</div>
<div class="item item1">
<image src="/common/images/nav07.png"></image>
<text class="title">丽人/美发</text>
</div>
<div class="item item2">
<image src="/common/images/nav08.png"></image>
<text class="title">周边游</text>
</div>
<div class="item item3">
<image src="/common/images/nav09.png"></image>
<text class="title">KTV</text>
</div>
<div class="item item4">
<image src="/common/images/nav10.png"></image>
```

```html
    <text class = "title">火锅</text>
  </div>
 </div>
 </div>
</div>
```

样式设置如代码示例 3.26 所示。

代码示例 3.26

```css
.container {
    width: 100%;
    height: 100%;
    background-color: #fff;
    justify-content: center;
    align-items: center;
    flex-direction: column;
}
.control {
    width: 100%;
    height: 200px;
    margin: 15px;
}
.control .wrapper {
    display: grid;
    grid-template-columns: repeat(5,1fr);
    grid-template-rows: repeat(2,1fr);
    grid-rows-gap: 5px;
    grid-columns-gap: 5px;
}
.item {
    width: 100%;
    height: 100%;
    justify-content: center;
    align-items: center;
    flex-direction: column;
}
.title {
    font-size: 13px;
    color: #000;
}
image {
    object-fit: contain;
}
```

效果如图 3.60 所示。

图 3.60　网格实现菜单导航

## 3.5　语法详细讲解

鸿蒙 HML 采用了目前流行的组件化开发思想,通过声明式的组件标记语言,使 UI 更加容易开发和复用。

### 3.5.1　HML 语法

HML(HarmonyOS Markup Language,鸿蒙标记语言)框架结构如图 3.61 所示,是一套类似 HTML 的标记语言,通过组件和事件构建出页面的内容。页面具备数据绑定、事件绑定、列表渲染、条件渲染和逻辑控制等高级能力。

下面介绍 HML 语法的特点,鸿蒙的 HML 采用了 MVVM(一种把视图与代码逻辑分离的界面层设计方法)设计模式,通过 Webpack(前端打包工具)对 HML 文件进行编译和打包。

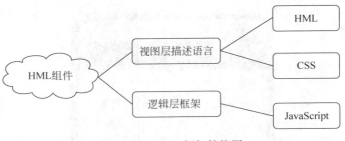

图 3.61　HML 框架结构图

### 1. HML

HML 是鸿蒙 JS 框架用来描述界面的标记语言,与 HTML 类似,如代码示例 3.27 所示。

代码示例 3.27　HML

```
<div class = "item-container">
<text class = "item-title">Image Show</text>
<div class = "item-content">
<image src = "/common/hm.png" class = "image"></image>
</div>
</div>
```

HML 只支持一个根节点,代码示例 3.28 这种写法目前是不支持的。

代码示例 3.28　不支持多标签

```
<div>
<text>hello</text>
</div>
<div>
<text>world</text>
</div>
```

### 2. 采用双向绑定机制

双向绑定机制是 MVVM 设计模式的核心部分,通过双向绑定机制彻底解决了界面与逻辑代码的耦合关系。鸿蒙的 ArkUI JavaScript 框架采用目前比较流行的双向绑定机制实现 UI 与代码逻辑分离,如代码示例 3.29 所示。

代码示例 3.29　双向绑定机制

```
<div class = "container">
<text class = "title">
    Hello {{title}}
</text>
<input type = "text" @change = "updateTitle"></input>
</div>
```

```
//xxx.js
export default {
    data: {
        title: 'World'
    },
    updateTitle(inputObj){
        this.title = inputObj.value
    }
}
```

当 input 输入内容时，change 事件会监听输入框中值的变化，把最新的值赋值给 title 属性，当 title 值发生变化时，双向绑定机制同步更新视图，如图 3.62 所示。

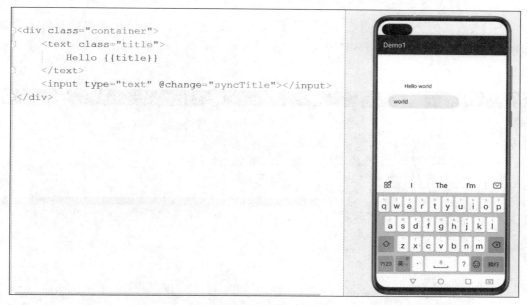

```
<div class="container">
    <text class="title">
        Hello {{title}}
    </text>
    <input type="text" @change="syncTitle"></input>
</div>
```

图 3.62 双向数据绑定

### 3. HML 界面事件绑定

在 HML 上通过事件绑定的回调函数实现接收一个事件对象参数，可以通过访问该事件对象获取事件信息，如代码示例 3.30 所示。

代码示例 3.30 界面事件响应

```
<!-- xxx.hml -->
<div>
<!-- 正常格式 -->
<div onclick = "clickfunc"></div>
<!-- 缩写 -->
<div @click = "clickfunc"></div>
```

```
</div>
//xxx.js
export default {
  data: {
    obj: '',
  },
  clickfunc: function(e) {
    this.obj = 'Hello World';
    console.log(e);
  },
}
```

**4. HML 列表渲染**

对列表数据的渲染可通过在 HML 上使用 for 指令实现,以此简化对界面标签元素的循环带来的复杂操作。for 循环指令大大减少了界面开发的工作量,如图 3.63 所示,如代码示例 3.31 所示。

代码示例 3.31　界面列表渲染

```
<!-- xxx.hml -->
<div class="array-container">
<!-- div 列表渲染 -->
<!-- 默认 $item 代表数组中的元素,$idx 代表数组中的元素索引 -->
<div for="{{array}}" tid="id">
<text>{{$idx}}.{{$item.name}}</text>
</div>
<!-- 自定义元素变量名称 -->
<div for="{{value in array}}" tid="id">
<text>{{$idx}}.{{value.name}}</text>
</div>
<!-- 自定义元素变量、索引名称 -->
<div for="{{(index, value) in array}}" tid="id">
<text>{{index}}.{{value.name}}</text>
</div>
</div>
//xxx.js
export default {
  data: {
    array: [
      {id: 1, name: 'jack', age: 18},
      {id: 2, name: 'tony', age: 18},
    ],
  },
}
```

tid 属性主要用来加速 for 循环的重新渲染,旨在列表中的数据有变更时,提高重新渲染的效率。tid 属性是用来指定数组中每个元素的唯一标识,如果未指定,则数组中每个元素的索引为该元素的唯一 id。例如上述 tid="id"表示数组中的每个元素的 id 属性为该元素的唯一标识。for 循环支持的写法如代码示例 3.32 所示。

**代码示例 3.32　for 循环支持的写法**

for = "array"：其中 array 为数组对象，array 的元素变量默认为 $ item
for = "v in array"：其中 v 为自定义的元素变量，元素索引默认为 $ idx
for = "(i, v) in array"：其中元素索引为 i，元素变量为 v，遍历数组对象 array

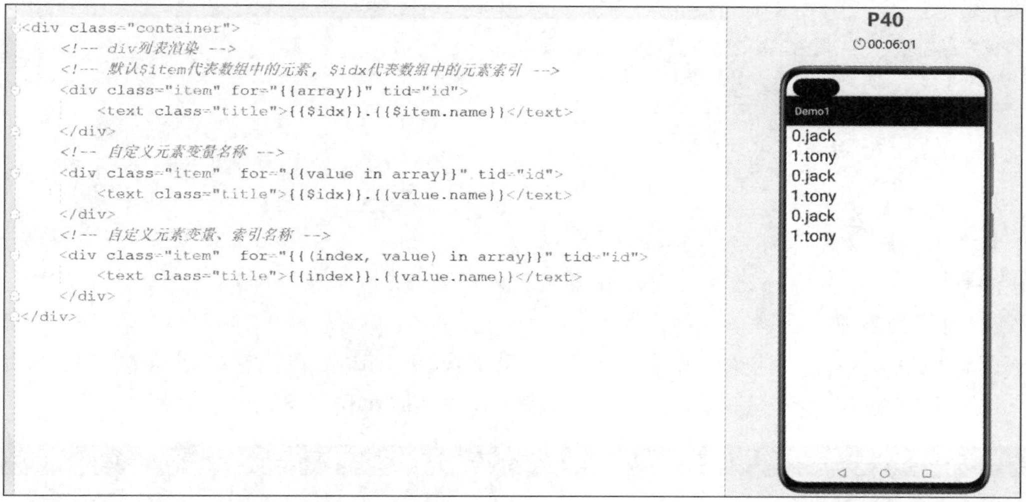

```
<div class="container">
    <!-- div列表渲染 -->
    <!-- 默认$item代表数组中的元素，$idx代表数组中的元素索引 -->
    <div class="item" for="{{array}}" tid="id">
        <text class="title">{{$idx}}.{{$item.name}}</text>
    </div>
    <!-- 自定义元素变量名称 -->
    <div class="item" for="{{value in array}}" tid="id">
        <text class="title">{{$idx}}.{{value.name}}</text>
    </div>
    <!-- 自定义元素变量、索引名称 -->
    <div class="item" for="{{(index, value) in array}}" tid="id">
        <text class="title">{{index}}.{{value.name}}</text>
    </div>
</div>
```

图 3.63　for 循环显示

**5. HML 条件渲染**

条件渲染分为两种：if/elif/else 和 show。两种写法的区别在于：第一种写法里当 if 为 false 时，组件不会在 vdom 中构建，也不会渲染，而第二种写法里当 show 为 false 时虽然也不渲染，但会在 vdom 中构建。

另外，当使用 if/elif/else 写法时，节点必须是兄弟节点，否则编译无法通过。实例如代码示例 3.33 所示。

**代码示例 3.33　界面条件渲染**

```
<!-- xxx.hml -->
<div>
<text if = "{{show}}"> Hello - TV </text>
<text elif = "{{display}}"> Hello - Wearable </text>
<text else> Hello - World </text>
</div>
//xxx.js
export default {
  data: {
    show: false,
```

```
      display: true,
   },
}
```

优化渲染优化：show 方法。当 show 为真时，节点可正常渲染；当 show 为假时，仅仅可将 display 样式设置为 none，如代码示例 3.34 所示。

**代码示例 3.34　优化渲染优化**

```
<!-- xxx.hml -->
<text show = "{{visible}}"> Hello World </text>
//xxx.js
export default {
  data: {
    visible: false,
  },
}
```

**注意**：禁止在同一个元素上同时设置 for 和 if 属性。

#### 6. 逻辑控制块

逻辑控制块使循环渲染和条件渲染变得更加灵活；block 在构建时不会被当作真实的节点编译。注意 block 标签只支持 for 和 if 属性，如代码示例 3.35 所示。

**代码示例 3.35　block 控制块**

```
<!-- xxx.hml -->
<list>
<block for = "glasses">
<list-item type = "glasses">
<text>{{ $item.name }}</text>
</list-item>
<block for = "$item.kinds">
<list-item type = "kind">
<text>{{ $item.color }}</text>
</list-item>
</block>
</block>
</list>
//xxx.js
export default {
  data: {
    glasses: [
      {name:'sunglasses', kinds:[{name:'XXX',color:'XXX'},{name:'XXX',color:'XXX'}]},
      {name:'nearsightedness mirror', kinds:[{name:'XXX',color:'XXX'}]},
    ],
  },
}
```

### 7. 模板引用

鸿蒙 ArkUI JS 框架支持直接在 HML 中引用其他的 HML 模板文件，这种做法的目的是更好地复用视图。这里每个独立的 HML 都是一个自定义的组件，如代码示例 3.36 所示。

代码示例 3.36 模板引用

```
<!-- template.hml -->
<div class = "item">
<text>Name: {{name}}</text>
<text>Age: {{age}}</text>
</div>
<!-- index.hml -->
<element name = 'comp' src = '../../common/template.hml'></element>
<div>
<comp name = "Tony" age = "18"></comp>
</div>
```

### 8. 页面生命周期

应用生命周期主要有两个：应用创建时调用的 onCreate 和应用销毁时触发的 onDestroy。

一个应用中可能会有多个页面，一个页面一般包括 onInit、onReady、onShow、onHide 和 onDestroy 等在页面创建、显示和销毁时会触发调用的事件。

(1) onInit：表示页面的数据已经准备好，可以使用 JS 文件中的 data 数据。

(2) onReady：表示页面已经编译完成，可以将界面显示给用户。

(3) onShow：JS UI 只支持应用同时运行并展示一个页面，当打开一个页面时，上一个页面就销毁了。当一个页面显示的时候，会调用 onShow。

(4) onHide：页面消失时被调用。

(5) onDestroy：页面销毁时被调用。

当应用从页面 A 跳转到页面 B 时，首先调用页面 A 的 onDestroy 函数。页面 A 销毁后，依次调用页面 B 的 onInit、onReady、onShow 函数来初始化和显示页面 B，如图 3.64 所示。

## 3.5.2 CSS 语法

CSS 是描述 HML 页面结构的样式语言。所有组件均存在系统默认样式，也可在页面 CSS 样式文件中对组件、页面自定义不同的样式。

下面从样式的几个重要方面来了解鸿蒙应用中 CSS 的用法。

### 1. 鸿蒙的尺寸单位

需要注意的是单位的使用场景：px 单位实际上是个弹性单位，帮助我们在不同分辨率下自动进行界面适配。

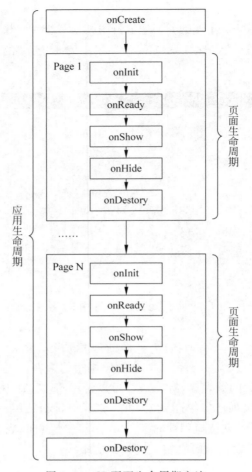

图 3.64　JS 页面生命周期方法

鸿蒙 ArkUI JS 支持两种常见的尺寸单位。

(1) 逻辑像素 px：

- 默认屏幕具有的逻辑宽度为 720px，实际显示时会将页面布局缩放至屏幕的实际宽度，如 100px 在实际宽度为 1440 物理像素的屏幕上，实际会被渲染为 200 物理像素（从 720px，所有尺寸放大 2 倍，到 1440px，所有尺寸放大 2 倍）。
- 当额外配置 autoDesignWidth 为 true 时，逻辑像素 px 将按照屏幕密度进行缩放，如 100px 在屏幕密度为 3 的设备上，实际会被渲染为 300 物理像素。当应用需要适配多种设备时，建议采用此方法。

(2) 百分比：表示该组件占父组件尺寸的百分比，如将组件的 width 设置为 50%，代表其宽度为父组件的 50%。

## 2. 样式引入方式

为了模块化管理和代码复用,CSS 样式文件支持@import 语句,以此导入 CSS 文件。

每个页面目录下存在一个与布局 HML 文件同名的 CSS 文件,用来描述该 HML 页面中组件的样式,决定组件应该如何显示。

(1) 内部样式,支持使用 style、class 属性来控制组件的样式,如代码示例 3.37 所示。

**代码示例 3.37　内部样式**

```
<!-- index.hml -->
<div class = "container">
<text style = "color:red">Hello World</text>
</div>
```

(2) 文件导入,合并外部样式文件。例如,在 common 目录中定义样式文件 style.css,并在 index.css 文件中进行导入,如代码示例 3.38 所示。

**代码示例 3.38　外部样式**

```
/* style.css */
.title {
  font-size: 50px;
}

/* index.css */
@import '../../common/style.css';
.container {
  justify-content: center;
}
```

## 3. 样式预编译

预编译提供了利用特有语法生成 CSS 的程序,可以提供变量、运算等功能,令开发者更便捷地定义组件样式,目前支持 less、sass 和 scss 的预编译。

**注意**:当使用样式预编译时,需要将原 CSS 文件的后缀改为 less、sass 或 scss,如将 index.css 改为 index.less、index.sass 或 index.scss。

当前文件使用样式预编译时,需要将原 index.css 改为 index.less,如代码示例 3.39 所示。

**代码示例 3.39　使用 less 预编译**

```
/* index.less */
/* 定义变量 */
@colorBackground: #000000;
.container {
  background-color: @colorBackground; /* 使用当前 less 文件中定义的变量 */
}
```

引用预编译文件,例如 common 中存在 style.scss 文件,需要将原 index.css 改为 index.scss,并引入 style.scss,如代码示例 3.40 所示。

**代码示例 3.40　使用 scss 预编译**

```
/* style.scss */
/* 定义变量 */
$colorBackground: #000000;
在 index.scss 中引用:
/* index.scss */
/* 引入外部 scss 文件 */
@import '../../common/style.scss';
.container {
  background-color: $colorBackground; /* 使用 style.scss 中定义的变量 */
}
```

### 4. 媒体查询

媒体查询(Media Query)在移动设备上应用十分广泛,开发者经常需要根据设备的大致类型或者特定的特征和设备参数(例如屏幕分辨率)来修改应用的样式。为此媒体查询提供了如下功能:

(1) 针对设备和应用的属性信息,可以设计出相匹配的布局样式。

(2) 当屏幕发生动态改变时(例如分屏、横竖屏切换),应用页面布局同步更新。

**注意**: media(媒体)属性值默认为设备的真实尺寸大小、物理像素和真实的屏幕分辨率。请勿与以 720px 为基准的项目配置宽度 px 混淆。

通用媒体特征如代码示例 3.41 所示。

**代码示例 3.41　媒体查询**

```
<!-- xxx.hml -->
<div>
<div class="container">
<text class="title">Hello World</text>
</div>
</div>

/* xxx.css */
.container {
  width: 300px;
  height: 600px;
  background-color: #008000;
}
@media screen and (device-type: tv) {
  .container {
    width: 500px;
    height: 500px;
    background-color: #fa8072;
```

```
    }
  }
  @media screen and (device-type: tv) {
    .container {
      width: 300px;
      height: 300px;
      background-color: #008b8b;
    }
  }
```

#### 5. 自定义字体样式

font-face 用于定义字体样式。应用可以在 style 中定义 font-face 来指定相应的字体名和字体资源，然后在 font-family 样式中引用该字体。

**注意**：自定义字体可以是从项目中的字体文件或网络字体文件中加载的字体，字体格式支持 TTF 和 OTF。网络字体文件：通过 URL 指定网络字体的地址，不支持设置多个 src。

这里通过一个简单案例介绍如何使用自定义样式，例如有以下的布局：

```
<div>
<text class = "demo-text">测试自定义字体</text>
</div>
```

TTF 文件通常放在 common 目录下。页面样式，如代码示例 3.42 所示。

**代码示例 3.42　自定义样式**

```
@font-face {
  font-family: HWfont;
  src: url("/common/HWfont.ttf");
}
.demo-text {
  font-family: HWfont;
}
```

#### 6. 动画样式

组件普遍支持的动画样式可以在 style 或 css 中进行设置，如动态的旋转、平移、缩放效果。

**注意**：@keyframes 的 from/to 不支持动态绑定。

对于不支持起始值或终止值缺省的情况，可以通过 from 和 to 显示指定起始和结束，如代码示例 3.43 所示。

**代码示例 3.43　关键帧动画**

```
@keyframes Go
{
    from {
```

```
        background-color: #f76160;
        transform:translate(0px) rotate(0deg) scale(1.0);
    }
    to {
        background-color: #09ba07;
        transform:translate(100px) rotate(180deg) scale(2.0);
    }
}
```

### 3.5.3  JS 逻辑

ArkUI JS 框架通过 MVVM 设计模式实现界面 UI 与 JS 逻辑的分离,这里介绍一下 HML 逻辑层的用法。JS 逻辑层支持 ES6 语法,但是需要通过引入 babel 进行处理。

#### 1. 支持 ES6 模块化标准

鸿蒙内置模块引入,所以不需要使用路径,内置模块通常以@system 开头:

```
import router from '@system.router';
```

自定义模块引入,可通过相对路径引入:

```
import utils from '../../common/utils.js';
```

#### 2. $refs 获取 DOM 元素

通过 $refs 获取 DOM 元素,如代码示例 3.44 所示。

**代码示例 3.44    $refs**

```
<!-- index.hml -->
<div class="container">
<image-animator
    class="image-player"
    ref="animator"
    images="{{images}}"
    duration="1s" onclick="handleClick"></image-animator>
</div>

//index.js
export default {
  data: {
    images: [
      { src: '/common/frame1.png' },
      { src: '/common/frame2.png' },
      { src: '/common/frame3.png' },
    ],
  },
  handleClick() {
```

```
      //获取 ref 属性为 animator 的 DOM 元素
      const animator = this.$refs.animator;
      const state = animator.getState();
      if (state === 'paused') {
        animator.resume();
      } else if (state === 'stopped') {
        animator.start();
      } else {
        animator.pause();
      }
    },
  };
```

### 3. $element 方法获取 HML 元素

$element 方法是鸿蒙 JS 内置的方法,用于获取 HML 元素,如代码示例 3.45 所示。

**代码示例 3.45　$element**

```
<!-- index.hml -->
<div class = "container">
  <image-animator
      class = "image-player"
      id = "animator"
      images = "{{images}}"
      duration = "1s"
      onclick = "handleClick"></image-animator>
</div>
//index.js
export default {
  data: {
    images: [
      { src: '/common/frame1.png' },
      { src: '/common/frame2.png' },
      { src: '/common/frame3.png' },
    ],
  },
  handleClick() {
      //获取 id 属性为 animator 的 DOM 元素
      const animator = this.$element('animator');
      const state = animator.getState();
    if (state === 'paused') {
      animator.resume();
    } else if (state === 'stopped') {
      animator.start();
    } else {
      animator.pause();
    }
  },
};
```

### 3.5.4 多语言支持

多语言支持,通过在文件组织中指定的 i18n 文件夹内放置每个语言地区下的资源定义文件即可,资源文件命名为"语言-地区.json"格式,例如英文(美国)的资源文件命名为 en-US.json。当开发框架无法在应用中找到系统语言的资源文件时,默认使用 en-US.json 中的资源内容。

资源文件用于存放应用在多种语言场景下的资源内容,开发框架使用 JSON 文件保存资源定义。

由于不同语言针对单复数有不同的匹配规则,在资源文件中的使用 zero、one、two、few、many、other 定义不同单复数场景下的词条内容。例如中文不区分单复数,仅存在 other 场景;英文存在 one、other 场景;阿拉伯语存在上述 6 种场景。

以 en-US.json 和 ar-AE.json 为例,资源文件内容格式如代码示例 3.46 所示。

代码示例 3.46　多语言定义

```
{
"strings": {
"hello": "Hello world!",
"object": "Object parameter substitution-{name}",
"array": "Array type parameter substitution-{0}",
"symbol": "@#$%^&*()_+-={}[]\\|:;\"'<>,./?",
"people": {
"one": "one person",
"other": "{count} people"
        }
    },
"files": {
"image": "image/en_picture.PNG"
        }
}
```

其他语言的配置文件,示例代码如下:

```
{
"strings": {
"plurals": {
"zero": "لا أحد",
"one": "وحده",
"two": "اثنان",
"few": "اشخاص ستة",
"many": "شخص خمسون",
"other": "شخص مائة"
            }
        }
}
```

上面介绍了如何定义多语言资源配置文件的方法,下面介绍如何引用资源。

在应用开发的页面中可使用多语言的语法,包含简单格式化和单复数格式化两种,都可以在 HML 或 JS 中使用。

**1. 简单格式化方法**

在应用中使用 $t 方法引用资源,$t 既可以在 HML 中使用,也可以在 JS 中使用。系统将根据当前语言环境和指定的资源路径(通过 $t 的 path 参数设置),显示对应语言的资源文件中的内容,如代码示例 3.47 所示。

**代码示例 3.47　简单格式化方法**

```html
<!-- xxx.hml -->
<div>
<!-- 不使用占位符,text 中显示"Hello world!" -->
<text>{{ $t('strings.hello') }}</text>
<!-- 具名占位符格式,运行时将占位符{name}替换为"Hello world" -->
<text>{{ $t('strings.object', { name: 'Hello world' }) }}</text>
<!-- 数字占位符格式,运行时将占位符{0}替换为"Hello world" -->
<text>{{ $t('strings.array', ['Hello world']) }}</text>
<!-- 先在 JS 中获取资源内容,再在 text 中显示"Hello world" -->
<text>{{ hello }}</text>
<!-- 先在 JS 中获取资源内容,并将占位符{name}替换为"Hello world",再在 text 中显示"Object
parameter substitution-Hello world" -->
<text>{{ replaceObject }}</text>
<!-- 先在 JS 中获取资源内容,并将占位符{0}替换为"Hello world",再在 text 中显示"Array type
parameter substitution-Hello world" -->
<text>{{ replaceArray }}</text>

<!-- 获取图片路径 -->
<image src = "{{ $t('files.image') }}" class = "image"></image>
<!-- 先在 JS 中获取图片路径,再在 image 中显示图片 -->
<image src = "{{ replaceSrc }}" class = "image"></image>
</div>


//xxx.js
//下面为在 JS 文件中的使用方法
export default {
  data: {
    hello: '',
    replaceObject: '',
    replaceArray: '',
    replaceSrc: '',
  },
  onInit() {
    //简单格式化
    this.hello = this.$t('strings.hello');
    this.replaceObject = this.$t('strings.object', { name: 'Hello world' });
```

```
            this.replaceArray = this.$t('strings.array', ['Hello world']);
            this.replaceSrc = this.$t('files.image');
    },
}
```

**2. 单复数格式化**

单复数格式化如代码示例3.48所示。

代码示例3.48 单复数格式化

```
<!-- xxx.hml -->
<div>
<!-- 当传递数值为 0 时:"0 people" 阿拉伯语中此处将匹配 key 为 zero 的词条 -->
<text>{{ $tc('strings.plurals', 0) }}</text>
<!-- 当传递数值为 1 时:"one person" 阿拉伯语中此处将匹配 key 为 one 的词条 -->
<text>{{ $tc('strings.plurals', 1) }}</text>
<!-- 当传递数值为 2 时:"2 people" 阿拉伯语中此处将匹配 key 为 two 的词条 -->
<text>{{ $tc('strings.plurals', 2) }}</text>
<!-- 当传递数值为 6 时:"6 people" 阿拉伯语中此处将匹配 key 为 few 的词条 -->
<text>{{ $tc('strings.plurals', 6) }}</text>
<!-- 当传递数值为 50 时:"50 people" 阿拉伯语中此处将匹配 key 为 many 的词条 -->
<text>{{ $tc('strings.plurals', 50) }}</text>
<!-- 当传递数值为 100 时:"100 people" 阿拉伯语中此处将匹配 key 为 other 的词条 -->
<text>{{ $tc('strings.plurals', 100) }}</text>
</div>
```

## 3.6 内置组件

组件(Component)结构如图3.65所示,是构建页面的核心,每个组件通过对数据和方法的简单封装可以实现独立的可视、可交互功能单元。组件之间相互独立,随取随用,也可以在需求相同的地方重复使用。开发者还可以通过组件间的合理搭配生成满足业务需求的新组件,从而减少开发量。

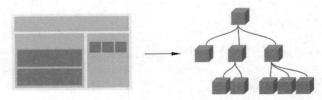

图3.65 组件结构图

### 3.6.1 容器组件

下面介绍几个常见的容器组件,这些组件有的保留了与HTML相同的命名,但是其实已经不再是 DOM 对象，它是鸿蒙操作系统内置的组件。

### 1. div 组件

div 组件用于页面结构的根节点或将内容进行分组，支持所有鸿蒙设备。div 组件主要用在界面的布局上，在鸿蒙布局小节详细介绍过。

div 组件在 HML 中的定义如代码示例 3.49 所示。

代码示例 3.49　div 组件

```
<div class = "container">
    <div class = "flex-box">
        <div class = "flex-item color-primary"></div>
        <div class = "flex-item color-warning"></div>
        <div class = "flex-item color-success"></div>
    </div>
</div>
```

### 2. list 组件

list 组件包含一系列相同宽度的列表项。适合连续、多行呈现同类数据，例如图片和文本，通常和子组件< list-item-group >和< list-item >一起使用。

实现城市索引列表，如图 3.66 所示。

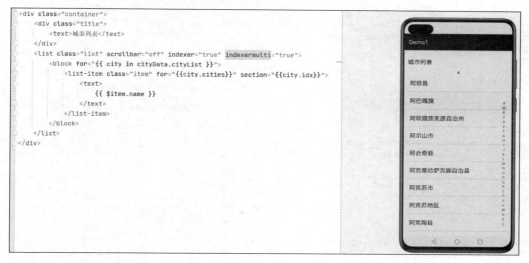

图 3.66　list 组件实现城市列表

list 组件可结合 list-item 组件一起使用，list 组件的 indexer 属性如果是 true，则表示开启右侧城市索引，同时需要结合 list-item 组件的 section 属性一起使用，section 属性的值是英语字母，如字母 A。

界面布局如代码示例 3.50 所示。

**代码示例 3.50　城市列表**

```html
<div class = "container">
<div class = "title">
<text>城市列表</text>
</div>
<list class = "list" scrollbar = "off" indexer = "true" indexermulti = "true">
<block for = "{{ city in cityData.cityList }}">
<list-item class = "item"
                  for = "{{city.cities}}"
                  section = "{{city.idx}}">
<text>
                  {{ $item.name }}
</text>
</list-item>
</block>
</list>
</div>
```

城市索引列表样式的实现如代码示例 3.51 所示。

**代码示例 3.51　城市索引列表样式**

```css
.container {
    display: flex;
    justify-content: center;
    align-items: center;
    flex-direction: column;
    padding: 10px;
}
.title {
    width:100%;
    height:60px;
    padding: 3px;
    justify-content: flex-start;
    align-items: center;
}

.title text {
    font-size: 20px;
}

.list {
    width:100%;
    height:100%;
}

.item {
    width:100%;
    height:60px;
```

```
        border-bottom: 1px;
        border-bottom-color: gainsboro;
        padding: 10px;
        justify-content: center;
        align-items:flex-start ;
        flex-direction: column;
        background-color: #fff;
}

text {
    font-size: 20px;
}
```

城市索引数据可通过 common 文件夹中的 cityIndex.json 数据导入。通过 ES6 模块化导入 JSON 数据，绑定到 data 对象中的 cityData 上，如代码示例 3.52 所示。

代码示例 3.52　导入城市 JSON 数据

```
//导入城市列表
import city from "../../common/cityIndex.json"

export default {
    data: {
        cityData:city
    }
}
```

城市索引列表的数据格式如图 3.67 所示。

```
"cityList": [
    {
        "idx": "A",
        "cities": [...]
    },
    {
        "idx": "B",
        "cities": [
            {
                "id": 2469,
                "name": "巴楚县",
                "pinyin": "bachu",
                "latitude": 39.783479,
                "longitude": 78.55041
            },
            {
                "id": 1693,
                "name": "巴东县",
                "pinyin": "badong",
                "latitude": 31.041403,
                "longitude": 110.336665
            },
            {
                "id": 223,
                "name": "白城市",
                "pinyin": "baicheng",
```

图 3.67　城市列表数据格式

通过 list 组件实现的城市索引列表如图 3.68 所示。

下面我们通过 list 组件实现一个商品的列表的例子，效果如图 3.69 所示。

图 3.68　城市列表　　　　　　图 3.69　商品列表

商品列表页面布局如代码示例 3.53 所示，在 list-item 组件中通过横向 Flex 布局，分成左右结构，左边的图片占整个横向的 200px，右边占横向的剩下所有空间。

代码示例 3.53　listdemo/index.html

```
< div class = "container">
< div class = "filter - wrap">
< toggle class = "margin" for = "{{ toggles }}">{{ $ item }}</toggle >
</div>
< list class = "list_product">
< list - item for = "{{ listData }}" class = "product">
< div class = "left">
< image src = "{{ $ item.pic}}">
</image>
</div>
< div class = "right">
< div class = "title">
```

```html
            <text>
                            {{$item.title}}
            </text>
        </div>
        <div class = "tags">
        <toggle class = "margin" for = "{{tags}}">{{$item}}</toggle>
        </div>
        <div class = "price">
        <text class = "m">¥ </text>
        <text class = "p">{{$item.price}}</text>
        </div>
        <div class = "promote">
        <image src = "/common/product/a1.png"/>
        <image src = "/common/product/a2.png"/>
        </div>
        <div class = "shopinfo">
        <text>{{$item.sale}}</text>
        <text>{{$item.shop}}</text>
        </div>
        </div>
        </list-item>
        </list>
        </div>
```

样式如代码示例 3.54 所示。

**代码示例 3.54　listdemo/index.css**

```css
.container {
    flex-direction: column;
    justify-content: center;
    align-items: center;
}

.margin {
    margin: 7px;
    color: #999;
}

.product {
    width: 100%;
    height: 150px;
    margin-top: 10px;
    display: flex;
    flex-direction: row;
}

.product > .left {
    width: 200px;
```

```css
    height: 100%;
    margin-left: 10px;
    display: flex;
}

.product > .left image {
    object-fit: contain;
}

.product > .right {
    flex: 1;
    height: 100%;
    border-bottom: 1px;
    border-bottom-color: #f2f2f2;
    flex-direction: column;
    margin-left: 10px;
    margin-right: 10px;
}

.product > .right > .title text {
    font-size: 14px;
    max-lines: 2;
    text-overflow: ellipsis;
    color: #333;
}

.product .right .price .m {
    color: #ff4142;
    font-size: 20px;
}

.product .right .price .p {
    color: #ff4142;
    font-size: 23px;
}

.promote {
    width: 100%;
    height: 25px;
    display: flex;
    flex-direction: row;
    justify-content: flex-start;
}

.promote image {
    margin: 5px;
    width: 60px;
    object-fit: fill;
}
```

```css
.shopinfo {
    display: flex;
    flex-direction: row;
}

.shopinfo text {
    font-size: 12px;
    margin-right: 5px;
    color: #999;
    padding: 5px;
}
```

JS 逻辑代码如代码示例 3.55 所示,这里定义了一个 listData 数组,页面通过 for 循环输出。

**代码示例 3.55    listdemo/index.js**

```js
export default {
    data: {
        toggles: ["京东物流","新品","品牌","系列"],
        tags:["Intel i5","内存 8GB","512GB"],
        listData:[
            {
"title":"【2021 新品】Morsiner 15.6 英寸酷睿 i7 笔记本电脑轻薄 2GB 独显游戏本金属超薄本学生手提 M5-Pro 套餐二",
"price":2948,
"pic":"/common/product/1.webp",
"sale":"已售 1700+件",
"shop":"Morsiner 拼购旗舰店"
            },
            {
"title":"15.6 英寸 超薄笔记本电脑 轻薄本 超极本 学生网课手提商务办公高速四核格莱富计算机笔记本分期免息 太空银 套餐三【8GB 内存+128GB 高速固态】",
"price":1388,
"pic":"/common/product/2.webp",
"sale":"已售 7800+件",
"shop":"格莱富计算机旗舰店"
            },
            {
"title":"15.6 英寸 超轻薄本 笔记本电脑轻薄超级本 学生网上课堂计算机 远程办公计算机 YEPO 笔记本电脑 高速四核 8GB 内存 128GB 固态(爆)",
"price":1588,
"pic":"/common/product/3.webp",
"sale":"已售 3.4 万件",
"shop":"YEPO 计算机旗舰店"
            },
            {
"title":"联想(Lenovo)小新 AIR14 2021 锐龙 R5 轻薄本手提商务办公学生学习超薄笔记本电脑 灰色 标配:锐龙 R5 六核 8GB 256GB 固态",
```

```
            "price":3999,
            "pic":"/common/product/4.webp",
            "sale":"已售 1.4 万件",
            "shop":"中柏平板京东自营旗舰店"
                },
                {
            "title":"惠普(HP)星 15 青春版 15.6 英寸轻薄窄边框笔记本电脑(R7 - 4700U 16G 512GSSD UMA FHD
        IPS)银",
            "price":4099,
            "pic":"/common/product/5.webp",
            "sale":"已售 3.9 万件",
            "shop":"惠普京东自营官方旗舰店"
                }
            ]
        }
}
```

### 3. swiper 组件

swiper 组件提供了切换子组件显示的能力。子组件支持除< list >之外的所有子组件。轮播图是最常见的页面结构，如图 3.70 所示。

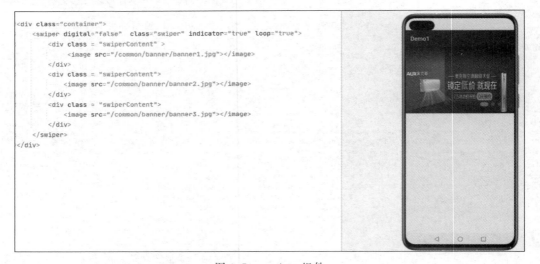

图 3.70　swiper 组件

swiper 组件的结构如代码示例 3.56 所示。indicator 属性表示是否启用导航点指示器，默认值为 true；loop 属性表示是否开启循环轮播。digital 属性表示是否启用数字导航点，默认值为 false。

**代码示例 3.56　swiper 用法**

```
< div class = "container">
```

```html
<swiper digital = "false" class = "swiper" indicator = "true" loop = "true">
    <div class = "swiperContent">
        <image src = "/common/banner/banner1.jpg"></image>
    </div>
    <div class = "swiperContent">
        <image src = "/common/banner/banner2.jpg"></image>
    </div>
    <div class = "swiperContent">
        <image src = "/common/banner/banner3.jpg"></image>
    </div>
</swiper>
</div>
```

样式如代码示例 3.57 所示。导航点指示器的位置需要通过样式控制，在.swiper 的样式中添加 indicator-bottom：20px 和 indicator-right：30px；将导航指示器的位置设置为右下角位置；indicator-color：#cf2411 用于设置导航点指示器的填充颜色；indicator-size：14px 用于设置导航点指示器的直径大小；indicator-selected-color 用于设置导航点指示器选中的颜色。

代码示例 3.57　swiper 样式

```css
.container {
    flex-direction: column;
    width: 100%;
    height: 100%;
    align-items: center;
}
.swiper {
    flex-direction: column;
    align-content: center;
    align-items: center;
    width: 100%;
    max-height: 230px;
    indicator-color: #cf2411;
    indicator-size: 14px;
    indicator-bottom: 20px;
    indicator-right: 30px;
    indicator-selected-color: #0f0;
}
.swiperContent {
    height: 100%;
    justify-content: center;
}
image{
    object-fit: fill;
}
```

**4. tabs 组件**

tabs 组件为子组件，仅支持一个 <tab-bar> 和支持一个 <tab-content>。

tabs 页签组件可配合 tab-bar 和 tab-content 组件一起使用，tab-bar 中的一个组件节点对应 tab-content 组件中的一个组件节点。当单击 tab-bar 中的一个元素的时候，tab-content 将自动显示对应索引的子节点。

实现图 3.71 和图 3.72 效果，这两个图的效果实际上是鸿蒙的适配效果，在 config.json 文件中将 deviceType 设置为[tv、phone]，这样就可以预览到下面的效果。

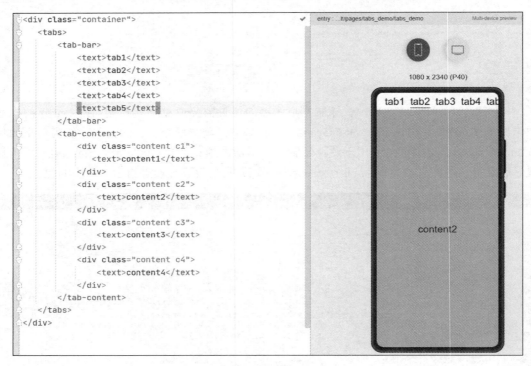

图 3.71　tabs 组件手机效果

TV 效果如图 3.72 所示。

tabs 页面结构如代码示例 3.58 所示。如果 tab-bar 中的元素超出横向范围，则可以自动拖动。

代码示例 3.58　tabs 页面结构

```
<div class = "container">
<tabs>
<tab-bar>
<text>tab1</text>
<text>tab2</text>
<text>tab3</text>
<text>tab4</text>
<text>tab5</text>
</tab-bar>
```

```html
<tab-content>
<div class = "content c1">
<text> content1 </text>
</div>
<div class = "content c2">
<text> content2 </text>
</div>
<div class = "content c3">
<text> content3 </text>
</div>
<div class = "content c4">
<text> content4 </text>
</div>
</tab-content>
</tabs>
</div>
```

图 3.72 tabs 组件 TV 效果

tabs 页面样式如代码示例 3.59 所示。

代码示例 3.59　tabs 页面样式

```css
.container {
    justify-content: center;
    align-items: center;
}
```

```css
.content {
    width: 100%;
    height: 100%;
    justify-content: center;
    align-items: center;
    background-color: gainsboro;
}

.c1 {
    background-color: red;
}

.c2 {
    background-color: greenyellow;
}

.c3 {
    background-color: goldenrod;
}

.c4 {
    background-color: darkblue;
}
```

tab-bar 切换可以触发 change 事件,如代码示例 3.60 所示。

**代码示例 3.60**

```js
export default {
    change: function(e) {
        console.log("Tab index: " + e.index);
    },
}
```

### 5. toolbar 组件

toolbar 组件放在界面底部,用于展示针对当前界面的操作选项。子组件支持< toolbar-item >子组件。工具栏最多可以展示 5 个 toolbar-item 子组件,如果存在 6 个及以上 toolbar-item 子组件,则保留前面 4 个子组件,后续的子组件将被收纳到工具栏上的更多项中,通过单击更多项弹窗可展示剩下的子组件,更多项展示的组件样式采用系统默认样式,toolbar-item 上设置的自定义样式不生效,如图 3.73 所示。

toolbar 组件的布局结构如代码示例 3.61 所示。如果省略 value 属性,则只显示图标。

**代码示例 3.61 toolbar 组件**

```html
<div class="container">
<toolbar style="position:fixed; bottom:0px;">
```

```
<toolbar-item icon='common/navs/f01.png'
              value='Option 1'></toolbar-item>
<toolbar-item icon='common/navs/f02.png'
              value='Option 2'></toolbar-item>
<toolbar-item icon='common/navs/f03.png'
              value='Option 3'></toolbar-item>
<toolbar-item icon='common/navs/f04.png'
              value='Option 4'></toolbar-item>
<toolbar-item icon='common/navs/f05.png'
              value='Option 5'></toolbar-item>
<toolbar-item icon='common/navs/f06.png'
              value='Option 6'></toolbar-item>
</toolbar>
</div>
```

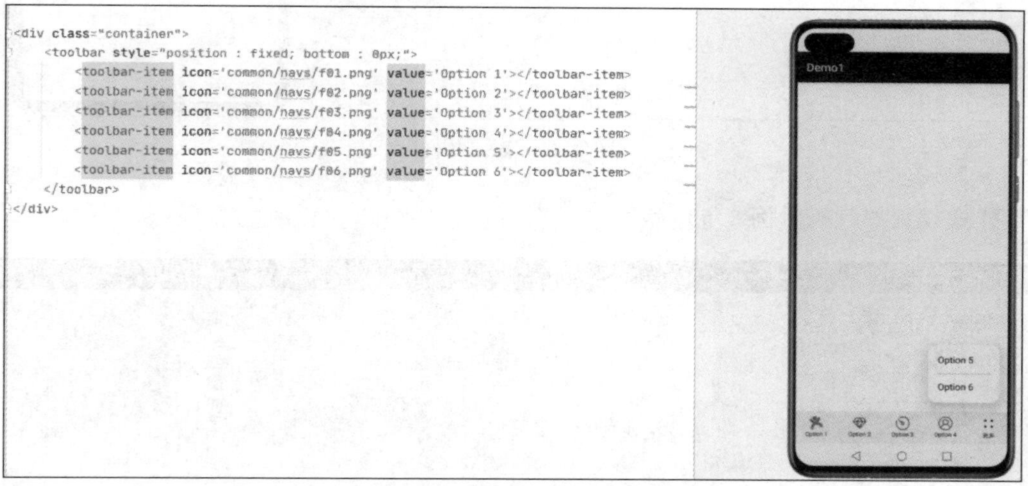

图 3.73 toolbar 组件

### 6. stack 组件

stack 组件按照顺序依次入栈,后一个子组件覆盖前一个子组件,如图 3.74 所示。stack 组件的布局结构如代码示例 3.62 所示。

**代码示例 3.62　stack 组件布局**

```
<stack class="stack-parent">
<div class="back-child bd-radius"></div>
<div class="positioned-child bd-radius"></div>
<div class="front-child bd-radius"></div>
</stack>
```

```html
<stack class="stack-parent">
    <div class="back-child bd-radius"></div>
    <div class="positioned-child bd-radius"></div>
    <div class="front-child bd-radius"></div>
</stack>
```

图 3.74　stack 组件

样式设置如代码示例 3.63 所示。

代码示例 3.63

```css
.stack-parent {
    width: 400px;
    height: 400px;
    background-color: #ccc;
    border-width: 1px;
    border-style: solid;
    justify-content: center;
    align-items: center;
    margin-top: 100px;
}
.back-child {
    width: 300px;
    height: 300px;
    background-color: #3f56ea;
}
.front-child {
    width: 150px;
    height: 150px;
    background-color: #00bfc9;
}
.positioned-child {
    width: 100px;
    height: 100px;
```

```
        left: 50px;
        top: 50px;
        background-color: #47cc47;
}
.bd-radius {
        border-radius: 16px;
}
```

#### 7. badge 组件

应用中如果有需要用户关注的新事件提醒,则可以采用新事件标记来标识,从 API Version 5 开始支持。仅支持单子组件节点,如果使用多子组件节点,则默认使用第 1 个子组件节点。

badge 组件的子组件大小不能超过 badge 组件本身的大小,否则子组件不会绘制,如图 3.75 所示。

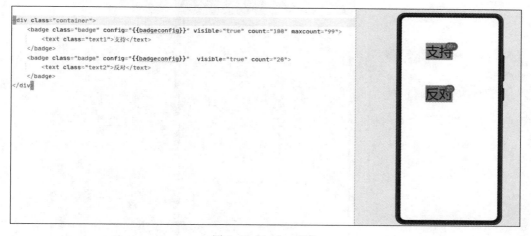

图 3.75　badge 组件

badge 组件的使用方式如代码示例 3.64 所示。

代码示例 3.64

```
<badge class="badge" config="{{badgeconfig}}" visible="true" count="100" maxcount="99">
<text class="text1">支持</text>
</badge>
```

通过代码可控制 badge 的样式,如代码示例 3.65 所示,config 属性的值可通过代码进行设置。textSize 用于标记数字文本的大小;badgeSize 表示圆点标记的默认大小。

代码示例 3.65

```
export default {
data:{
```

```
            badgeconfig:{
                badgeColor:"#0a59f7",
                textColor:"#ffffff",
            }
        }
    }
```

#### 8. refresh 组件

refresh 组件如图 3.76 所示，该容器不支持在智慧屏上使用。可以通过 refresh 组件的 type 属性设置组件刷新时的动效，共有两个可选值，不支持动态修改，refresh 组件的方法如表 3.12 所示。

图 3.76　refresh 组件

表 3.12　refresh 组件的方法

| 项目类型 | 参数 | 描述 |
| --- | --- | --- |
| refresh | { refreshing: refreshingValue } | 下拉刷新状态变化时触发。可能值为 false：当前处于下拉刷新过程中 true：当前未处于下拉刷新过程中 |

续表

| 项目类型 | 参 数 | 描 述 |
|---|---|---|
| pulldown | {state: string} | 下拉开始和松手时触发。可能值为<br>start：表示开始下拉<br>end：表示结束下拉 |

auto：默认效果，列表界面拉到顶部后，列表不移动，下拉后有转圈弹出。

pulldown：列表界面拉到顶部后，可以继续往下滑动一段距离触发刷新，刷新完成后有回弹效果（如果子组件含有 list，为了防止下拉效果冲突，则需将 list 的 scrolleffect 设置为 no）。

如代码示例 3.66 所示，属性 onrefresh 的值是刷新时回调的方法，refreshing 属性用于标识刷新组件当前是否正在刷新，默认值为 false。

**代码示例 3.66　refresh_demo.hml**

```html
<refresh onrefresh = "refresh"
        refreshing = "{{ this.isRefreshing }}"
        lasttime = "true"
        type = "pulldown"
        friction = "42">
<div class = "container">
<text class = "title">
        下拉更新数据：
</text>
<text class = "content">
        {{ title }}
</text>
</div>
</refresh>
```

样式如代码示例 3.67 所示。

**代码示例 3.67　refresh_demo.css**

```css
.container {
    display: flex;
    justify-content: center;
    align-items: center;
    left: 0px;
    top: 0px;
    width: 454px;
    height: 454px;
    flex-direction: column;
}

.title {
    font-size: 20px;
    text-align: center;
```

```css
    width: 200px;
    height: 100px;
}

.content {
    font-size: 20px;
    color: red;
}
```

处理刷新请求的代码如代码示例 3.68 所示。

**代码示例 3.68　refresh_demo.js**

```js
const injectRef = Object.getPrototypeOf(global) || global;
injectRef.regeneratorRuntime = require('@babel/runtime/regenerator');

export default {
    data: {
        title: 'HarmonyOS 2.0',
        isRefreshing:null
    },
    requestData(delay) {
        return new Promise((resolve) => {
            setTimeout(() => {
                this.isRefreshing = false;
                //Simulate refresh results
                const messageNum = Math.round(Math.random());
                console.log("requestData:" + messageNum);
                resolve(messageNum);
            }, delay);
        });
    },
    async showResult(){
        try {
            const messageNum = await this.requestData(5000);
            console.log("showResult:" + messageNum);
            this.title = "更新为HarmonyOS 2.0." + messageNum
        }catch(e){
            console.error('requestData caught exception' + e);
        }
    },
    refresh: function (refreshingValue) {
        this.showResult();
        this.isRefreshing = refreshingValue.refreshing;
    }
}
```

### 9. panel 组件

panel 组件如图 3.77 所示,提供了一种轻量的内容展示窗口,可方便地在不同尺寸中切换,属于弹出式组件。可滑动面板暂不支持渲染属性,包括 for、if 和 show。

可以设置可滑动面板的类型 type,不可动态变更,可选值有以下 3 种,通常和滑动面板的 3 种可选模式一起使用。

(1) minibar:提供 minibar 和全屏展示切换效果。

(2) foldable:内容永久展示,提供大(全屏)、中(半屏)、小 3 种尺寸展示切换效果,如图 3.77 所示。

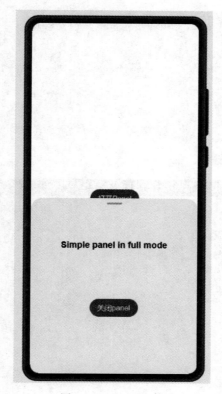

图 3.77　panel 组件

(3) temporary:内容临时展示区,提供大(全屏)、中(半屏)2 种尺寸展示切换效果。

滑动面板有 3 种状态,mode 参数可选值有以下 3 种。

(1) mini:当类型为 minibar 和 foldable 时,为最小状态;如果类型为 temporary,则不生效。

(2) half:当类型为 foldable 和 temporary 时,为类半屏状态;如果类型为 minibar,则不生效。

(3) full:类全屏状态。

单击"关闭 panel"按钮,通过 this.$element('simplepanel').show()方法弹出 panel,如代码示例 3.69 所示。

**代码示例 3.69    panel_demo.hml**

```html
<div class = "doc-page">
<div class = "btn-div">
<button type = "capsule" value = "打开 Panel" onclick = "showPanel"></button>
</div>
<panel id = "simplepanel" type = "foldable" mode = "half" onsizechange = "changeMode" miniheight = "200px">
<div class = "panel-div">
<div class = "inner-txt">
<text class = "txt">Simple panel in {{modeFlag}} mode</text>
</div>
<div class = "inner-btn">
<button type = "capsule" value = "关闭 panel" onclick = "closePanel"></button>
</div>
</div>
</panel>
</div>
```

样式如代码示例 3.70 所示。

**代码示例 3.70    panel_demo.css**

```css
.doc-page {
    flex-direction: column;
    justify-content: center;
    align-items: center;
}
.btn-div {
    width: 100%;
    height: 200px;
    flex-direction: column;
    align-items: center;
    justify-content: center;
}
.txt {
    color: #000000;
    font-weight: bold;
    font-size: 19px;
}
.panel-div {
    width: 100%;
    flex-direction: column;
    align-items: center;
}
.inner-txt {
    width: 100%;
    height: 160px;
    flex-direction: column;
    align-items: center;
```

```css
        justify-content: center;
    }
    .inner-btn {
        width: 100%;
        height: 120px;
        justify-content: center;
        align-items: center;
    }
```

modeFlag 属性用于设置 panel 的弹出方式,打开和关闭可通过 show、close 方法实现,如代码示例 3.71 所示。

代码示例 3.71　panel_demo.js

```js
export default {
    data: {
        modeFlag: "half"
    },
    showPanel(e) {
        this.$element('simplepanel').show()
    },
    closePanel(e) {
        this.$element('simplepanel').close()
    },
    changeMode(e) {
        this.modeFlag = e.mode
    }
}
```

### 10. stepper 组件

当完成一个任务需要多个步骤时,可以使用 stepper 组件展示当前进展。子组件仅支持<stepper-item>子组件。步骤导航器内的步骤顺序按照子组件<stepper-item>的顺序进行排序,如图 3.78 所示。

stepper 案例的布局结构如代码示例 3.72 所示。

代码示例 3.72　stepper_demo.hml

```html
<div class = "container">
<stepper class = "stepper" id = "mystepper" index = "0" onnext = "nextclick" onback = "backclick">
<stepper-item class = "stepperItem" label = "{{label_1}}">
<div class = "stepperItemContent">
<text class = "text">第 1 页</text>
</div>
<button type = "capsule" class = "button" value = "设置右边按钮的状态" onclick = "setRightButton"></button>
</stepper-item>
```

```
<stepper-item class = "stepperItem" label = "{{label_2}}">
    <div class = "stepperItemContent">
        <text class = "text">第 2 页</text>
    </div>
    <button type = "capsule" class = "button" value = "设置右边按钮的状态" onclick = "setRightButton"></button>
</stepper-item>
<stepper-item class = "stepperItem" label = "{{label_3}}">
    <div class = "stepperItemContent">
        <text class = "text">第 3 页</text>
    </div>
    <button type = "capsule" class = "button" value = "设置右边按钮的状态" onclick = "setRightButton"></button>
</stepper-item>
</stepper>
</div>
```

图 3.78 stepper 组件

样式如代码示例 3.73 所示。

代码示例 3.73 stepper_demo.css

```
.container {
    margin-top: 20px;
```

```css
    flex-direction: column;
    justify-content: center;
    align-items: center;
    height: 300px;
}
.stepperItem {
    height: 100%;
    flex-direction: column;
    justify-content: center;
    align-items: center;

}
.stepperItemContent {
    color: #0000ff;
    font-size: 50px;
    justify-content: center;
    align-items: center;
}
.button {
    width: 60%;
    margin-top: 30px;
    justify-content: center;
}
```

逻辑处理如代码示例3.74所示。

**代码示例3.74　stepper_demo.js**

```js
export default {
    data: {
        label_1:
        {
            prevLabel: '返回',
            nextLabel: '下一步',
            status: 'normal'
        },
        label_2:
        {
            prevLabel: '返回',
            nextLabel: '下一步',
            status: 'normal'
        },
        label_3:
        {
            prevLabel: '返回',
            nextLabel: '下一步',
            status: 'normal'
        },
    },
```

```
        setRightButton(e) {
            this.$element('mystepper').setNextButtonStatus({status: 'skip', label: '跳过'});
        },
        nextclick(e) {
            var index = {
                pendingIndex: e.pendingIndex
            }
            return index;
        },
        backclick(e) {
            var index = {
                pendingIndex: e.pendingIndex
            }
            return index;
        },
    }
```

### 11. dialog 组件

dialog 组件如图 3.79 所示,子组件支持单个子组件。dialog 属性、样式均不支持动态更新。

图 3.79  dialog 组件效果

案例布局如代码示例 3.75 所示。

代码示例 3.75 dialog_demo.hml
```html
<div class = "doc-page">
<div class = "btn-div">
<button type = "capsule" value = "单击弹出 dialog 窗口" class = "btn" onclick = "showDialog">
</button>
</div>
<dialog id = "simpledialog" class = "dialog-main" oncancel = "cancelDialog">
<div class = "dialog-div">
<div class = "inner-txt">
<text class = "txt">请确认填写的信息！</text>
</div>
<div class = "inner-btn">
<button type = "capsule" value = "再改改" onclick = "cancelSchedule" class = "btn-txt btn_cancel"></button>
<button type = "capsule" value = "提交" onclick = "setSchedule" class = "btn-txt"></button>
</div>
</div>
</dialog>
</div>
```

样式如代码示例 3.76 所示。

代码示例 3.76 dialog_demo.css
```css
.doc-page {
    flex-direction: column;
    justify-content: center;
    align-items: center;
}
.btn-div {
    width: 100%;
    height: 200px;
    flex-direction: column;
    align-items: center;
    justify-content: center;
}
.btn_cancel {
    background-color: #ccc;
}
.txt {
    color: #000000;
    font-weight: bold;
    font-size: 19px;
}
.dialog-main {
    width: 300px;
}
```

```css
.dialog-div {
    flex-direction: column;
    align-items: center;
}
.inner-txt {
    width: 300px;
    height: 80px;
    flex-direction: column;
    align-items: center;
    justify-content: space-around;
}
.inner-btn {
    width: 300px;
    height: 80px;
    justify-content: space-around;
    align-items: center;
}
```

逻辑处理如代码示例 3.77 所示。

**代码示例 3.77　dialog_demo.js**

```js
import prompt from '@system.prompt';

export default {
    showDialog(e) {
        this.$element('simpledialog').show()
    },
    cancelDialog(e) {
        prompt.showToast({
            message: 'Dialog cancelled'
        })
    },
    cancelSchedule(e) {
        this.$element('simpledialog').close()
        prompt.showToast({
            message: 'Successfully cancelled'
        })
    },
    setSchedule(e) {
        this.$element('simpledialog').close()
        prompt.showToast({
            message: 'Successfully confirmed'
        })
    }
}
```

## 3.6.2 基础组件

### 1. button 组件

button 组件如图 3.80 所示,包括胶囊按钮、圆形按钮、文本按钮、弧形按钮、下载按钮。如果 icon 使用云端路径,则需要开启权限列表 ohos.permission.INTERNET,不支持子组件。

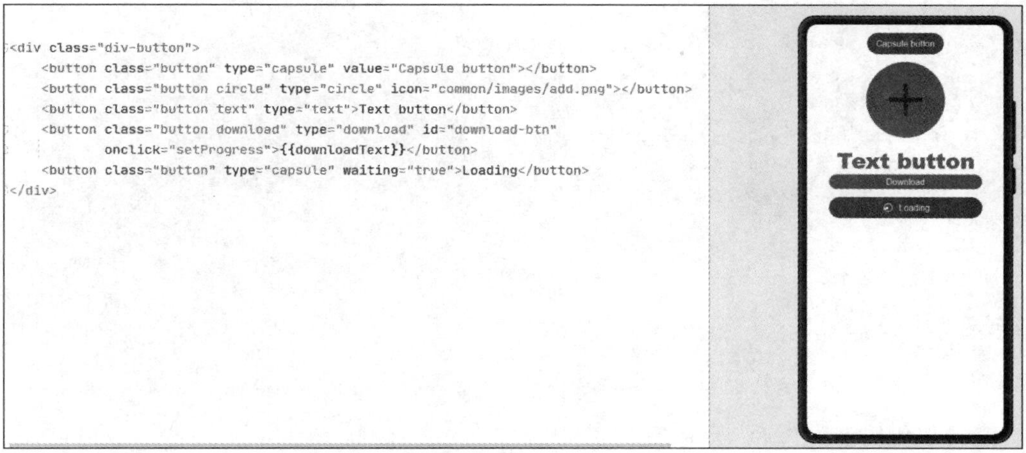

图 3.80 button 组件效果

图 3.80 所示的布局如代码示例 3.78 所示。

**代码示例 3.78  button_demo.hml**

```html
<div class = "div-button">
<button class = "button" type = "capsule" value = "Capsule button"></button>
<button class = "button circle" type = "circle" icon = "common/images/add.png"></button>
<button class = "button text" type = "text">Text button</button>
<button class = "button download" type = "download" id = "download-btn"
        onclick = "setProgress">{{downloadText}}</button>
<button class = "button" type = "capsule" waiting = "true">Loading</button>
</div>
```

样式如代码示例 3.79 所示。

**代码示例 3.79  button_demo.css**

```css
.div-button {
    flex-direction: column;
    align-items: center;
}
.button {
    margin-top: 15px;
}
```

```css
.button:waiting {
    width: 280px;
}
.circle {
    background-color: #007dff;
    radius: 72px;
    icon-width: 72px;
    icon-height: 72px;
}
.text {
    text-color: red;
    font-size: 40px;
    font-weight: 900;
    font-family: sans-serif;
    font-style: normal;
}
.download {
    width: 280px;
    text-color: white;
    background-color: #007dff;
}
```

逻辑如代码示例 3.80 所示。

**代码示例 3.80　button_demo.js**

```js
export default {
    data: {
        progress: 5,
        downloadText: "Download"
    },
    setProgress(e) {
        this.progress += 10;
        this.downloadText = this.progress + "%";
        this.$element('download-btn').setProgress({ progress: this.progress });
        if (this.progress >= 100) {
            this.downloadText = "Done";
        }
    }
}
```

### 2. input 组件

input 组件包括单选框、复选框、按钮和单行文本输入框。不支持子组件，如图 3.81 所示。

input 组件类型的可选值为 text、email、date、time、number、password、button、checkbox 和 radio。其中 text、email、date、time、number 和 password 这 6 种类型之间支持动态切换修改，button、checkbox 和 radio 不支持动态修改。可选值定义如下。

(1) text：定义一个单行的文本字段。
(2) email：定义用于 e-mail 地址的字段。
(3) date：定义用于输入日期的控件(包括年、月、日,但不包括时间)。
(4) time：定义用于输入时间的控件(不带时区)。
(5) number：定义用于输入数字的字段。
(6) password：定义密码字段(字段中的字符会被遮蔽)。
(7) button：定义可单击的按钮。
(8) checkbox：定义复选框。
(9) radio：定义单选按钮,允许在多个拥有相同 name 值的选项中选中其中一个。
智能穿戴仅支持 button、radio、checkbox 类型。

图 3.81 input 组件效果

案例布局如代码示例 3.81 所示。

代码示例 3.81　input_demo.hml

```
< div class = "container">
< text >
```

```html
<span>用户登录</span>
</text>
<input class="input" type="text" value="" maxlength="20" enterkeytype="send"
        headericon="/common/search.png" placeholder="输入用户名或者邮箱" onchange="change"
        onenterkeyclick="enterkeyClick">
</input>
<input class="input" type="password" value="" maxlength="20"
enterkeytype="send"
        headericon="/common/mima.png" placeholder="输入密码"
onchange="change"
        onenterkeyclick="enterkeyClick">
</input>
<div class="check">
<input onchange="checkboxOnChange" checked="true" type="checkbox"></input>
<text>记住用户名</text>
</div>
<input class="button" type="button" value="登 录" onclick="buttonClick"></input>
</div>
```

样式如代码示例3.82所示。

**代码示例3.82　input_demo.css**

```css
.container {
    width:100%;
    height: 500px;
    flex-direction: column;
    justify-content: center;
    align-items: center;
}
.input {
    placeholder-color: gray;
    width: 300px;
    margin-top: 10px;
}
.button {
    background-color: gray;
    margin-top: 20px;
    width: 150px;
}

.check {
    justify-content: center;
    align-items: center;
}

.check text {
    font-size: 15px;
}
```

逻辑处理如代码示例 3.83 所示。

代码示例 3.83　input_demo.js
```js
import prompt from '@system.prompt'
export default {
    change(e){
        prompt.showToast({
            message: "value: " + e.value,
            duration: 3000,
        });
    },
    enterkeyClick(e){
        prompt.showToast({
            message: "enterkey clicked",
            duration: 3000,
        });
    },
    buttonClick(e){
        this.$element("input").showError({
            error: 'error text'
        });
    },
    checkboxOnChange(e) {
        prompt.showToast({
            message:'checked: ' + e.checked,
            duration: 3000,
        });
    }
}
```

3. toggle 组件

toggle 组件用于从一组选项中进行选择，并可能在界面上实时显示选择后的结果。通常这一组选项都由状态按钮构成，如图 3.82 所示。

上面案例的布局如代码示例 3.84 所示。

代码示例 3.84　toggle_demo.hml
```html
<div class = "container">
<text class = "margin">1. 多选</text>
<div style = "flex-wrap: wrap">
<toggle class = "margin" for = "{{toggles}}">{{ $item }}</toggle>
</div>
<divider class = "divider" vertical = "false"></divider>
<text class = "margin">2. 单选</text>
<div style = "flex-wrap: wrap">
<toggle class = "margin"
            for = "{{toggle_list}}"
```

```
                    id = "{{ $ item.id}}"
                    checked = "{{ $ item.checked}}"
                    value = "{{ $ item.name}}" @change = "allchange"
@click = "allclick({{ $ item.id}})">
</toggle>
</div>
</div>
```

图 3.82  toggle 组件效果

样式如代码示例 3.85 所示。

代码示例 3.85  toggle_demo.css

```css
.container {
    justify-content: flex-start;
    align-items: flex-start;
    flex-direction: column;
    margin-top: 10px;
}
.margin {
    margin: 7px;
```

```
    }
    .divider {
        margin: 10px;
        color: #ccc;
        stroke-width: 3px;
        line-cap: round;
    }
```

逻辑如代码示例3.86所示。

**代码示例3.86　toggle_demo.js**
```
export default {
    data: {
        toggle_list: [
                    { "id":"1001", "name":"客厅", "checked":true },
                    { "id":"1002", "name":"卧室", "checked":false },
                    { "id":"1003", "name":"次卧", "checked":false },
                    { "id":"1004", "name":"厨房", "checked":false },
                    { "id":"1005", "name":"书房", "checked":false },
                    { "id":"1006", "name":"花园", "checked":false },
                    { "id":"1007", "name":"洗澡间", "checked":false },
                    { "id":"1008", "name":"阳台", "checked":false },
        ],
        toggles:["客厅","卧室","厨房","书房"],
        idx: ""
    },
    allclick(arg) {
        this.idx = arg
    },
    allchange(e) {
        if (e.checked === true) {
            for (var i = 0; i < this.toggle_list.length; i++) {
                if (this.toggle_list[i].id === this.idx) {
                    this.toggle_list[i].checked = true
                } else {
                    this.toggle_list[i].checked = false
                }
            }
        }
    }
}
```

**4. menu 组件**

menu组件作为临时性弹出窗口，用于展示用户可执行的操作，如图3.83所示。案例布局如代码示例3.87所示。

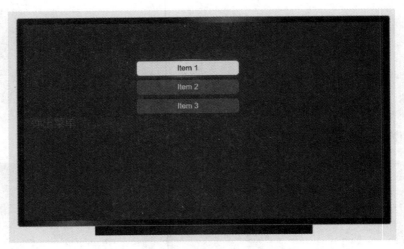

图 3.83  menu 组件效果

代码示例 3.87  menu_demo.hml

```html
<div class = "container">
<text onclick = "onTextClick" class = "title-text">弹出菜单</text>
<menu id = "apiMenu" onselected = "onMenuSelected">
<option value = "Item 1">Item 1</option>
<option value = "Item 2">Item 2</option>
<option value = "Item 3">Item 3</option>
</menu>
</div>
```

通过 show 方法，可以指定显示的坐标位置，如代码示例 3.88 所示。

代码示例 3.88  menu_demo.js

```js
import prompt from '@system.prompt';
export default {
    onMenuSelected(e) {
        prompt.showToast({
            message: e.value
        })
    },
    onTextClick() {
        this.$element("apiMenu").show({x:280,y:120});
    }
}
```

5. marquee 组件

跑马灯组件，用于展示一段单行滚动的文字，如图 3.84 所示。
案例布局如代码示例 3.89 所示。

图 3.84 marquee 组件效果

代码示例 3.89 marquee_demo.hml

```
<div class = "container">
<marquee id = "customMarquee"
         class = "customMarquee"
         scrollamount = "{{scrollAmount}}"
         loop = "{{loop}}"
         direction = "{{marqueeDir}}"
         onbounce = "onMarqueeBounce"
         onstart = "onMarqueeStart"
         onfinish = "onMarqueeFinish">
    {{marqueeCustomData}}
</marquee>
<div class = "content">
<button class = "controlButton" onclick = "onStartClick">开始播放</button>
<button class = "controlButton" onclick = "onStopClick">停止播放</button>
</div>
</div>
```

样式如代码示例 3.90 所示。

代码示例 3.90 marquee_demo.css

```
.container {
    flex-direction: column;
    justify-content: center;
    align-items: center;
    background-color: deepskyblue;
}
.customMarquee {
    width: 100%;
```

```css
        height: 80px;
        padding: 10px;
        margin: 20px;
        border: 4px solid #ff8888;
        border-radius: 20px;
        font-size: 40px;
        color: #ff8888;
        font-weight: bolder;
        font-family: serif;
        background-color: #ffdddd;
}
.content {
        flex-direction: row;
        justify-content: center;
        align-items: center;
}
.controlButton {
        width: 100px;
        height: 30px;
        margin-right: 10px;
        background-color: deepskyblue;
}
```

start 方法可以启动走马灯效果，如代码示例 3.91 所示。

**代码示例 3.91　marquee_demo.js**

```js
export default {
    data: {
        scrollAmount: 30,
        loop: 3,
        marqueeDir: 'left',
        marqueeCustomData: 'HarmonyOS 2.0',
    },
    onMarqueeBounce: function() {
        console.log("onMarqueeBounce");
    },
    onMarqueeStart: function() {
        console.log("onMarqueeStart");
    },
    onMarqueeFinish: function() {
        console.log("onMarqueeFinish");
    },
    onStartClick (evt) {
        this.$element('customMarquee').start();
    },
    onStopClick (evt) {
        this.$element('customMarquee').stop();
    }
}
```

### 6. image 组件

image 组件用来渲染展示的图片。

本组件需要访问网络，因此需要申请 ohos.permission.INTERNET 权限（如果使用云端路径）。

首先，需要在配置文件 config.json 下的 module 下添加 reqPermissions，如代码示例 3.92 所示。

**代码示例 3.92　申请权限**

```
"module": {
"abilities": [],
"reqPermissions": [
    {
"name": "ohos.permission.INTERNET"
    }
  ]
}
```

现在可以在 HML 中使用 URL 的方式引用网络图片地址了。

```
<image src = "http://blog.51itcto.com/wp-content/uploads/2020/11/1062-220x150.jpeg">
</image>
```

object-fit 设置图片的缩放类型，如表 3.13 所示。

表 3.13　缩放类型

| 缩放类型 | 描述 |
| --- | --- |
| cover | 保持宽高比进行缩小或者放大，使图片两边都大于或等于显示边界，居中显示 |
| contain | 保持宽高比进行缩小或者放大，使图片完全显示在显示边界内，居中显示 |
| fill | 不保持宽高比进行放大缩小，使图片填充满显示边界 |
| none | 保持原有尺寸进行居中显示 |
| scale-down | 保持宽高比居中显示，图片缩小或者保持不变 |

### 7. chart 组件

chart 组件用于呈现线图、柱状图、量规图界面。下面分别介绍 chart 组件的几种图表的用法。

线图如图 3.85 所示，使用方法如代码示例 3.93 所示。

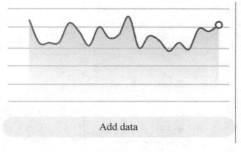

图 3.85　线图

代码示例 3.93　线图

```html
<!-- xxx.hml -->
<div class="container">
  <stack class="chart-region">
    <image class="chart-background" src="common/background.png"></image>
    <chart class="chart-data" type="line" ref="linechart" options="{{lineOps}}" datasets="{{lineData}}"></chart>
  </stack>
  <button value="Add data" onclick="addData"></button>
</div>
```

```css
/* xxx.css */
.container {
  flex-direction: column;
  justify-content: center;
  align-items: center;
}
.chart-region {
  height: 400px;
  width: 700px;
}
.chart-background {
  object-fit: fill;
}
.chart-data {
  width: 700px;
  height: 600px;
}
```

```js
//xxx.js
export default {
  data: {
    lineData: [
      {
        strokeColor: '#0081ff',
        fillColor: '#cce5ff',
        data: [763, 550, 551, 554, 731, 654, 525, 696, 595, 628, 791, 505, 613, 575, 475, 553, 491, 680, 657, 716],
        gradient: true,
      }
    ],
    lineOps: {
      xAxis: {
        min: 0,
        max: 20,
        display: false,
      },
      yAxis: {
        min: 0,
        max: 1000,
        display: false
```

```
      },
      series: {
        lineStyle: {
          width: "5px",
          smooth: true,
        },
        headPoint: {
          shape: "circle",
          size: 20,
          strokeWidth: 5,
          fillColor: '#ffffff',
          strokeColor: '#007aff',
          display: true
        },
        loop: {
          margin: 2,
          gradient: true,
        }
      }
    },
  },
  addData() {
    this.$refs.linechart.append({
      serial: 0,
      data: [Math.floor(Math.random() * 400) + 400]
    })
  }
}
```

柱状图如图3.86所示,使用方法如代码示例3.94所示。

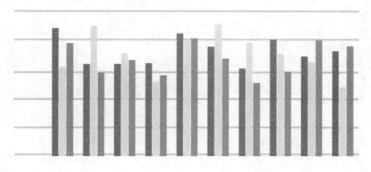

图3.86 柱状图

**代码示例3.94 柱状图**

```
<!-- xxx.hml -->
<div class = "container">
```

```
<stack class = "data-region">
<image class = "data-background" src = "common/background.png"></image>
<chart class = "data-bar" type = "bar" id = "bar-chart" options = "{{barOps}}" datasets = "{{barData}}"></chart>
</stack>
</div>
/* xxx.css */
.container {
  flex-direction: column;
  justify-content: center;
  align-items: center;
}
.data-region {
  height: 400px;
  width: 700px;
}
.data-background {
  object-fit: fill;
}
.data-bar {
  width: 700px;
  height: 400px;
}
//xxx.js
export default {
  data: {
    barData: [
      {
        fillColor: '#f07826',
        data: [763, 550, 551, 554, 731, 654, 525, 696, 595, 628],
      },
      {
        fillColor: '#cce5ff',
        data: [535, 776, 615, 444, 694, 785, 677, 609, 562, 410],
      },
      {
        fillColor: '#ff88bb',
        data: [673, 500, 574, 483, 702, 583, 437, 506, 693, 657],
      },
    ],
    barOps: {
      xAxis: {
        min: 0,
        max: 20,
        display: false,
        axisTick: 10
      },
      yAxis: {
        min: 0,
```

```
            max: 1000,
            display: false
        },
    },
}
```

量规图如图 3.87 所示,使用方法如代码示例 3.95 所示。

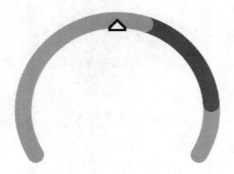

图 3.87 量规图

代码示例 3.95 量规图

```
<!-- xxx.hml -->
<div class = "container">
<div class = "gauge-region">
<chart class = "data-gauge" type = "gauge" percent = "50"></chart>
</div>
</div>
/* xxx.css */
.container {
  flex-direction: column;
  justify-content: center;
  align-items: center;
}
.gauge-region {
  height: 400px;
  width: 400px;
}
.data-gauge {
  colors: #83f115, #fd3636, #3bf8ff;
  weights: 4, 2, 1;
}
```

### 8. picker 组件

picker 组件支持的类型包括普通选择器、日期选择器、时间选择器、时间日期选择器和多列文本选择器。

时间选择器如图3.88所示。将type值设置为time,如代码示例3.96所示。

图3.88 时间选择器

代码示例3.96 picker_demo.hml

```html
<div class="container">
<button @click="showPicker"
        class="button"
        type="capsule"
        value="时间选择器(type=time)">

</button>
<picker type="time" id="picker"></picker>
</div>
```

代码中通过show方法弹出picker组件,如代码示例3.97所示。

代码示例3.97 picker_demo.js

```js
export default {
    data: {
```

```
            title: 'World'
        },
        showPicker(){
            this.$element("picker").show()
        }
    }
```

日期选择器如图 3.89 所示,将 type 值设置为 date,如代码示例 3.98 所示。

图 3.89 日期选择器

代码示例 3.98 picker_demo.hml

```
<button @click = "showPicker"
        class = "button"
        type = "capsule"
        value = "日期选择器(type = date)">
</button>

<picker type = "date" id = "picker"></picker>
```

日期时间选择器如图 3.90 所示,将 type 值设置为 datetime,代码如下:

```
<picker type = "datetime" id = "picker"></picker>
```

图 3.90　日期时间选择器

普通选择器：type=text，如代码示例 3.99 所示，如图 3.91 所示。

**代码示例 3.99　picker_demo.hml**

```
<button @click = "showPicker"
        class = "button"
        type = "capsule"
        value = "普通选择器:type = text">
</button>

<picker range = "{{singleData}}" type = "text" id = "picker"></picker>
```

使用时需要使用数据绑定的方式，如 range={{data}}，还需要在 JS 中声明相应变量：singleData:[100,200,300,400,500]，如代码示例 3.100 所示。

**代码示例 3.100　picker_demo.js**

```
export default {
    data: {
```

```
            singleData:[100,200,300,400,500]
    },
    showPicker(){
        this.$element("picker").show()
    }
}
```

图 3.91　普通文本选择器

多列文本选择器: type=multi-text,如代码示例 3.101 所示,如图 3.92 所示。

代码示例 3.101　picker_demo.hml

```
<button @click = "showPicker"
        class = "button"
        type = "capsule"
        value = "多列文本选择器(type = multi - text)">
</button>
<picker range = "{{data}}" type = "multi - text" id = "picker"></picker>
```

设置多列文本选择器的选择项,其中 range 为二维数组。长度表示多少列,数组的每项表示每列的数据,如 [["a","b"],["c","d"]],如代码示例 3.102 所示。

代码示例3.102　picker_demo.js

```js
export default {
    data: {
        data:[["a","b"],["c","d"]]
    },
    showPicker(){
        this.$element("picker").show()
    }
}
```

图 3.92　多列文本选择器

#### 9. picker-view 组件

嵌入页面的滑动选择器。如图 3.93 所示，和 picker 组件类似，可以设置滑动选择器的类型，type 属性不支持动态修改，可选项如下。

(1) text：文本选择器。

(2) time：时间选择器。

(3) date：日期选择器。

(4) datetime：日期时间选择器。

(5) multi-text：多列文本选择器。

图 3.93　picker-view 组件

picker-view 组件，如代码示例 3.103 所示。

代码示例 3.103　picker_view_demo.hml

```
<div class = "container" @swipe = "handleSwipe">
<text class = "title">
    选择的时间:{{time}}
</text>
<picker-view class = "time-picker" type = "time" selected = "{{defaultTime}}" @change = "handleChange"></picker-view>
</div>
```

样式如代码示例 3.104 所示。

代码示例 3.104　picker_view_demo.css

```
.container {
    flex-direction: column;
    justify-content: center;
    align-items: center;
```

```css
        left: 0px;
        top: 0px;
        width: 454px;
        height: 454px;
}
.title {
    font-size: 20px;
    text-align: center;
}
.time-picker {
    width: 320px;
    height: 210px;
    margin-top: 20px;
}
```

逻辑如代码示例 3.105 所示。

**代码示例 3.105　picker_view_demo.js**

```javascript
export default {
    data: {
        defaultTime: "",
        time: "",
    },
    onInit() {
        this.defaultTime = now();
    },
    handleChange(data) {
        this.time = concat(data.hour, data.minute);
    }
};

function now() {
    const date = new Date();
    const hours = date.getHours();
    const minutes = date.getMinutes();
    return concat(hours, minutes);
}


function fill(value) {
    return (value > 9 ? "" : "0") + value;
}


function concat(hours, minutes) {
    return `${fill(hours)}:${fill(minutes)}`;
}
```

### 10. piece 组件

piece 组件是一种块状的入口,可包含图片和文本。常用于展示收件人,例如:邮件收件人或信息收件人。如果使用云端路径,则需要申请 ohos.permission.INTERNET 权限,如图 3.94 所示。

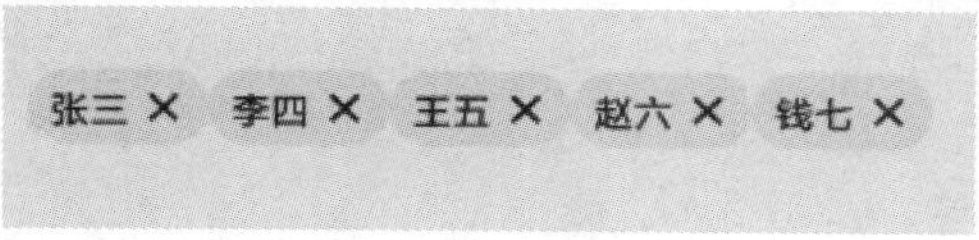

图 3.94 piece 组件

piece 组件的布局如代码示例 3.106 所示。

```
代码示例 3.106    piece_demo.hml
<div class = "container">
<piece for = "{{tags}}"
        content = "{{ $item }}"
        closable = "true"
        onclose = "closeSecond">
</piece>
</div>
```

循环 tags 输出一组 piece 组件,如代码示例 3.107 所示。

```
代码示例 3.107    piece_demo.js
export default {
    data: {
        tags:["张三","李四","王五","赵六","钱七"]
    },
    closeSecond(e) {
        console.log(JSON.stringify(e))
    }
}
```

### 11. qrcode 组件

qrcode 组件生成并显示二维码,如图 3.95 所示。

当二维码的 width 和 height 不一致时,以二者最小值作为二维码的边长,并且最终生成的二维码居中显示。

当二维码的 width 和 height 只设置一个时,取设置的值作为二维码的边长。当都不设置时,使用默认值 200px 作为边长,代码如下:

```
<qrcode value = "https://huawei.com"></qrcode>
```

可以通过 color 设置二维码的颜色,并且可以通过 background-color 设置二维码的背景颜色，如代码示例 3.108 所示。

代码示例 3.108　qrcode_demo.css

```css
.qrcode {
    color: red;
    background-color: darkcyan;
    width: 300px;
    height: 300px;
}
```

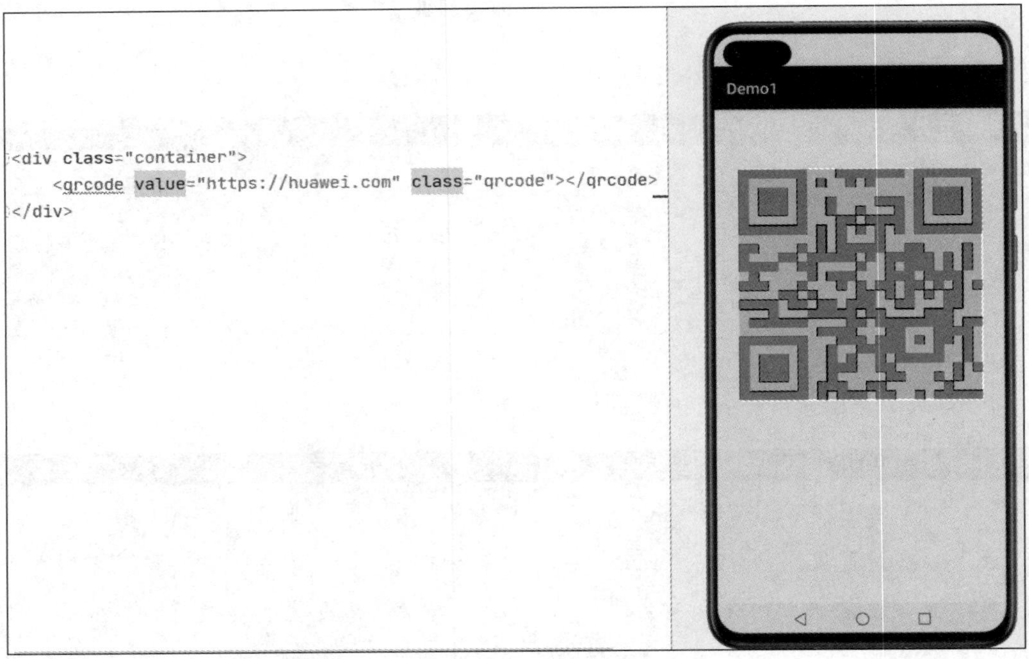

```html
<div class="container">
    <qrcode value="https://huawei.com" class="qrcode"></qrcode>
</div>
```

图 3.95　qrcode 组件

### 3.6.3　媒体组件

#### 1. video 组件

video 组件需要访问网络，因此需要申请 ohos.permission.INTERNET 权限（如果使用云端路径）。用法如代码示例 3.109 所示。

代码示例 3.109　video_deom.hml

```html
< video id = 'id'
    src = '/common/mmmp4'
    muted = 'false'
    autoplay = 'false'
    poster = '/common/m.png'
```

```
        controls = "true"
        onprepared = 'preparedCallback'
        onstart = 'startCallback'
        onpause = 'pauseCallback'
        onfinish = 'finishCallback'
        onerror = 'errorCallback'
        onseeking = 'seekingCallback'
        onseeked = 'seekedCallback'
        ontimeupdate = 'timeupdateCallback'
        onlongpress = 'change_fullscreenchange'
        onclick = "change_start_pause">
</video>
```

效果如图 3.96 所示。

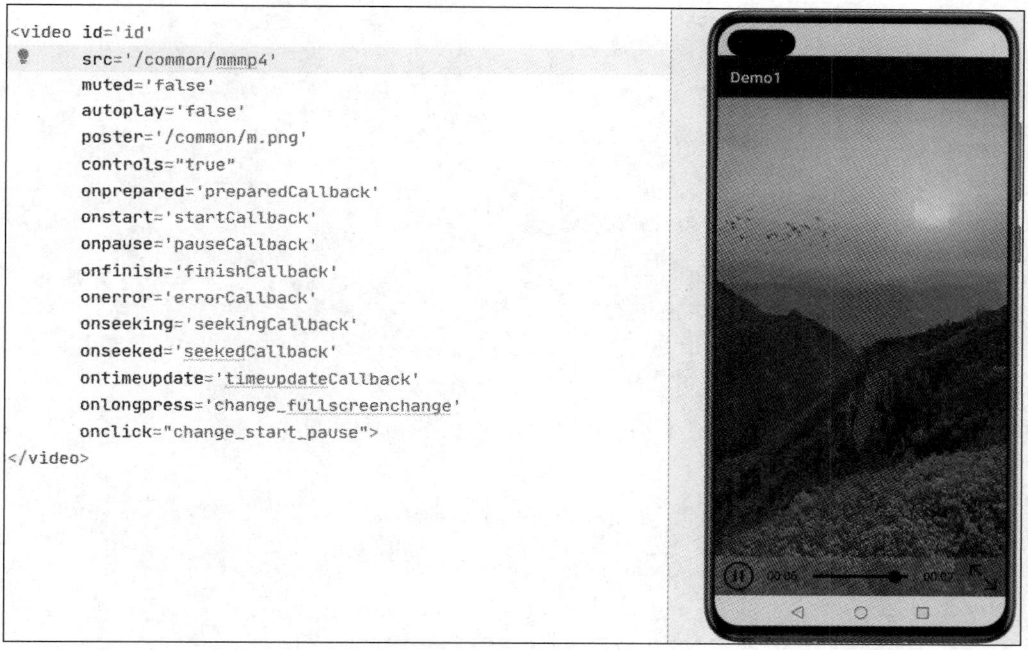

```
<video id='id'
    src='/common/mmmp4'
    muted='false'
    autoplay='false'
    poster='/common/m.png'
    controls="true"
    onprepared='preparedCallback'
    onstart='startCallback'
    onpause='pauseCallback'
    onfinish='finishCallback'
    onerror='errorCallback'
    onseeking='seekingCallback'
    onseeked='seekedCallback'
    ontimeupdate='timeupdateCallback'
    onlongpress='change_fullscreenchange'
    onclick="change_start_pause">
</video>
```

图 3.96 video 组件

通过 API 提供的方法控制视频,如代码示例 3.110 所示。

代码示例 3.110　video_deom.js

```js
export default {
    data: {
        event: '',
        seekingtime: '',
        timeupdatetime: '',
```

```
            seekedtime: '',
            isStart: true,
            isfullscreenchange: false,
            duration: '',
        },
        preparedCallback: function (e) {
            this.event = '视频连接成功';
            this.duration = e.duration;
        },
        startCallback: function () {
            this.event = '视频开始播放';
        },
        pauseCallback: function () {
            this.event = '视频暂停播放';
        },
        finishCallback: function () {
            this.event = '视频播放结束';
        },
        errorCallback: function () {
            this.event = '视频播放错误';
        },
        seekingCallback: function (e) {
            this.seekingtime = e.currenttime;
        },
        timeupdateCallback: function (e) {
            this.timeupdatetime = e.currenttime;
        },
        change_start_pause: function () {
            if (this.isStart) {
                this.$element('videoId').pause();
                this.isStart = false;
            } else {
                this.$element('videoId').start();
                this.isStart = true;
            }
        },
        change_fullscreenchange: function () { //全屏
            if (!this.isfullscreenchange) {
                this.$element('videoId').requestFullscreen({
                    screenOrientation: 'default'
                });
                this.isfullscreenchange = true;
            } else {
                this.$element('videoId').exitFullscreen();
                this.isfullscreenchange = false;
            }
        }
    }
}
```

## 2. 短视频 App 实战

下面通过 video 组件模仿短视频 App 的效果,短视频整个页面是一个播放的视频,如图 3.97 所示。

图 3.97 video 组件实现短视频

首先,实现页面布局,可以把页面分为 3 部分:上部、中部和下部。中部用于显示 video,右下为社交区,包括红心、留言、转发按钮。视频上部为视频标题,视频下部为视频描述。

社交区域采用 fixed 定位方式固定在右边的位置上。

视频标题和描述同样采用 fixed 定位方式固定在左下边的位置上。

实现如代码示例 3.111 所示。

代码示例 3.111 video_dy_demo.hml

```
<div class = "container">
<video ref = "avplayer"
        class = "video"
        controls = "false"
```

```html
            autoplay = "true"
            onclick = "resumePlayer"
            src = "{{ videos[currentVideoIndex].url }}"
            onswipe = "nextVideo"
            onFinish = "pausePlayer">
</video>
<image src = "/res/img/icon_play_pause.png" style = "object-fit : none;"
            show = "{{ showPauseBtn }}"></image>

<div class = "socialBar">
<button type = "circle"
                icon = "{{ likeImgUrl }}"
                style = "background-color : transparent;"
                onclick = "likeit"></button>
<button type = "circle" icon = "/res/img/icon_comment.png"
                style = "background-color : transparent;"></button>
<button type = "circle" icon = "/res/img/icon_transfer.png"
                style = "background-color : transparent;"
                onclick = "transfer"></button>
</div>

<div class = "video-title">
<text style = "color : white; font-size : 20px; font-weight : bold;">
            {{ videos[currentVideoIndex].title }}
</text>
<text style = "color : white; font-size : 16px;">
            {{ videos[currentVideoIndex].desc }}
</text>
</div>
</div>
```

样式如代码示例 3.112 所示。

**代码示例 3.112　video_dy_demo.css**

```css
.container {
    flex-direction: column;
    justify-content: center;
    align-items: center;
}

.video {
    width : 100%;
    height : 100%;
    object-fit : cover;
}

.socialBar {
    position: fixed;
```

```css
    flex-direction: column;
    justify-content: space-between;
    align-items: center;
    height: 200px;
    bottom: 150px;
    right: 30px;
}

.video-title {
    position: fixed;
    flex-direction: column;
    justify-content: center;
    align-items: flex-start;
    left:30px;
    bottom: 20px;
    width: 220px;
    height: 200px;
}

.pauseButton {
    background-image: url('res/video/icon_play_pause.png');
}
```

逻辑实现如代码示例 3.113 所示,这里当在视频上向上滑动的时候,播放下一个视频,通过在 video 组件上的 onswiper 方法监听向上的滑动事件并调用 nextVideo 方法,以此获取下一个视频的索引。

代码示例3.113 video_dy_demo.js

```js
export default {
    data: {
        btnClicked: false,
        showPauseBtn: false,
        playing: true,
        likeBtnClicked: false,
        likeImgUrl: '/res/img/icon_like_before.png',
        videos: [
            {
                url: "/res/video/tianlongbabu.mp4",
                title: "看跳舞的女生",
                desc: "是不是男生都爱看跳舞的女生"
            },
            {
                url: "/res/video/driver.mp4",
                title: "@杏子的精彩视频",
                desc: "我怀疑你在开车,可是,我没有证据啊"
            }
        ],
```

```
        currentVideoIndex: 0
    },
    likeit() {
        this.likeBtnClicked = !this.likeBtnClicked

        if (this.likeBtnClicked) {
            this.likeImgUrl = '/res/img/icon_like_after.png'
        } else {
            this.likeImgUrl = '/res/img/icon_like_before.png'
        }
    }
,
    pausePlayer() {
        this.$refs.avplayer.pause()
        this.showPauseBtn = false
    },
    nextVideo() {
        if (this.currentVideoIndex < this.videos.length - 1) {
            this.currentVideoIndex += 1
        } else {
            this.currentVideoIndex = 0
        }
    },
    resumePlayer() {
        let player = this.$refs.avplayer
        if (this.playing) {
            player.pause()
        } else {
            player.start()
        }
        this.playing = !this.playing
        this.showPauseBtn = !this.showPauseBtn
    },
}
```

### 3.6.4 画布组件

画布组件，用于自定义绘制图形。

**注意**：canvas 对象的获取，只能在页面生命周期 onShow()方法中获取 Context。

canvas 画布的方法如表 3.14 所示，需要注意的是 getContext()方法不支持在 onInit 和 onReady 中进行调用，只能在 onShow()方法中调用。

表 3.14 canvas 的方法

| 项目类型 | 参数 | 描述 |
| --- | --- | --- |
| getContext | String | 不支持在 onInit 和 onReady 中进行调用 |

下面我们介绍一下 canvas 的用法，代码如下：

```
<canvas id="board" class="board"></canvas>
```

在 onShow() 方法中获取 Context 对象，代码如下：

```
var heroCanvas = this.$element("heroCanvas")
this.heroCxt = heroCanvas.getContext("2d")
```

下面通过一个自由涂鸦画板的例子来实现涂鸦效果，如代码示例 3.114 所示。

canvas 实现涂鸦效果，先在 canvas 上绑定 ontouchend、ontouchstart、ontouchmove 这 3 个监听事件，这些监听事件的事件对象可以获取监听的 touch 事件的坐标位置信息，以此实现涂鸦功能。

**代码示例 3.114　canvas_demo.hml**

```
<div class="container">
<canvas id="board"
        ontouchend="paintEnd"
        ontouchstart="painStart"
        ontouchmove="paint"
        class="board">
</canvas>
</div>
```

涂鸦效果如图 3.98 所示。

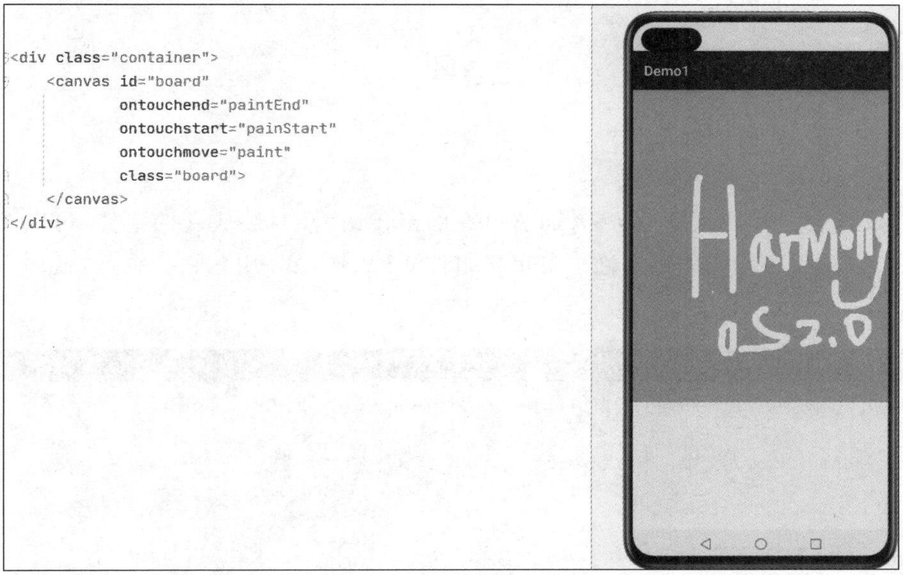

图 3.98　canvas 组件涂鸦

涂鸦实现,如代码示例 3.115 所示。

代码示例 3.115　canvas_demo.js
```js
export default {
    data: {
        cxt: {}
    },
    onInit() {
    },
    onShow() {
        this.cxt = this.$element("board").getContext("2d");
    },
    painStart(e) {
        this.cxt.beginPath();
        this.cxt.strokeStyle = "white";
        this.cxt.lineWidth = 10
        this.cxt.lineCap = "round"
        this.cxt.lineJoin = "round"
        //绘制起点
        this.cxt.moveTo(e.touches[0].localX, e.touches[0].localY)
    },
    paint(e) {
        console.error(e.touches[0].localX);
        this.cxt.lineTo(e.touches[0].localX, e.touches[0].localY);
        this.cxt.stroke();
    },
    paintEnd() {
        this.cxt.closePath();
    },
}
```

## 3.7　自定义组件

自定义组件是用户根据业务需求,将已有的组件组合,封装成新的组件,可以在工程中多次调用,以此提高代码的可读性。自定义组件通过 element 引入宿主页面,使用方法如代码示例 3.116 所示。

代码示例 3.116　自定义组件的引入方式
```
<element name='comp' src='../../common/component/comp.hml'></element>
<div>
<comp prop1='xxxx' @child1="bindParentVmMethod"></comp>
</div>
```

element 标签的属性说明:
(1) name 属性指自定义组件名称(非必填),组件名称对大小写不敏感,默认使用小写。

src 属性指自定义组件 HML 文件路径(必填),若没有设置 name 属性,则默认使用 HML 文件名作为组件名。

(2) 事件绑定:自定义组件中绑定子组件事件可使用(on|@)child1 语法,子组件中通过 this.\$emit('child1',{params:'传递参数'})触发事件并进行传值,父组件执行 bindParentVmMethod 方法并接收子组件传递的参数。

**注意**:子组件中使用驼峰命名法命名的事件,在父组件中绑定时需要使用短横线分隔命名形式,例如:@children-event 表示绑定子组件的 childrenEvent 事件,如@children-event="bindParentVmMethod"。

自定义组件中的内置对象说明,如图 3.99 所示。

| 属性 | 类型 | 描述 |
| --- | --- | --- |
| data | Object/Function | 页面的数据模型,类型是对象或者函数,如果类型是函数,返回值必须是对象。属性名不能以\$或_开头,不要使用保留字for、if、show和tid。<br>data与private、public不能重合使用 |
| props | Array/Object | props用于组件之间的通信,可以通过<tag xxxx='value'>的方式传递给组件;props名称必须用小写,不能以\$或_开头,不要使用保留字for、if、show和tid。目前props的数据类型不支持Function |
| computed | Object | 用于在读取或设置中进行预先处理,计算属性的结果会被缓存。计算属性名不能以\$或_开头,不要使用保留字 |

图 3.99 自定义组件的内置对象说明

## 3.7.1 自定义组件定义

**注意**:自定义组件与 page 中的组件结构是一样的,包含 HML、CSS、JS 文件。

下面具体介绍创建组件的步骤。

步骤 1:在 common 目录下存放着常见的公开组件,一个目录就是一个组件,每个组件分为三部分:index.css、index.hml、index.js,如图 3.100 所示。

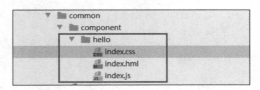

图 3.100 自定义组件

步骤 2:编写好自定义组件后,通过 element 在需要引用的地方引入,如代码示例 3.117 所示。

**代码示例 3.117 自定义组件的引入**

```
<!-- 通过 element 引用自定义组件: name 组件名 -->
<element name = "hello"
         src = "../../common/component/hello/index.hml">
</element>
```

```
<div class = "container">
<hello></hello> //组件声明
</div>
```

### 3.7.2 自定义组件事件与交互

组件创建好后,可以给一个组件添加输入和输出属性,如图 3.101 所示,这样就可以复用组件的逻辑了。

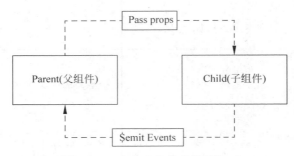

图 3.101 自定义组件事件与交互

自定义组件可以通过 props 声明属性,父组件通过设置属性向子组件传递参数。camelCase(驼峰命名法)的 prop 名,在外部父组件传递参数时需要使用 kebab-case(短横线分隔命名)形式,即当属性 compProp 在父组件引用时需要转换为 comp-prop。给自定义组件添加 props,通过父组件向下传递参数,如代码示例 3.118 所示。

代码示例 3.118 props 属性

```
<!-- comp.hml -->
<div class = "item">
<text class = "title-style">{{compProp}}</text>
</div>
//comp.js
export default {
    props: ['compProp'],
}
<!-- xxx.hml -->
<element name = 'comp' src = '../../common/component/comp/comp.hml'></element>
<div class = "container">
<comp comp-prop = "{{title}}"></comp>
</div>
```

子组件可以通过固定值 default 设置默认值,当父组件没有设置该属性时,将使用其默认值。此情况下 props 属性必须为对象形式,不能用数组形式,如代码示例 3.119 所示。

代码示例 3.119 props 默认值

```
<!-- comp.hml -->
```

```
<div class = "item">
<text class = "title-style">{{title}}</text>
</div>
//comp.js
export default {
  props: {
    title: {
      default: 'title',
    },
  },
}
```

本示例中加入了一个 text 组件,用于显示标题,标题的内容是一个自定义属性,用于显示用户设置的标题内容,当用户没有设置时显示默认值 title。在引用该组件时可添加该属性的设置,代码如下:

```
<!-- xxx.hml -->
<element name = 'comp' src = '../../common/component/comp/comp.hml'></element>
<div class = "container">
<comp title = "自定义组件"></comp>
</div>
```

父子组件之间数据的传递是单向的,只能从父组件传递给子组件,子组件不能直接修改父组件传递下来的值,可以将 props 传入的值用 data 接收后作为默认值,再对 data 的值进行修改,如代码示例 3.120 所示。

**代码示例 3.120　props 的单向传递**

```
//comp.js
export default {
  props: ['defaultCount'],
  data() {
    return {
      count: this.defaultCount,
    };
  },
  onClick() {
    this.count = this.count + 1;
  },
}
```

### 1. $watch 监控数据改变

如果需要观察组件中属性的变化,则可以通过 $watch 方法增加属性变化回调。使用方法如代码示例 3.121 所示。

**代码示例 3.121　$watch**

```
//comp.js
```

```js
export default {
  props: ['title'],
  onInit() {
    this.$watch('title', 'onPropertyChange');
  },
  onPropertyChange(newV, oldV) {
    console.info('title 属性变化 ' + newV + '' + oldV);
  },
}
```

#### 2. computed 计算属性

自定义组件中经常需要在读取或设置某个属性时进行预处理,以此提高开发效率,此种情况需要使用 computed 字段。computed 字段中可通过设置属性的 get() 和 set() 方法在属性读写的时候进行触发,使用方式如代码示例 3.122 所示。

**代码示例 3.122　computed 计算属性**

```js
//comp.js
export default {
  props: ['title'],
  data() {
    return {
      objTitle: this.title,
      time: 'Today',
    };
  },
  computed: {
    message() {
      return this.time + '' + this.objTitle;
    },
    notice: {
      get() {
        return this.time;
      },
      set(newValue) {
        this.time = newValue;
      },
    },
  },
  onClick() {
    console.info('get click event ' + this.message);
    this.notice = 'Tomorrow';
  },
}
```

这里声明的第 1 个计算属性 message 默认只有 get() 函数,message 的值会取决于 objTitle 的值的变化。get() 函数只能读取而不能改变值,当需要赋值给计算属性的时候可以提供一个 set() 函数,如示例中的 notice。

**3. 组件的输出属性**

输出属性是在组件的声明标签上定义的,格式如下:

```
<hello title = "hello world" @out = "callbackfunc"></hello>
```

**注意**:这里 title 是组件的输入属性,@out 是输出属性名,@out 的值是在父组件中定义的方法,可以通过这个函数从组件内将数据传出去。

子组件也可以通过绑定的事件向上传递参数,在自定义事件上添加传递参数的示例如代码示例 3.123 所示。

**代码示例 3.123 组件的输出属性**

```html
<!-- comp.hml -->
<div class = "item">
<text class = "text-style" onclick = "childClicked">单击这里查看隐藏文本</text>
<text class = "text-style" if = "{{showObj}}">hello world</text>
</div>
//comp.js
export default {
    childClicked () {
       this. $ emit('eventType1', {text: '收到子组件参数'});
       this.showObj = !this.showObj;
    },
}
```

子组件向上传递参数 text,父组件接收时通过 e.detail 获取参数,如代码示例 3.124 所示。

**代码示例 3.124 获取参数**

```html
<!-- xxx.hml -->
<div class = "container">
<text>父组件:{{text}}</text>
<comp @event-type1 = "textClicked"></comp>
</div>
//xxx.js
export default {
    data: {
      text: '开始',
    },
    textClicked (e) {
      this.text = e.detail.text;
    },
}
```

## 3.8 本章小结

本章通过 7 个小节,分别介绍了 ArkUI JS 应用框架的语法,以及接口调用。ArkUI JS 框架适用于富设备,如手机、TV、手表。

# 第 4 章 ArkUI JS 与 Java 混合开发

本章讲解 ArkUI JS 与 Java UI 混合编程，通过混合编程可以充分发挥不同框架的优点，让项目开发更加容易。

## 4.1 JavaScript 调用 Service Ability

JS UI 框架提供了 JS FA（Feature Ability）调用 Java PA（Particle Ability）的机制。该机制在 HarmonyOS 引擎内提供了一种通道来传递方法调用、数据返回、事件上报，开发者可根据需要自行实现 FA 和 PA 两端的对应接口，以此完成对应的功能逻辑。

FA（JS API）调用 PA（Java API）的机制，包含远端调用 Ability 和本地调用 Internal Ability 两种方式。

FA 提供了以下 3 个 JS 接口：

（1）FeatureAbility.callAbility(OBJECT)：调用 PA 能力。

（2）FeatureAbility.subscribeAbilityEvent(OBJECT，Function)：订阅 PA 能力。

（3）FeatureAbility.unsubscribeAbilityEvent(OBJECT)：取消订阅 PA 能力。

PA 端提供的两个接口：

（1）boolean IRemoteObject.onRemoteRequest（int code，MessageParcel data，MessageParcel reply，MessageOption option）：Ability 方式，与 FA 通过 rpc 方式通信，该方式的优点在于 PA 可以被不同的 FA 调用。

（2）boolean onRemoteRequest(int code，MessageParcel data，MessageParcel reply，MessageOption option)：Internal Ability 方式，集成在 FA 中，适用于与 FA 业务逻辑关联性强，响应时延要求高的服务。该方式仅支持本 FA 访问调用。

### 4.1.1 JS 端调用远端 Service Ability

JS 端与 Java 端通过接口扩展机制进行通信，通过 bundleName 和 abilityName 进行关联。在 FeatureAbility Plugin 收到 JS 调用请求后，系统根据开发者在 JS 指定的 abilityType、Ability 或 Internal Ability 来选择对应的方式进行处理。开发者在 onRemoteRequest()方法中实现 PA

提供的业务逻辑,不同的业务通过业务码来区分。

下面介绍在JavaScript项目中,如何创建一个提供计算服务的例子:

第1步:在项目的java目录下的包上右击鼠标,创建一个空的Service Ability,如图4.1所示。

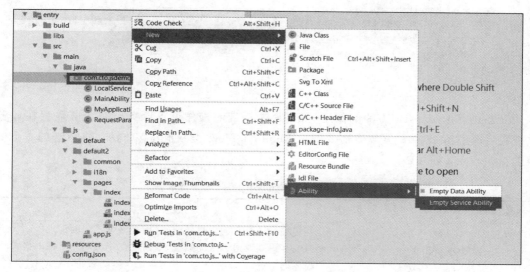

图 4.1 创建 Service Ability

第2步:编写Service Ability,FA在请求PA服务时会调用Ability connectAbility连接PA,连接成功后,需要通过onConnect返回一个远程代理remote对象,如代码示例4.1所示。

首先,创建一个服务器端的代理类MyRemoteObject。这个类必须是一个RemoteObject类的子类,同时实现IRemoteBroker接口,重写接口中的onRemoteRequest()方法,该方法用于获取从客户端发送过来的请求,处理后返回给客户端,如代码示例4.1所示。

**代码示例4.1 MyRemoteObject**

```
class MyRemoteObject extends RemoteObject implements IRemoteBroker {

    public MyRemoteObject() {
        super("MyRemoteObject");
    }

    @Override
    public IRemoteObject asObject() {
        return this;
    }

    @Override
    public boolean onRemoteRequest ( int code, MessageParcel data, MessageParcel reply, MessageOption option) throws RemoteException {
```

```java
        //读取客户端发送过来的数据,并把发送的字符串转换成对象
        String zsonStr = data.readString();
        RequestParam param = ZSONObject.stringToClass(zsonStr, RequestParam.class);
        //返回结果仅支持可序列化的 Object 类型
        Map<String, Object> zsonResult = new HashMap<String, Object>();
        zsonResult.put("code", 200);
        zsonResult.put("abilityResult", param.getFirstNum() + param.getSecondNum());
        reply.writeString(ZSONObject.toZSONString(zsonResult));
        return true;
    }
}
```

onRemoteRequest()方法的作用是把 JS 端携带的操作请求业务码及业务数据在业务执行完后返回给 JS 端。开发者需要继承 RemoteObject 类并重写该方法。该方法的 4 个参数说明如表 4.1 所示。

表 4.1 onRemoteRequest()方法的参数说明

| 参数名 | 类型 | 非空 | 说明 |
| --- | --- | --- | --- |
| code | number | 是 | 客户端发送过来的随机数字码,用来区分不同的请求 |
| data | MessageParcel | 否 | 封装客户端发送过来的数据,当前仅支持 JSON 字符串格式 |
| reply | MessageParcel | 否 | 服务器端处理后的结果,通过 reply 包装返回,当前仅支持 String 格式。一般 MessageParcel 的第 1 个值是处理成功或者失败的数字码,第 2 个值才是真正处理的结果 |
| option | MessageOption | 否 | 指示操作是同步还是异步的方式 |

在 Service Ability 中通过 onConnect()方法返回上面定义的远程代理对象,如代码示例 4.2 所示。onConnect()方法的作用是开发者的 PA 首次被 FA 连接时回调,并返回 IRemoteObject 对象,用于后续的业务通信。开发者需要继承 Ability 类并重写该方法。

代码示例 4.2 Service Ability

```java
import ohos.aafwk.ability.Ability;
import ohos.aafwk.content.Intent;
import ohos.rpc.*;
import ohos.hiviewdfx.HiLog;
import ohos.hiviewdfx.HiLogLabel;
import java.util.HashMap;
import java.util.Map;
import ohos.utils.zson.ZSONObject;

public class RemoteServiceAbility extends Ability {

    private MyRemoteObject remoteObject = new MyRemoteObject();

    @Override
```

```
    public IRemoteObject onConnect(Intent intent) {
        return remoteObject.asObject();
    }

    @Override
    public void onDisconnect(Intent intent) {
    }

}
```

第 3 步：在组件中调用编写好的计算服务 Service Ability，如代码示例 4.3 所示。

**代码示例 4.3　JavaScript 调用 PA**

```
export default {
    data: {
        num: 0
    },
    onInit() {

    },
    plus: async function(){
        var actionData = {};
        actionData.firstNum = 1024;
        actionData.secondNum = 2048;

        var action = {};
        action.bundleName = 'com.example.demo3';
        action.abilityName = 'com.example.demo3.RemoteServiceAbility';
        action.messageCode = 101;
        action.data = actionData;
        action.abilityType = 0;
        action.syncOption = 0;

        var result = await FeatureAbility.callAbility(action);
        var ret = JSON.parse(result);
        if (ret.code == 200) {
            //对每次返回的结果进行追加
            this.num += ret.abilityResult;
            console.info('plus result is:' + JSON.stringify(ret.abilityResult));
        } else {
            console.error('plus error code:' + JSON.stringify(ret.code));
        }
    }
}
```

FeatureAbility.callAbility(OBJECT)可调用 PA(Particle Ability)提供的能力。该方法的参数如表 4.2 所示。

表 4.2  FeatureAbility.callAbility 方法的参数说明

| 参数名 | 类型 | 非空 | 说明 |
| --- | --- | --- | --- |
| bundleName | String | 是 | Ability 的包名称,需要与 PA 端匹配,区分大小写 |
| abilityName | String | 是 | Ability 名称,需要与 PA 端匹配,区分大小写 |
| messageCode | number | 否 | Ability 操作码(操作码用于定义 PA 的业务功能,需要与 PA 端约定) |
| data | Object | 否 | 发送到 Ability 的数据(根据不同的业务携带相应的业务数据,数据字段名称需要与 PA 端约定) |
| abilityType | number | 否 | Ability 类型,对应 PA 端不同的实现方式。<br>0: Ability,拥有独立的 Ability 生命周期,FA 使用远端进程通信拉起并请求 PA 服务,适用于提供基本服务供多 FA 调用或者在后台独立运行的场景,具体 Java 侧接口定义见 Ability 模块接口(Java 语言,Ability 方式)。<br>1: Internal Ability,与 FA 共进程,采用内部函数调用的方式和 FA 通信,适用于对 PA 响应时延要求较高的场景,不支持其他 FA 访问调用能力,具体 Java 侧接口定义见 ArkUIInternalAbility 类(Java 语言,Internal Ability 方式) |
| syncOption | number | 否 | PA 侧请求消息处理同步/异步选项,非必填,默认使用同步方式。当前异步方式仅支持 AbilityType 为 Internal Ability 类型。<br>0: 同步方式,默认方式。<br>1: 异步方式 |

## 4.1.2  JS 端订阅远端 Service Ability

JS 端可以订阅 Service Ability,订阅 PA 后,PA 会不断地向订阅端推送数据变化。订阅端可以在适当的时候取消订阅,如图 4.2 所示。

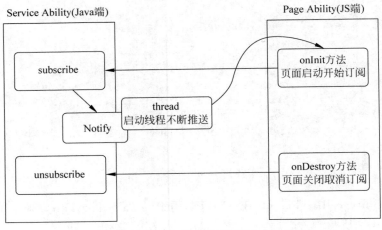

图 4.2  JS 端订阅 PA

创建远程代码类 MySubRemoteObject，如代码示例 4.4 所示。

代码示例 4.4　demo3/java/MySubRemoteObject.java

```java
package com.example.demo3;

import ohos.rpc.*;
import ohos.utils.zson.ZSONObject;
import java.util.HashMap;
import java.util.HashSet;
import java.util.Map;
import java.util.Set;

public class MySubRemoteObject extends RemoteObject implements IRemoteBroker {

    //支持多 FA 订阅,如果仅支持单 FA 订阅,则可直接使用变量存储:private IRemoteObject
    //remoteObjectHandler;
    private Set<IRemoteObject> remoteObjectHandlers = new HashSet<IRemoteObject>();

    public MySubRemoteObject() {
        super("可订阅的 Service Ability");
    }

    @Override
    public IRemoteObject asObject() {
        return this;
    }

    @Override
    public boolean onRemoteRequest(int code, MessageParcel data, MessageParcel reply, MessageOption option) throws RemoteException {
        switch (code){
            case 1000: {
                remoteObjectHandlers.add(data.readRemoteObject());
                startNotify();
                //return result, the key field should be negotiated with JS side
                Map<String, Object> zsonResult = new HashMap<String, Object>();
                zsonResult.put("code", 200);
                reply.writeString(ZSONObject.toZSONString(zsonResult));
                break;
            }
            case 1001:{
                remoteObjectHandlers.remove(data.readRemoteObject());
                //return result, the key field should be negotiated with JS side
                Map<String, Object> zsonResult = new HashMap<String, Object>();
                zsonResult.put("code", 200);
                reply.writeString(ZSONObject.toZSONString(zsonResult));
                break;
            }
            default:{
```

```
                    reply.writeString("service not defined");
                    return false;
                }
            }
            return true;
        }

        public void startNotify() {
            System.out.println("------ startNotify ------");
            new Thread(() -> {
                while (true) {
                    try {
                        Thread.sleep(5 * 100);
                        reportEvent();
                    } catch (RemoteException | InterruptedException e) {
                        break;
                    }
                }
            }).start();
        }

        private void reportEvent() throws RemoteException {
            MessageParcel data = MessageParcel.obtain();
            MessageParcel reply = MessageParcel.obtain();
            MessageOption option = new MessageOption();
            Map<String, Object> zsonEvent = new HashMap<String, Object>();
            zsonEvent.put("abilityEvent", "test event!");
            System.out.println("------------");
            data.writeString(ZSONObject.toZSONString(zsonEvent));
            for (IRemoteObject item : remoteObjectHandlers) {
                item.sendRequest(100, data, reply, option);
            }
            reply.reclaim();
            data.reclaim();
        }
    }
}
```

接下来,创建 Service Ability,如代码示例 4.5 所示。

**代码示例 4.5　demo3/java/RemoteEventServiceAbility.java**

```
import ohos.aafwk.ability.Ability;
import ohos.aafwk.content.Intent;
import ohos.rpc.IRemoteObject;
import ohos.hiviewdfx.HiLog;
import ohos.hiviewdfx.HiLogLabel;

public class RemoteEventServiceAbility extends Ability {

    private MySubRemoteObject mySubRemoteObject = new MySubRemoteObject();
```

```
@Override
public void onStart(Intent intent) {
    super.onStart(intent);
}

@Override
public IRemoteObject onConnect(Intent intent) {
    super.onConnect(intent);
    return mySubRemoteObject.asObject();
}
}
```

JS 端订阅和取消订阅,如代码示例 4.6 所示。

FeatureAbility.subscribeAbilityEvent 订阅的推送结果通过回调方法返回最新推送的结果,这里主要尽量使用箭头函数的写法,保持 this 指向。

**代码示例 4.6 demo3/index.js**

```
export default {
    data: {
        num: 0,
        msg:""
    },
    onInit() {
        this.subscribe();
    },
    subscribe: async function() {
        var action = {};
        action.bundleName = 'com.example.demo3';
        action.abilityName = 'com.example.demo3.RemoteEventServiceAbility';
        action.messageCode = 1000;
        action.abilityType = 0;
        action.syncOption = 0;

        var result = await FeatureAbility.subscribeAbilityEvent(action, (callbackData) => {
            var callbackJson = JSON.parse(callbackData);
            console.info('eventData is: ' + JSON.stringify(callbackJson.data));
            this.eventData = JSON.stringify(callbackJson.data.abilityEvent);
            this.msg += this.eventData + "!"
        });
        var ret = JSON.parse(result);
        if (ret.code == 200) {
            console.info('subscribe success, result:' + result);
            this.msg = ret.code + ":订阅成功"
        } else {
            console.error('subscribe error, result:' + result);
        }
    },
```

```
unsubscribe: async function() {
    var action = {};
    action.bundleName = 'com.example.demo3';
    action.abilityName = 'com.example.demo3.RemoteEventServiceAbility';
    action.messageCode = 1001;
    action.abilityType = 0;
    action.syncOption = 0;

    var result = await FeatureAbility.unsubscribeAbilityEvent(action);
    var ret = JSON.parse(result);
    if (ret.code == 200) {
        console.info('unsubscribe success, result: ' + result);
    } else {
        console.error('unsubscribe error, result: ' + result);
    }
}
}
```

## 4.2　JS 端调用音乐播放 Service Ability

本节完成一个分布式控制 MP3 播放器功能，效果如图 4.3 所示。前台的界面采用 JS UI 开发，音乐播放控制通过 Java 端 Service Ability 实现 MP3 播放、暂停、继续播放和调节音量功能。播放控制器采用 Java UI 实现。

图 4.3　JS 端音乐播放器＋Java 远程服务

JS UI 音乐播放器页面与 Java 端 MP3 播放服务之间的服务调用效果如图 4.4 所示。

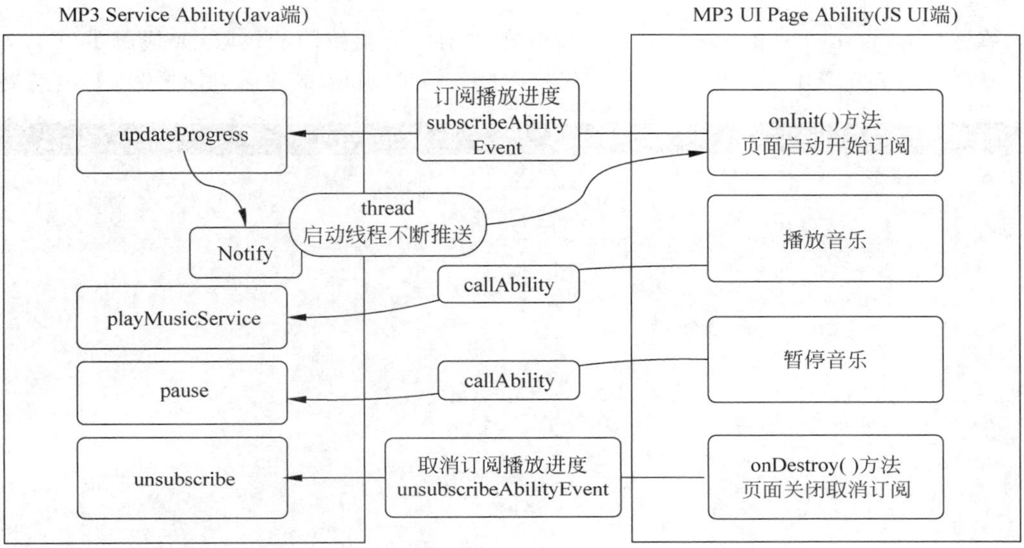

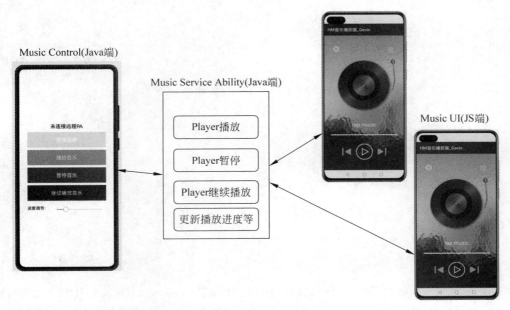

图 4.4　服务调用关系图

Java 端音乐控制器可以通过 Java 端音乐 Service Ability 控制多个 JS 端音乐播放器播放，如图 4.5 所示。

图 4.5　Java 端控制多端

## 4.2.1 申请分布式使用权限

在 config.json 文件的 module 中配置分布式调用需要使用的权限,如代码示例 4.7 所示。这里主要配置 DISTRIBUTED_DATASYNC、servicebus 的权限,其他可以暂时忽略。

**代码示例 4.7　HmMusicPlayer/config.json**

```json
"reqPermissions": [
  {
    "name": "ohos.permission.WRITE_USER_STORAGE"
  },
  {
    "name": "ohos.permission.MICROPHONE"
  },
  {
    "name": "ohos.permission.READ_USER_STORAGE"
  },
  {
    "name": "ohos.permission.DISTRIBUTED_DATASYNC"
  },
  {
    "name": "ohos.permission.INTERNET"
  },
  {
    "name": "ohos.permission.servicebus.ACCESS_SERVICE"
  },
  {
    "name": "ohos.permission.servicebus.BIND_SERVICE"
  },
  {
    "name": "ohos.permission.DISTRIBUTED_DEVICE_STATE_CHANGE"
  },
  {
    "name": "ohos.permission.GET_DISTRIBUTED_DEVICE_INFO"
  },
  {
    "name": "ohos.permission.GET_BUNDLE_INFO"
  },
  {
    "name": "ohos.permission.LOCATION"
  }
]
```

同时需要在 MainAbility 中获取权限,获取的权限与 config.json 文件中的权限申请保持一致,如代码示例 4.8 所示。MainAbility 是 JS 端 Ability,默认管理 js 目录下的 default component。

代码示例 4.8　HmMusicPlayer/java/MainAbility

```java
public class MainAbility extends AceAbility {
    @Override
    public void onStart(Intent intent) {
        super.onStart(intent);
        requestPermission();
    }

    //获取权限
    private void requestPermission() {
        String[] permission = {
"ohos.permission.CAMERA",
"ohos.permission.READ_USER_STORAGE",
"ohos.permission.WRITE_USER_STORAGE",
"ohos.permission.DISTRIBUTED_DATASYNC",
"ohos.permission.MICROPHONE",
"ohos.permission.GET_DISTRIBUTED_DEVICE_INFO",
"ohos.permission.KEEP_BACKGROUND_RUNNING",
"ohos.permission.NFC_TAG"};
        List<String> applyPermissions = new ArrayList<>();
        for (String element : permission) {
            if (verifySelfPermission(element) != 0) {
                if (canRequestPermission(element)) {
                    applyPermissions.add(element);
                }
            }
        }
        requestPermissionsFromUser(applyPermissions.toArray(new String[0]), 0);
    }

    @Override
    public void onStop() {
        super.onStop();
    }
}
```

## 4.2.2　创建 Java 端 Service Ability

创建 Java 端音乐播放远程调用的 Service Ability，该 Service Ability 依赖以下几个重要的类：

（1）AudioManager 类用于管理声频、控制声频设备、调整音量等，这里以这个类控制远程设备的音量。

（2）AVMetadataHelper 类用于描述媒体元数据，即描述多媒体数据，例如媒体标题、媒体时长、媒体的帧数据等。

（3）Player 类用于视频播放，包括播放控制、播放设置和播放查询，如播放的开始/停止、播放速度设置和是否循环播放等。

创建 MusicPlayerServiceAbility.java 文件，如代码示例 4.9 所示。

**代码示例 4.9　HmMusicPlayer/Java/MusicPlayerServiceAbility**

```java
public class MusicPlayerServiceAbility extends Ability {
    private static final HiLogLabel LABEL_LOG = new HiLogLabel(3, 0xD001100, "Demo");
    //支持多 FA 订阅,如果仅支持单 FA 订阅,则可直接使用变量存储:private IRemoteObject
    //remoteObjectHandler;
    private Set<IRemoteObject> remoteObjectHandlers = new HashSet<IRemoteObject>();

    private Audio audio;
    private static Source sVideoSource;
    private boolean isPlay = false;
    private static AVMetadataHelper sAvMetadataHelper = new AVMetadataHelper();
    public Player sPlayer;

    @Override
    protected void onStart(Intent intent) {
        super.onStart(intent);
        audio = new Audio(getContext());
    }

    public IRemoteObject onConnect(Intent intent) {
        return null;
    }

    public boolean preparePlayer() {
        if (sPlayer == null) {
            sPlayer = new Player(MusicPlayerServiceAbility.this);
        }
        return true;
    }

    private void playMP3File(String mediaName) {
        preparePlayer();
        FileDescriptor fileDescriptor = null;
        try {
            File mp3Path = getExternalFilesDir(Environment.DIRECTORY_MUSIC);
            if (!mp3Path.exists()) {
                mp3Path.mkdirs();
            }
            File mp3File = new File(mp3Path.getAbsolutePath() + "/" + "1.mp3");
            Resource res = getResourceManager()
                    .getRawFileEntry("resources/rawfile/1.mp3").openRawFile();
            Byte[] buf = new Byte[4096];
            int count = 0;
            FileOutputStream fos = new FileOutputStream(mp3File);
            while ((count = res.read(buf)) != -1) {
```

```
                fos.write(buf, 0, count);
            }
            fileDescriptor = new FileInputStream(mp3File).getFD();
            sVideoSource = new Source(fileDescriptor);
            sPlayer.setSource(sVideoSource);
            sPlayer.prepare();
            isPlay = true;
            sPlayer.play();
            sAvMetadataHelper.setSource(fileDescriptor);
        } catch (IOException e) {
            e.printStackTrace();
        }
    }

    public int getProgress() {
        int duration = sPlayer.getDuration();
        return duration;
    }
}
```

上面的 onConnect 方法()返回了一个 IRemoteObject 接口的实现子类。

这里我们直接把 MusicPlayerRemoteObjecty 类作为 MusicPlayerServiceAbilit 类的内部类,如代码示例 4.10 所示。

**代码示例 4.10　HmMusicPlayer/Java/MusicPlayerServiceAbility.java**

```
class MusicPlayerRemoteObject extends RemoteObject implements IRemoteBroker {

    public MusicPlayerRemoteObject() {
        super("远程音乐播放控制 Ability");
    }

    @Override
    public IRemoteObject asObject() {
        return this;
    }

    @Override
    public boolean onRemoteRequest ( int code, MessageParcel data, MessageParcel reply,
MessageOption option) throws RemoteException {
        System.out.println("收到命令======" + code);
        switch (code) {
            case 1000: {
                preparePlayer();
                if (sPlayer.isNowPlaying()) {
                    break;
                }
                if (isPlay) {
```

```
                    sPlayer.play();
                    break;
                }
                //播放
                playMP3File("1.mp3");
                break;
            }
            case 1001: {
                //暂停
                preparePlayer();
                if (sPlayer.isNowPlaying()) {
                    sPlayer.pause();
                    isPlay = false;
                }
                break;
            }
            case 1002: {
                //继续播放
                preparePlayer();
                //这个地方有 bug
                sPlayer.play();
                isPlay = false;
                break;
            }
            case 1003:
                audio.vloumnUp();
                break;
            case 1004:
                audio.vloumnDown();
                break;
        }
        return true;
    }
}
```

这里设置了几个 Service Ability 端,用于响应客户端发送的请求编码:播放代码 1000、暂停代码 1001、继续播放代码 1002、控制音量代码 1003 和 1004。

调节音量的高低使用的是 AudioManager 类,为了调用方便,封装了 Audio.java 类,该类提供了两种方法,即调高音量和调低音量,如代码示例 4.11 所示。

**代码示例 4.11　HmMusicPlayer/Java/utils/Audio.java**

```java
public class Audio {

    private AudioManager audioManager;
    public static final int VLOMUN_STEP = 1;

    public Audio(Context context){
```

```java
        audioManager = new AudioManager(context);
    }

    public boolean vloumnUp(){
        try {
            int volume = audioManager.getVolume(AudioManager.AudioVolumeType.STREAM_MUSIC);
            return audioManager.setVolume(AudioManager.AudioVolumeType.STREAM_MUSIC, volume + VLOMUN_STEP);

        } catch (AudioRemoteException e) {
            e.printStackTrace();
        }
        return false;
    }
    public boolean vloumnDown(){
        try {
            int volume = audioManager.getVolume(AudioManager.AudioVolumeType.STREAM_MUSIC);
            return audioManager.setVolume(AudioManager.AudioVolumeType.STREAM_MUSIC, volume - VLOMUN_STEP);

        } catch (AudioRemoteException e) {
            e.printStackTrace();
        }
        return false;
    }
}
```

### 4.2.3 音乐播放器前端的 UI

音乐播放器的界面效果如图 4.3 所示,代码如代码示例 4.12 所示。

**代码示例 4.12　HmMusicPlayer/js/index.hml**

```html
<div class = "container">
<div class = "cd">
<image id = "poster" class = "poster" src = "/common/icons/a1.png">
</image>
<div class = "circle"></div>
</div>
<div class = "lyrics">
<text> rap music </text>
</div>
<div class = "control">
<image class = "next" src = "/common/icons/last.png">
</image>
<image if = "{{ status == 'start' }}" class = "play" @click = "play" src = "/common/icons/play.png">
</image>
```

```html
<image elif="{{ status == 'pause' }}" class="play" @click="pause" src="/common/icons/pause.png">
</image>
<image class="next" src="/common/icons/next.png">
</image>
</div>
<progress class="min-progress" type="horizontal" percent="10" secondarypercent="50">
</progress>
<div class="stick" id="stick">
<image class="pointer" src="/common/icons/stick.png">
</image>
</div>
<image class="qianyi" src="/common/icons/qianyi.png"></image>
</div>
```

这里需要处理两个动画效果: 音乐唱片 CD 随音乐播放转动和唱片磁针的动画, 这两个动画效果都是通过代码来处理的, 如代码示例 4.13 所示。

动画的初始化需要放在 onShow() 方法中, 通过 this.$element().animate() 方法创建。

**代码示例 4.13　HmMusicPlayer/js/index.js**

```js
onShow(){
    var options = {
        duration: 3000,
        easing: 'linear',
        fill: 'forwards',
        iterations: 'Infinity',
    };
    var frames = [
        {transform: {rotate: '0deg'}},
        {transform: {rotate: '360deg'}}
    ];
    this.aniPoster = this.$element('poster').animate(frames, options);

    var options2 = {
        duration: 3000,
        easing: 'linear',
        fill: 'forwards',
        iterations: 1,
    };
    var frames2 = [
        {transform: {rotate: '0deg'}},
        {transform: {rotate: '18deg'}}
    ];
    this.aniStick = this.$element('stick').animate(frames2, options2);
},
```

aniPoster 和 aniStick 是创建好的动画对象，开启动画和关闭动画只需调用对象的 play() 和 cancel() 方法。

页面样式如代码示例 4.14 所示。

**代码示例 4.14　HmMusicPlayer/js/index.css**

```css
.container {
    flex-direction: column;
    justify-content: flex-start;
    align-items: center;
    flex-direction: column;
    background-image: url("/common/icons/bg.png");
    background-repeat: no-repeat;
}

.cd {
    width: 500px;
    height: 500px;
    margin-top: 220px;
    position: relative;
}

.circle {
    width:100px;
    height:100px;
    border-radius: 50px;
    background-color: #ccc;
    position: absolute;
    left: 200px;
    top:200px;
}

.poster {
    width: 500px;
    height: 500px;
    border-radius: 250px;
}

.stick {
    width:38px;
    height: 300px;
    position: fixed;
    right:60px ;
    top:220px;
}

.lyrics {
    flex-direction: column;
    justify-content: center;
```

```css
    align-items: center;
    text-align: center;
    margin-top: 100px;
}

.lyrics text {
    font-size: 40px;
}

.control {
    flex-direction: row;
    justify-content: space-between;
    align-items: center;
    width: 400px;
    margin-top: 50px;
}

.control .play {
    width:150px;
    height:150px;
    object-fit: contain;
}

.control .next {
    width:80px;
    height:80px;
    object-fit: contain;
}

.min-progress {
    margin-top: 30px;
}

.qianyi {
    position: fixed;
    top:100px;
    left:50px;
    width: 60px;
    height: 60px;
    object-fit: contain;
}
```

## 4.2.4　封装 JS 前端调用 Service Ability 的方法

在 js→default→common 目录下创建 utils.js 文件,封装调用 Service Ability 的调用方法,如代码示例 4.15 所示。

代码示例 4.15　HmMusicPlayer/js/default/common/utils.js

```js
export default {
    subscribe: async function(code,callbackData) {
        var action = {};
        action.bundleName = 'com.charjedu.ptgamebook';
        action.abilityName = 'com.charjedu.ptgamebook.hmmusicplayer.MusicPlayerServiceAbility';
        action.messageCode = code;
        action.abilityType = 0;
        action.syncOption = 0;

        var result = await FeatureAbility.subscribeAbilityEvent(action, callbackData);
        var ret = JSON.parse(result);
        return ret;
    },
    unsubscribe: async function(code) {
        var action = {};
        action.bundleName = 'com.charjedu.ptgamebook';
        action.abilityName = 'com.charjedu.ptgamebook.hmmusicplayer.MusicPlayerServiceAbility';
        action.messageCode = code;
        action.abilityType = 0;
        action.syncOption = 0;

        var result = await FeatureAbility.unsubscribeAbilityEvent(action);
        var ret = JSON.parse(result);
        return ret;
    },
    callAbility:async function(code){
        var action = {};
        action.bundleName = 'com.charjedu.ptgamebook';
        action.abilityName = 'com.charjedu.ptgamebook.hmmusicplayer.MusicPlayerServiceAbility';
        action.messageCode = code;
        action.data = {};
        action.abilityType = 0;
        action.syncOption = 0;
        var result = await FeatureAbility.callAbility(action);
        var ret = JSON.parse(result);
        if (ret.code == 0) {
            console.info('plus result is:' + JSON.stringify(ret.abilityResult));
        } else {
            console.error('plus error code:' + JSON.stringify(ret.code));
        }
    },
    startAbility: async function(target){
        let result = await FeatureAbility.startAbility(target);
        let ret = JSON.parse(result);
        if (ret.code == 0) {
            console.log('success');
```

```
            } else {
                console.log('cannot start browing service, reason: ' + ret.data);
            }
        },
    }
```

### 4.2.5　JS 端调用 Service Ability 的方法

在页面逻辑代码中引入 utils.js 类,单击按钮调用对应的 Service Ability 的代码,如代码示例 4.16 所示。

**代码示例 4.16　HmMusicPlayer/js/index.js**

```
import utils from "../../common/utils.js"

export default {
    data: {
        status:'start',
        aniPoster: '',
        aniStick:''
    },
    onInit() {

    },

    onShow(){
        var options = {
            duration: 3000,
            easing: 'linear',
            fill: 'forwards',
            iterations: 'Infinity',
        };
        var frames = [
            {transform: {rotate: '0deg'}},
            {transform: {rotate: '360deg'}}
        ];
        this.aniPoster = this.$element('poster').animate(frames, options);

        var options2 = {
            duration: 3000,
            easing: 'linear',
            fill: 'forwards',
            iterations: 1,
        };
        var frames2 = [
            {transform: {rotate: '0deg'}},
            {transform: {rotate: '18deg'}}
        ];
```

```
            this.aniStick = this.$element('stick').animate(frames2, options2);
        },
        play: async function(){
            this.status = 'pause'
            this.aniPoster.play();
            this.aniStick.play();
            await utils.callAbility(1000)
        },
        pause: async function(){
            this.status = 'start'
            this.aniPoster.cancel();
            this.aniStick.cancel();
            await utils.callAbility(1001)
        },
        continuePlay: async function(){
            await utils.callAbility(1002)
        },
        vloumnUp: async function(){
            await utils.callAbility(1003)
        },
        vloumnDown: async function(){
            await utils.callAbility(1004)
        },
        changePage: async function(){
            let actionData = {
                uri: 'www.huawei.com'
            };
            let target = {
                bundleName: "com.charjedu.ptgamebook",
                abilityName: "com.charjedu.ptgamebook.hmmusicplayer.MainAbility",
                data: actionData
            };
            await utils.startAbility(target);
        },
}
```

## 4.2.6 音乐播放器遥控 UI

音乐播放器遥控页面可采用 Java UI 开发，控制远程调用 Service Ability，如代码示例 4.17 所示。

**代码示例 4.17　HmMusicPlayer/resources/layout/ability_player_main.xml**

```xml
<?xml version = "1.0" encoding = "utf-8"?>
<DirectionalLayout
    xmlns:ohos = "http://schemas.huawei.com/res/ohos"
    ohos:height = "match_parent"
    ohos:alignment = "vertical_center"
```

```xml
        ohos:width = "match_parent"
        ohos:orientation = "vertical">

<Text
        ohos:text = "未连接远程 PA"
        ohos:layout_alignment = "horizontal_center"
        ohos:id = " $ + id:txtDevice"
        ohos:height = "match_content"
        ohos:width = "match_content"
        ohos:text_size = "20fp">
</Text>

<Button
        ohos:id = " $ + id:btn_link_remote"
        ohos:height = "60vp"
        ohos:width = "300vp"
        ohos:text_color = " # fff"
        ohos:background_element = " # ccff"
        ohos:top_margin = "10vp"
        ohos:layout_alignment = "horizontal_center"
        ohos:text = "连接设备"
        ohos:text_size = "20fp"
        />

<Button
        ohos:id = " $ + id:btn_play"
        ohos:height = "60vp"
        ohos:width = "300vp"
        ohos:background_element = " # 0f0"
        ohos:layout_alignment = "horizontal_center"
        ohos:text_color = " # fff"
        ohos:text = "播放音乐"
        ohos:text_size = "20fp"
        ohos:top_margin = "10vp"
        />

<Button
        ohos:id = " $ + id:btn_pause"
        ohos:height = "60vp"
        ohos:width = "300vp"
        ohos:text_color = " # fff"
        ohos:background_element = " # c00"
        ohos:layout_alignment = "horizontal_center"
        ohos:text = "暂停音乐"
        ohos:top_margin = "10vp"
        ohos:text_size = "20fp"
        />

<Button
```

```xml
        ohos:id = " $ + id:btn_continuePlay"
        ohos:height = "60vp"
        ohos:width = "300vp"
        ohos:text_color = " # fff"
        ohos:background_element = "blue"
        ohos:layout_alignment = "horizontal_center"
        ohos:text = "继续播放音乐"
        ohos:top_margin = "10vp"
        ohos:text_size = "20fp"
        />


<DirectionalLayout
        ohos:below = " $ id:btn_continuePlay"
        ohos:top_margin = "20vp"
        ohos:id = " $ + id:remote_progress_seek_layout"
        ohos:width = "match_parent"
        ohos:height = "match_content"
        ohos:alignment = "horizontal_center"
        ohos:orientation = "horizontal">

<Text
        ohos:id = " $ + id:remote_progress_seek_title"
        ohos:width = "match_content"
        ohos:height = "match_content"
        ohos:text_size = "16fp"
        ohos:text = "进度调节:"/>
<Slider
        ohos:id = " $ + id:remote_progress_seek"
        ohos:height = "30vp"
        ohos:left_margin = "20vp"
        ohos:orientation = "horizontal"
        ohos:min = "0"
        ohos:max = "100"
        ohos:progress = "20"
        ohos:width = "200vp"/>
</DirectionalLayout>
</DirectionalLayout>
```

界面效果如图 4.6 所示，当界面启动后，自动检测附近在线可连接的鸿蒙操作系统设备，检测到可连接设备后自动连接，并组网。

## 4.2.7 音乐播放器遥控逻辑实现

创建 PlayerMainAbility.java 文件，音乐播放器控制器功能如下：
(1) 连接远程 Service Ability。
(2) 将音乐播放命令发送给 Service Ability，让远程连接设备播放音乐。

图 4.6　Java 端遥控器效果图

（3）将音乐暂停播放命令发送给 Service Ability，让远程连接设备暂停播放音乐。
（4）将音乐继续播放命令发送给 Service Ability，让远程连接设备继续播放音乐。
（5）将播放进度命令发送给 Service Ability，让远程连接设备按指定的进度播放音乐。
（6）调节遥控器的实体音量按键，实现远程连接设备同步调节。

打开 slice→PlayerMainAbilitySlice.java 文件，在 onStart()方法中获取界面中定义的按钮，设置按钮的单击事件。

连接远程音乐 Service Ability，界面中有很多命令需要发送，所以采用 connectAbility 的方法调用远程的 Service Ability，这样只需连接成功，就可以不断地将命令发送给 Service Ability 进行处理，如代码示例 4.18 所示。

代码示例 4.18　HmMusicPlayer/java/PlayerMainAbilitySlice.java

```java
@Override
public void onStart(Intent intent) {
    super.onStart(intent);
    super.setUIContent(ResourceTable.Layout_ability_player_main);

    txtRemoteDeviceInfo = (Text) findComponentById(ResourceTable.Id_txtDevice);
```

```java
//跑马灯效果
txtRemoteDeviceInfo.setTruncationMode(Text.TruncationMode.AUTO_SCROLLING);
//始终处于自动滚动状态
txtRemoteDeviceInfo.setAutoScrollingCount(Text.AUTO_SCROLLING_FOREVER);
//启动跑马灯效果
txtRemoteDeviceInfo.startAutoScrolling();
List < DeviceInfo > remoteDevice = DeviceUtils.getRemoteDevice();
txtRemoteDeviceInfo.setText("正在扫描附近的设备");
if(remoteDevice != null && remoteDevice.size()> 0){
    txtRemoteDeviceInfo.setText("连接设备:" + remoteDevice.get(0).getDeviceName());
}

//查找连接远程设备
Button btn_link_remote = (Button) findComponentById(ResourceTable.Id_btn_link_remote);
btn_link_remote.setClickedListener(component -> {
    Intent connectPAIntent = new Intent();
    String bundleName = "com.charjedu.ptgamebook";
    String abilityName = "com.charjedu.ptgamebook.hmmusicplayer.MusicPlayerServiceAbility";
    Operation operation = new Intent.OperationBuilder()
            .withDeviceId(DeviceUtils.getDeviceId())
            .withBundleName(bundleName)
            .withAbilityName(abilityName)
            .withFlags(Intent.FLAG_ABILITYSLICE_MULTI_DEVICE)
            .build();
    connectPAIntent.setOperation(operation);
    connectAbility(connectPAIntent, mConn);
});
}
```

单击连接远程设备按钮,连接成功后,在界面顶端会弹出连接成功信息。当控制器界面启动后,自动查找附近可用的鸿蒙设备,并且自动连接组网,组网成功后通过界面的跑马灯提示连接的设备名称。

这里需要查找附近的鸿蒙在线设备,DeviceUtils.java 文件封装的是获取附近设备信息的包装类,如代码示例 4.19 所示。

**代码示例 4.19　HmMusicPlayer/java/utils/DeviceUtils.java**

```java
public class DeviceUtils {
    private static final String TAG = DeviceUtils.class.getSimpleName();

    private DeviceUtils() {
    }

    /**
     * Get group id.
     *
     * @return group id.
     */
```

```java
    public static String getGroupId() {
        AccountAbility account = AccountAbility.getAccountAbility();
        DistributedInfo distributeInfo = account.queryOsAccountDistributedInfo();
        return distributeInfo.getId();
    }

    public static List<String> getAvailableDeviceId() {
        List<String> deviceIds = new ArrayList<>();

        List<DeviceInfo> deviceInfoList = DeviceManager.getDeviceList(DeviceInfo.FLAG_GET_ONLINE_DEVICE);
        if (deviceInfoList == null) {
            return deviceIds;
        }

        if (deviceInfoList.size() == 0) {
            System.out.println("did not find other device");
            return deviceIds;
        }

        for (DeviceInfo deviceInfo : deviceInfoList) {
            deviceIds.add(deviceInfo.getDeviceId());
        }

        return deviceIds;
    }

    /**
     * Get available id
     *
     * @return available device ids
     */
    public static List<String> getAllAvailableDeviceId() {
        List<String> deviceIds = new ArrayList<>();

        List<DeviceInfo> deviceInfoList = DeviceManager.getDeviceList(DeviceInfo.FLAG_GET_ALL_DEVICE);
        if (deviceInfoList == null) {
            return deviceIds;
        }
        System.out.println("deviceInfoList size " + deviceInfoList.size());
        if (deviceInfoList.size() == 0) {
            System.out.println("did not find other device");
            return deviceIds;
        }

        for (DeviceInfo deviceInfo : deviceInfoList) {
            deviceIds.add(deviceInfo.getDeviceId());
        }
```

```java
        return deviceIds;
    }

    /**
     * Get remote device info
     *
     * @return Remote device info list.
     */
    public static List<DeviceInfo> getRemoteDevice() {
        List<DeviceInfo> deviceInfoList = DeviceManager.getDeviceList(DeviceInfo.FLAG_GET_ONLINE_DEVICE);
        return deviceInfoList;
    }

    /**
     * @return java.lang.String
     * @Param []
     * @description 获取当前组网下可迁移的设备 id
     */
    public static String getDeviceId() {
        String deviceId = "";
        List<String> outerDevices = DeviceUtils.getAvailableDeviceId();
        System.out.println("getDeviceId DeviceUtils.getRemoteDevice() = " + outerDevices);
        if (outerDevices == null || outerDevices.size() == 0) {
            System.out.println("no other device to continue");
        } else {
            for (String item : outerDevices) {
                System.out.println("item deviceId = " + item);
            }
            deviceId = outerDevices.get(0);
        }
        System.out.println("continueAbility to deviceId = " + deviceId);
        return deviceId;
    }
}
```

这里通过 connectAbility(connectPAIntent, mConn) 方法连接远程 Service Ability，连接的结果通过 mConn 接口方法返回，如代码示例 4.20 所示。MyRemoteProxy 用于调用由端定义的用来操作远程 Service 的代理类，当客户端与 Service 端连接成功后，通过 onAbilityConnectDone() 方法实现客户端调用代理的创建。

**代码示例 4.20　HmMusicPlayer/java/slice/PlayerMainAbilitySlice.java**

```java
//当连接完成时,用来提供管理已连接 PA 的能力
private MyRemoteProxy mProxy = null;
//用于管理连接关系
private IAbilityConnection mConn = new IAbilityConnection() {
```

```java
    @Override
    public void onAbilityConnectDone(ElementName element, IRemoteObject remote, int resultCode) {
        //跨设备 PA 连接完成后,会返回一个序列化的 IRemoteObject 对象
        //通过该对象得到控制远端服务的代理
        mProxy = new MyRemoteProxy(remote);
        new ToastDialog(getContext())
                .setText("成功连接远程音乐服务")
                .setAlignment(LayoutAlignment.TOP)
                .show();
    }

    @Override
    public void onAbilityDisconnectDone(ElementName element, int resultCode) {
        //当已连接的远端 PA 关闭时,会触发该回调
        //支持开发者按照返回的错误信息进行 PA 生命周期管理
        disconnectAbility(mConn);
    }
};
```

MyRemoteProxy 类的定义如代码示例 4.21 所示。

**代码示例 4.21　HmMusicPlayer/java/pa/MyRemoteProxy.java**

```java
public class MyRemoteProxy implements IRemoteBroker {

    private final IRemoteObject remote;

    public MyRemoteProxy (IRemoteObject remote) {
        this.remote = remote;
    }

    @Override
    public IRemoteObject asObject() {
        return remote;
    }

    public int sendCommand(int code) throws RemoteException {
        MessageParcel data = MessageParcel.obtain();
        MessageParcel reply = MessageParcel.obtain();
        //option 不同的取值,决定采用同步或异步方式跨设备控制 PA
        //本例需要同步获取对端 PA 执行加法的结果,因此采用同步的方式,即 MessageOption.TF_SYNC
        //具体 MessageOption 的设置,可参考相关 API 文档
        MessageOption option = new MessageOption(MessageOption.TF_SYNC);

        try {
            remote.sendRequest(code, data, reply, option);
            int errCode = reply.readInt();
            if (errCode != 200) {
```

```
                throw new RemoteException();
            }
            int result = reply.readInt();
            return result;
        } finally {
            data.reclaim();
            reply.reclaim();
        }
    }
}
```

通过上面的步骤创建好客户端调用远程 Service 的代理后,可以通过这个代理向服务器端发送命令了,MyRemoteProxy 类中 sendCommand()方法用于在客户端向服务器端发送执行命令,在这种方法中使用客户端代理对象的 sendRequest()方法发送与服务器端约定的代码。

接下来,就可以在客户端向服务器端发送播放、暂停、继续等命令,以此供服务器端执行,如代码示例 4.22 所示。

**代码示例 4.22** HmMusicPlayer/java/slice/PlayerMainAbilitySlice.java

```
//播放
Button btn_play = (Button) findComponentById(ResourceTable.Id_btn_play);
btn_play.setClickedListener(component -> {
    System.out.println("---------- 连接播放 ---------");
    if (mProxy != null) {
        try {
            mProxy.sendCommand(1000);
        } catch (RemoteException e) {
            e.printStackTrace();
        }
    }else{
        new ToastDialog(getContext())
                .setText("请先单击连接远程设备")
                .setAlignment(LayoutAlignment.CENTER)
                .show();
    }
});

//暂停
Button btn_pause = (Button) findComponentById(ResourceTable.Id_btn_pause);
btn_pause.setClickedListener(component -> {
    if (mProxy != null) {
        try {
            mProxy.sendCommand(1001);
        } catch (RemoteException e) {
            e.printStackTrace();
```

```
        }
    }else{
        new ToastDialog(getContext())
                .setText("请先单击连接远程设备")
                .setAlignment(LayoutAlignment.CENTER)
                .show();
    }
});

//重新播放
Button btn_continueplay = (Button) findComponentById(ResourceTable.Id_btn_continuePlay);
btn_continueplay.setClickedListener(component -> {
    if (mProxy != null) {
        try {
            mProxy.sendCommand(1002);
        } catch (RemoteException e) {
            e.printStackTrace();
        }
    }else{
        new ToastDialog(getContext())
                .setText("请先单击连接远程设备")
                .setAlignment(LayoutAlignment.CENTER)
                .show();
    }
});

//进度控制
Slider remote_progress_seek = (Slider) findComponentById(ResourceTable.Id_remote_progress_seek);
remote_progress_seek.setValueChangedListener(new Slider.ValueChangedListener() {
    @Override
    public void onProgressUpdated(Slider slider, int i, boolean b) {

    }

    @Override
    public void onTouchStart(Slider slider) {

    }

    @Override
    public void onTouchEnd(Slider slider) {
        System.out.println(slider.getProgress());
        if (mProxy != null) {
            try {
                mProxy.sendCommand(1005);
            } catch (RemoteException e) {
                e.printStackTrace();
            }
        }else{
```

```
            new ToastDialog(getContext())
                    .setText("请先单击连接远程设备")
                    .setAlignment(LayoutAlignment.CENTER)
                    .show();
        }
    }
});
```

### 4.2.8 通过实体音量键控制远程设备音量

打开音乐播放器的控制器界面后,可以通过调节手机的实体音量键来控制远程已经连接成功的手机的音量。

这里通过重写 onKeyDown()方法,判断不同的类型按键,将控制音量的代码发送给 ServiceAbility,如代码示例 4.23 所示。

**代码示例 4.23**　HmMusicPlayer/java/slice/PlayerMainAbilitySlice.java

```java
@Override
public boolean onKeyDown(int keyCode, KeyEvent keyEvent) {
    if(mProxy == null){
        return false;
    }
    if (keyCode == KeyEvent.KEY_VOLUME_UP) {
        //mHandler.sendEvent(KEY_VOLUME_UP);

        new Thread(new Runnable() {
            @Override
            public void run() {
                try {
                    mProxy.sendCommand(1003);
                } catch (RemoteException e) {
                    e.printStackTrace();
                }
            }
        }).start();
        return true;
    } else if (keyCode == KeyEvent.KEY_VOLUME_DOWN) {
        //mHandler.sendEvent(KEY_VOLUME_DOWM);

        new Thread(new Runnable() {
            @Override
            public void run() {
                try {
                    mProxy.sendCommand(1004);
                } catch (RemoteException e) {
                    e.printStackTrace();
                }
```

```
            }
        }).start();
        return true;
    }
    return true;
}
```

### 4.2.9　JS 端订阅 Service Ability 中的播放状态

4.2.8 节实现了控制器端控制远程端的音乐播放,但是无法实现界面的效果与 Service Ability 的播放状态同步,因此需要在 JS 端订阅 Service Ability 的状态。

JS 端通过订阅 Service Ability 中的 Player 的状态,来改变页面的展现。例如获取 Player 的播放进度和播放状态等信息。

在 index.js 文件的 onInit()方法中发起订阅,通过订阅的回调方法接收 Service Ability 推送的状态消息,如代码示例 4.24 所示。

代码示例 4.24　HmMusicPlayer/js/default/index.js

```js
data: {
        status: 'start',
        aniPoster: '',
        aniStick: '',
        cacheStatus:new Set(),
        ctime:0,
        total:0,
        progress:0
},
onInit() {
    this.subPlayerService();
},
subPlayerService: async function(){
    await utils.subscribe(100, (data) => {

        //service 推送的结果
        var callbackJson = JSON.parse(data);
        console.info('eventData is: ' + JSON.stringify(callbackJson.data));
        //判断播放状态,status:1 表示正在播放,status:-1 表示暂停
        if(!this.cacheStatus.has(callbackJson.data.status)){
            this.cacheStatus.add(callbackJson.data.status)
            if (callbackJson.data.status == 1) {
                this.cacheStatus.delete(-1)
                this.play();
            } else if (callbackJson.data.status == -1) {
                this.cacheStatus.delete(1)
                this.pause()
            }
```

```
            }
            //计算播放进度
            if(callbackJson.data.total > 0){
                this.progress = (callbackJson.data.ctime/callbackJson.data.total) * 100;
            }

        })//订阅播放

    },
```

这里数字 100 是 Service 端协商的编码,当服务器端接收到 100 编码时,每隔 500ms 推送一次 Player 的状态。修改 MusicPlayerServiceAbility 类,添加 100 编码的订阅方法,如代码示例 4.25 所示。

代码示例 4.25　HmMusicPlayer/java/MusicPlayerServiceAbility

```java
public class MusicPlayerServiceAbility extends Ability {
    private static final HiLogLabel LABEL_LOG = new HiLogLabel(3, 0xD001100, "Demo");
    //支持多 FA 订阅,如果仅支持单 FA 订阅,则可直接使用变量存储:private IRemoteObject remoteObjectHandler;
    private Set<IRemoteObject> remoteObjectHandlers = new HashSet<IRemoteObject>();

    private Audio audio;
    private static Source sVideoSource;
    private boolean isPlay = false;
    private static AVMetadataHelper sAvMetadataHelper = new AVMetadataHelper();
    public Player sPlayer;

    @Override
    protected void onStart(Intent intent) {
        super.onStart(intent);
        audio = new Audio(getContext());
    }

    public IRemoteObject onConnect(Intent intent) {
        return new MusicPlayerRemoteObject().asObject();
    }

    public boolean preparePlayer() {
        if (sPlayer == null) {
            sPlayer = new Player(MusicPlayerServiceAbility.this);
        }
        return true;
    }

    private void playMP3File(String mediaName) {
        preparePlayer();
```

```java
            FileDescriptor fileDescriptor = null;
            try {
                File mp3Path = getExternalFilesDir(Environment.DIRECTORY_MUSIC);
                if (!mp3Path.exists()) {
                    mp3Path.mkdirs();
                }
                File mp3File = new File(mp3Path.getAbsolutePath() + "/" + "1.mp3");
                Resource res = getResourceManager()
                        .getRawFileEntry("resources/rawfile/1.mp3").openRawFile();
                Byte[] buf = new Byte[4096];
                int count = 0;
                FileOutputStream fos = new FileOutputStream(mp3File);
                while ((count = res.read(buf)) != -1) {
                    fos.write(buf, 0, count);
                }
                fileDescriptor = new FileInputStream(mp3File).getFD();
                sVideoSource = new Source(fileDescriptor);
                sPlayer.setSource(sVideoSource);
                sPlayer.prepare();
                isPlay = true;
                sPlayer.play();
                sAvMetadataHelper.setSource(fileDescriptor);
            } catch (IOException e) {
                e.printStackTrace();
            }
    }

    public int getProgress() {
        int duration = sPlayer.getDuration();
        return duration;
    }

    public void updateProgress() {
        new Thread(() -> {
            while (true) {
                try {
                    Thread.sleep(5 * 100);
                    //播放状态
                    int status = 0;

                    //当前播放的时间
                    int cTime = 0;

                    //总时间
                    int total = 0;

                    if(sPlayer != null){
                        System.out.println("------ sPlayer ------" + sPlayer.isNowPlaying());
                        if(sPlayer.isNowPlaying()){
```

```java
                        status = 1;
                        cTime = sPlayer.getCurrentTime();
                        total = sPlayer.getDuration();
                    }else{
                        status = -1;
                    }

                }
                reportEvent(status,cTime,total);
            } catch (RemoteException | InterruptedException e) {
                break;
            }
        }
    }).start();
}

private void reportEvent(int status, int ctime, int total) throws RemoteException {
    MessageParcel data = MessageParcel.obtain();
    MessageParcel reply = MessageParcel.obtain();
    MessageOption option = new MessageOption();
    Map<String, Object> zsonEvent = new HashMap<String, Object>();
    zsonEvent.put("status",status);
    zsonEvent.put("ctime",ctime);
    zsonEvent.put("total",total);
    System.out.println("------ reportEvent ------ " + status + " ----- ctime" + ctime
+ " ---- total" + total);
    data.writeString(ZSONObject.toZSONString(zsonEvent));
    //这段代码再确定是否必须推送
    for (IRemoteObject item : remoteObjectHandlers) {
        item.sendRequest(0, data, reply, option);
    }
    reply.reclaim();
    data.reclaim();
}

class MusicPlayerRemoteObject extends RemoteObject implements IRemoteBroker {

    public MusicPlayerRemoteObject() {
        super("远程音乐播放控制 Ability");
    }

    @Override
    public IRemoteObject asObject() {
        return this;
    }

    @Override
    public boolean onRemoteRequest(int code, MessageParcel data, MessageParcel reply,
MessageOption option) throws RemoteException {
```

```java
System.out.println("收到命令======" + code);
switch (code) {
    case 100:{
        remoteObjectHandlers.add(data.readRemoteObject());
        //异步推送
        updateProgress();
        //返回一个同步信息
        Map<String, Object> zsonResult = new HashMap<String, Object>();
        zsonResult.put("code", 200);
        reply.writeString(ZSONObject.toZSONString(zsonResult));
        break;
    }
    case 1000: {
        preparePlayer();
        if (sPlayer.isNowPlaying()) {
            break;
        }
        if (isPlay) {
            sPlayer.play();
            break;
        }
        //播放
        playMP3File("1.mp3");
        break;
    }
    case 1001: {
        //暂停
        preparePlayer();
        if (sPlayer.isNowPlaying()) {
            sPlayer.pause();
            isPlay = false;
        }
        break;
    }
    case 1002: {
        //继续播放
        preparePlayer();
        //这个地方有 Bug
        sPlayer.play();
        isPlay = false;
        break;
    }
    case 1003:
        audio.vloumnUp();
        break;
    case 1004:
        audio.vloumnDown();
        break;
    case 1005:
```

```
                        //微秒 μs,1000μs = 1ms
                        sPlayer.rewindTo(1000 * 1000);
                        break;
                }
                return true;
        }
    }
}
```

当 JS 端订阅 Service Ability 的状态时,Service 端每隔 500ms 发送一次状态结果,因此在 JS 端的回调方法也会每隔 500ms 执行一次,这会造成界面逻辑每隔 500ms 调用一次的问题,为了解决这个问题,可在 data 中定义一个 Set 类型的 cacheStatus 对象,用来缓存发送过来的状态,如果两次的状态一样,则不再触发页面逻辑,如代码示例 4.26 所示。

**代码示例 4.26**

```
var callbackJson = JSON.parse(data);
if(!this.cacheStatus.has(callbackJson.data.status)){
     this.cacheStatus.add(callbackJson.data.status)
     if (callbackJson.data.status == 1) {
          this.cacheStatus.delete(-1)
          this.play();
     } else if (callbackJson.data.status == -1) {
          this.cacheStatus.delete(1)
          this.pause()
     }
}
```

## 4.2.10 本节小结

本节通过鸿蒙分布式调度控制音乐播放的例子,介绍了如何在 Java 端和 JS 端连接 Service Ability,以及通过创建可订阅的 Service Ability 实现页面的状态订阅。

## 4.3 JavaScript 项目混合 Java UI 开发

可以在 JavaScript UI 项目中,同时使用 Java UI 混合开发项目,这种混合开发的好处是,对于一些 JavaScript UI 无法实现的功能,可以通过 Java UI 实现,鸿蒙 Dev Eco Studio 支持混合打包应用,这样可以更好地帮助开发者根据不同的需求选择不同的 UI 框架。

这里先创建 JavaScript 项目,在创建好的 JavaScript 项目中添加 Java UI 页面,如图 4.7 所示。

创建好 JavaScript 项目后,右击 java 目录下的包名,新建一个空的 Java Page Ability,如图 4.8 所示。

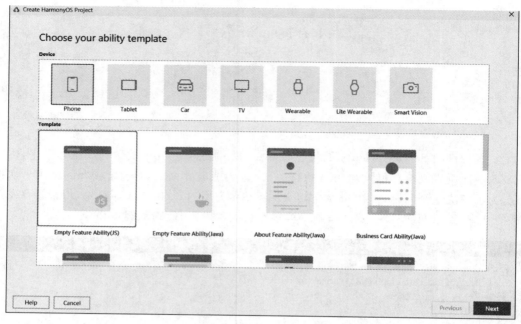

图 4.7 创建 JavaScript 项目

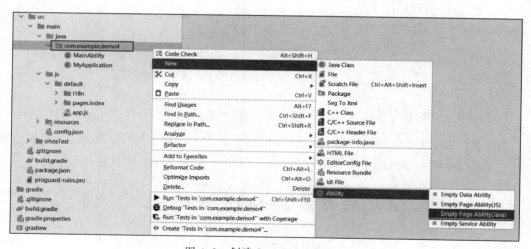

图 4.8 创建 Java Page Ability

单击空的 Java Page Ability，在目录结果中添加了 layout 和 slice 目录，用来编写 xml 和 slice 代码，如图 4.9 所示。

### 4.3.1 JS Ability 和 Java Ability 跳转

搭建好混合项目后，我们来修改 resources→layout→ability_java_main.xml 布局，效果如图 4.10 所示。

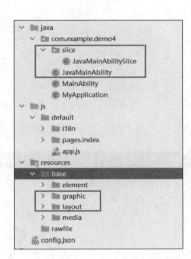

图 4.9 JS+Java 混合目录结构　　图 4.10 Java layout 效果

如代码示例 4.27 所示。

**代码示例 4.27　demo4/resource/layout/ability_java_main.xml**

```xml
<?xml version = "1.0" encoding = "utf-8"?>
<DirectionalLayout
    xmlns:ohos = "http://schemas.huawei.com/res/ohos"
    ohos:height = "match_parent"
    ohos:width = "match_parent"
    ohos:alignment = "center"
    ohos:background_element = "#00FFFF"
    ohos:orientation = "vertical">

<Text
    ohos:id = "$ + id:text_helloworld"
    ohos:height = "match_content"
    ohos:width = "match_content"
    ohos:layout_alignment = "horizontal_center"
    ohos:text = "Java UI 首页"
    ohos:text_size = "30fp"
    ohos:text_color = "#fff"
    />

<Button
    ohos:id = "$ + id:btn_change"
```

```xml
            ohos:top_margin = "30fp"
            ohos:text = "跳转到 JS Component 首页"
            ohos:background_element = "#000"
            ohos:text_color = "#ffffff"
            ohos:text_size = "23fp"
            ohos:padding = "5fp"
            ohos:height = "match_content"
            ohos:width = "match_content">
</Button>

</DirectionalLayout>
```

在 slice 代码中添加从 Java Ability 跳转到 JS Ability 的代码,如代码示例 4.28 所示。

**代码示例 4.28   demo4/java/slice/JavaMainAbilitySlice.java**

```java
@Override
public void onStart(Intent intent) {
    super.onStart(intent);
    super.setUIContent(ResourceTable.Layout_ability_java_main);

    Button btn_change = (Button) findComponentById(ResourceTable.Id_btn_change);
    btn_change.setClickedListener(component -> {
        Intent secondIntent = new Intent();
        //指定待启动 FA 的 bundleName 和 abilityName
        Operation operation = new Intent.OperationBuilder()
                .withDeviceId("")
                .withBundleName("com.example.demo4")
                .withAbilityName("com.example.demo4.MainAbility")
                .build();
        secondIntent.setOperation(operation);
        //通过 AbilitySlice 的 startAbility 接口实现启动另一个页面
        startAbility(secondIntent);
    });
}
```

接下来,修改 JavaScript 项目页面,打开 js→default→page→index→index.html,修改后如代码示例 4.29 所示。

**代码示例 4.29   demo4/js/default/index.html**

```html
<div class = "container">
<text class = "title">
        JavaScript UI 页面
</text>
<button type = "capsule" @click = "changePage">
        跳转到 Java UI 首页
</button>
</div>
```

修改 index.js 文件的代码,添加 JS 跳转到 Java 的逻辑,如代码示例 4.30 所示。

代码示例 4.30　demo4/js/default/index.js

```js
export default {
    data: {
        title: ""
    },
    onInit() {

    },
    changePage:async function(){
        let actionData = {
            uri: 'www.huawei.com'
        };
        let target = {
            bundleName: "com.example.demo4",
            abilityName: "com.example.demo4.JavaMainAbility",
            data: actionData
        };

        let result = await FeatureAbility.startAbility(target);
        let ret = JSON.parse(result);
        if (ret.code == 0) {
            console.log('success');
        } else {
            console.log('cannot start browing service, reason: ' + ret.data);
        }
    }
}
```

现在,就可以实现在 JS Ability 和 Java Ability 之间进行跳转了,如图 4.11 所示。

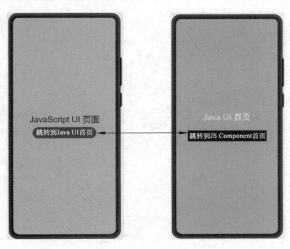

图 4.11　Java Ability 与 JS Ability 之间相互跳转

### 4.3.2 JS 端调用相机拍照功能

对于一些 JS UI 中没有提供的功能,可以使用 Java UI 实现。本节中,在 JS 项目中使用 Java UI Slice 实现拍照功能,拍照完成后跳转到 JS UI 页面,如图 4.12 所示。

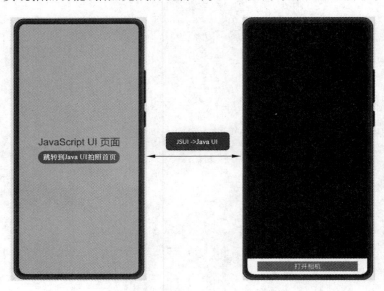

图 4.12　JS 项目中使用 Java UI 完成相机拍照功能

在 JS 项目中创建 CameraMainAbility,在 ability_camera_main.xml 文件中编写布局,如代码示例 4.31 所示。

```
代码示例 4.31    demo5_camera/resource/layout/ability_camera_main.xml
<?xml version = "1.0" encoding = "utf - 8"?>
<DirectionalLayout
    xmlns:ohos = "http://schemas.huawei.com/res/ohos"
    ohos:height = "match_parent"
    ohos:width = "match_parent"
    ohos:alignment = "vertical_center"
    ohos:orientation = "vertical">

<DirectionalLayout
        ohos:id = " $ + id:surface_container"
        ohos:height = "0vp"
        ohos:width = "match_parent"
        ohos:background_element = " # 000000"
        ohos:weight = "29"/>

<Button
        ohos:id = " $ + id:btn_open_camera"
```

```xml
        ohos:height = "30vp"
        ohos:width = "300vp"
        ohos:background_element = "#f00"
        ohos:layout_alignment = "horizontal_center"
        ohos:text = "打开相机"
        ohos:text_color = "#fff"
        ohos:text_size = "20fp"
        ohos:top_margin = "10vp"
        ohos:bottom_margin = "10vp"
        />
</DirectionalLayout>
```

打开相机逻辑,如代码示例 4.32 所示。

**代码示例 4.32  demo5_camera/java/slice/CameraMainAbilitySlice.java**

```java
import static ohos.media.camera.device.Camera.FrameConfigType.FRAME_CONFIG_PREVIEW;

public class CameraMainAbilitySlice extends AbilitySlice {

    private static final int SCREEN_WIDTH = 1080;
    private static final int SCREEN_HEIGHT = 2340;
    private static final int IMAGE_RCV_CAPACITY = 9;
    private ImageReceiver imageReceiver;
    private Surface previewSurface;
    private SurfaceProvider surfaceProvider;
    private boolean isCameraRear;
    private Camera cameraDevice;
    //执行回调的 EventHandler
    EventHandler eventHandler = new EventHandler(EventRunner.create("CameraCb"));

    @Override
    public void onStart(Intent intent) {
        super.onStart(intent);
        super.setUIContent(ResourceTable.Layout_ability_camera_main);

        //相机预览
        initSurface();

        //测试打开相机
        Button btn_open_camera = (Button) findComponentById(ResourceTable.Id_btn_open_camera);
        btn_open_camera.setClickedListener(component -> {
            openCamera();
        });
    }

    private void initSurface() {
        surfaceProvider = new SurfaceProvider(this);
        DirectionalLayout.LayoutConfig params = new DirectionalLayout.LayoutConfig(
```

```java
                            ComponentContainer.LayoutConfig.MATCH_PARENT, ComponentContainer.
LayoutConfig.MATCH_PARENT);
        surfaceProvider.setLayoutConfig(params);
        surfaceProvider.pinToZTop(true);    //设置为true才可以看见
        surfaceProvider.getSurfaceOps().get().addCallback(new SurfaceCallBack());
        ((ComponentContainer)
                findComponentById(ResourceTable.Id_surface_container)).addComponent
(surfaceProvider);
    }

    class SurfaceCallBack implements SurfaceOps.Callback {
        @Override
        public void surfaceCreated(SurfaceOps callbackSurfaceOps) {
            if (callbackSurfaceOps != null) {
                    callbackSurfaceOps.setFixedSize(surfaceProvider.getHeight(),
surfaceProvider.getWidth());
            }
            openCamera();
        }

        @Override
        public void surfaceChanged(SurfaceOps callbackSurfaceOps, int format, int width, int
height) {
        }

        @Override
        public void surfaceDestroyed(SurfaceOps callbackSurfaceOps) {
        }
    }

    private void openCamera() {
        imageReceiver = ImageReceiver.create(SCREEN_WIDTH, SCREEN_HEIGHT, ImageFormat.
JPEG, IMAGE_RCV_CAPACITY);

        //获取CameraKit对象
        CameraKit cameraKit = CameraKit.getInstance(getApplicationContext());
        if (cameraKit == null) {
            //处理cameraKit获取失败的情况
        }
        try {
            //获取当前设备的逻辑相机列表
            String[] cameraIds = cameraKit.getCameraIds();
            if (cameraIds.length <= 0) {
                System.out.println("cameraIds size is 0");
            }
            //相机创建和相机运行时的回调
            CameraStateCallbackImpl cameraStateCallback = new CameraStateCallbackImpl();
            //创建相机设备
```

```java
            cameraKit.createCamera(cameraIds[0], cameraStateCallback, eventHandler);
        } catch (IllegalStateException e) {
            //处理异常
        }
    }

    private final class CameraStateCallbackImpl extends CameraStateCallback {
        @Override
        public void onCreated(Camera camera) {
            previewSurface = surfaceProvider.getSurfaceOps().get().getSurface();
            if (previewSurface == null) {
                //HiLog.info(TAG, "create camera filed, preview surface is null");
                return;
            }

            try {
                Thread.sleep(200);
            } catch (InterruptedException exception) {
                //HiLog.info(TAG, "Waiting to be interrupted");
            }

            CameraConfig.Builder cameraConfigBuilder = camera.getCameraConfigBuilder();
            cameraConfigBuilder.addSurface(previewSurface);
            cameraConfigBuilder.addSurface(imageReceiver.getRecevingSurface());
            camera.configure(cameraConfigBuilder.build());
            cameraDevice = camera;

            //e/nableImageGroup();
        }

        @Override
        public void onConfigured(Camera camera) {
            //配置相机设备
            FrameConfig.Builder framePreviewConfigBuilder = camera.getFrameConfigBuilder(FRAME_CONFIG_PREVIEW);
            framePreviewConfigBuilder.addSurface(previewSurface);
            try {
                //启动循环帧捕获
                camera.triggerLoopingCapture(framePreviewConfigBuilder.build());
            } catch (IllegalArgumentException e) {
                // HiLog.error(TAG, "Argument Exception");
            } catch (IllegalStateException e) {
                //HiLog.error(TAG, "State Exception");
            }
        }

        @Override
```

```java
        public void onPartialConfigured(Camera camera) {
            //当使用 addDeferredSurfaceSize 配置相机后,会接到此回调
        }

        @Override
        public void onReleased(Camera camera) {
            //释放相机设备
        }
    }

    @Override
    public void onActive() {
        super.onActive();
    }

    private void releaseCamera() {
        if (cameraDevice != null) {
            cameraDevice.release();
            cameraDevice = null;
        }

        if (imageReceiver != null) {
            imageReceiver.release();
            imageReceiver = null;
        }

        if (eventHandler != null) {
            eventHandler.removeAllEvent();
            eventHandler = null;
        }
    }

    @Override
    protected void onStop() {
        super.onStop();
        releaseCamera();
    }

    @Override
    public void onForeground(Intent intent) {
        super.onForeground(intent);
    }
}
```

# 第 5 章　ArkUI JS 游戏开发案例

本章介绍使用 ArkUI JS 框架开发一款基于鸿蒙操作系统的飞机大战游戏。这款飞机大战游戏是在普通单机游戏的基础上添加了鸿蒙操作系统分布式支持,使游戏可以同时使用多台鸿蒙操作系统设备,为游戏玩家提供分布式场景的游戏体验。

## 5.1　飞机大战游戏介绍

鸿蒙飞机大战游戏,在传统的飞机大战游戏的基础上,增加了鸿蒙操作系统的分布式支持,实现为不同的游戏功能提供最佳的鸿蒙设备展现。例如使用鸿蒙智慧屏展示游戏界面、鸿蒙手机模拟游戏的游戏手柄、鸿蒙音箱设备播放游戏声音,不同的游戏功能通过不同算力的设备来展示,为游戏提供分布式立体的体验效果,如图 5.1 所示。

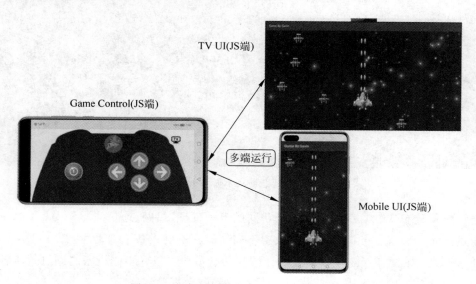

图 5.1　根据不同能力显示不同的游戏 UI

游戏涉及多种角色,例如太空、主角飞机、敌机、子弹,还涉及爆炸效果在内的多个游戏角色的变化。同时,游戏中还添加了鸿蒙分布式操作系统的技术元素,这些元素的功能如下。

(1) 可以流转的游戏界面:在手机上运行的游戏可以流转到电视上,而且可以继续在电视上运行。

(2) 设备转换能力:手机秒变游戏控制器,把游戏场景流转到鸿蒙智慧屏上,游戏的声音通过音箱播放,鸿蒙操作系统可以把不同设备连接起来形成一个虚拟的超级终端。

(3) 可以有多台设备申请加入游戏中:多台设备可以申请加入游戏中,此时游戏将从单主角的游戏变成多主角的游戏。

(4) 游戏界面可以根据不同的设备进行适配显示,充分体现鸿蒙的一套代码多设备部署的能力。

通过搭载鸿蒙操作系统的智慧屏,能够更好地展现游戏的画面,让游戏玩家可以更好地体验游戏的画面感,如图 5.2 所示。

图 5.2　智慧屏展示游戏的效果图

通过鸿蒙分布式软总线实现鸿蒙设备间的自动发现、组网、通信。
通过鸿蒙分布式任务调度实现游戏中不同功能在不同设备上启动、运行。
通过鸿蒙分布式 FA、PA 展现技术让游戏功能模块可以在不同设备之间进行流转。
通过鸿蒙分布式数据库系统,提供不同设备间的数据共享。
鸿蒙分布式飞机大战游戏交互效果如图 5.3 所示。

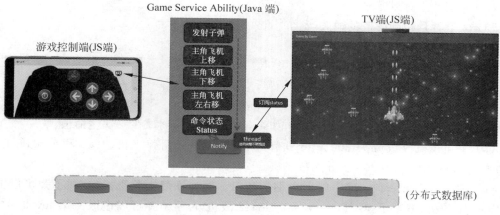

图 5.3　分布式游戏的交互图

## 5.2　飞机大战游戏分析

本游戏案例采用 JS 端的 Canvas 实现游戏效果绘制，因为 Canvas 在不同的鸿蒙设备上运行的性能不同，因此需要特别注意性能问题。

### 5.2.1　游戏性能问题分析

目前鸿蒙的原生 Canvas 并不能很好地支持游戏开发，鸿蒙 JS UI 中提供的 Canvas 主要用来绘制图表。对于飞机大战游戏场景，会涉及多个游戏角色的动画效果，鸿蒙的 Canvas 渲染能力比较有限，同时对于多游戏角色动画的实现目前虽然可以通过 JS 中的 setInterval 实现，但是鸿蒙 JS 运行时对多 setInterval 的执行效率同样非常低，而且多 setInterval 会造成屏幕闪屏的问题，因此需要尽量避免由于对 setInterval 的调用太多而造成内存泄漏问题和清除画布带来的闪屏问题。

本案例中的飞机大战需要在不同的硬件设备上运行，因此如何处理鸿蒙 Canvas 的渲染不足问题和 setInterval 效率低下问题非常关键。

为了解决 Canvas 绘制的性能问题，这里通过在整个游戏逻辑中只调用一个 setInterval 定时器的方式来处理游戏中的多角色动画，这样可以极大地减少因为清除画布带来的闪屏和内存泄漏问题。

### 5.2.2　游戏角色分析

本游戏案例中的游戏角色包括太空背景、主角飞机、敌人飞机、子弹、爆炸效果。具体的角色分析如下。

**1. 游戏太空背景**

（1）实现太空滚动背景。

（2）实现太空背景音乐。
（3）太空背景跟随游戏的通关变化。

### 2．主角飞机

（1）游戏玩家可以选择不同的主角飞机。
（2）主角飞机可以控制移动和发射子弹。
（3）主角飞机不可以超出屏幕移动。
（4）主角飞机生命值可以自由设置，如果飞机被撞一次将减少一次生命值。
（5）主角飞机等级分为 300 积分铁牌飞机、800 积分铜牌飞机、1000 积分银牌飞机、2000 积分金牌飞机和 10000 积分无敌飞机。
（6）主角飞机可一颗子弹击毁同级别的敌机。例如铁牌主角飞机，一颗子弹可以击毁一架 level1 敌机，三颗子弹可以击毁一架 level2 敌机，5 颗子弹可以击毁一架 level3 敌机，10 颗子弹可以击毁 boss 飞机。
（7）主角飞机被敌机撞击三次后会爆炸销毁，结束游戏。
（8）主角飞机爆炸效果。

### 3．敌人飞机

（1）敌机分 4 个级别：level1、level2、level3、boss，它们会随机出现，出现频率最高的是 level1，出现频率最低的是 level3，boss 飞机出现在最后。
（2）敌机出现在屏幕最上方，随机位置。
（3）敌机左右来回运动下落，增加游戏难度。
（4）敌机可以发射子弹。
（5）敌机被主角子弹击中后会爆炸销毁。
（6）敌机运动出屏幕最下边后会销毁敌机对象。

### 4．子弹

（1）子弹随飞机发射，碰撞爆炸。
（2）子弹运动方向向上或者向下。
（3）子弹运动的坐标超出屏幕后会销毁。
（4）子弹碰撞敌机会爆炸，并销毁子弹。

### 5．爆炸效果

（1）发生碰撞，爆炸效果绘制。
（2）爆炸位置为子弹或者敌机位置。
（3）爆炸声音。
（4）爆炸效果随后消失。

## 5.3　飞机大战核心算法

该游戏中涉及几个常见的算法，如碰撞检测算法和子弹发射运动算法等。

## 5.3.1 碰撞检测算法

碰撞检测就是判定两个物体是否在同一时间内占用一块空间,从数学的角度来看,就是两个物体有没有交集。

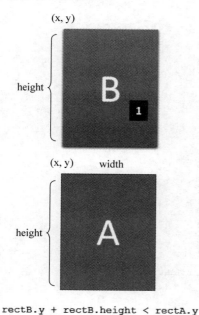

图 5.4  B 在 A 的上面

检测碰撞的方法有很多,一般使用以下两种:

(1) 从几何图形的角度来检测,就是判断一个物体是否与另一个有重叠,可以用物体的矩形边界来判断。

(2) 检测距离,就是判断两个物体是否足够近到发生碰撞,需要计算距离和判断两个物体是否足够近。

下面从几何图形的角度来检测 Canvas 中的对象是否发生碰撞,该检测方法如下。

**1. 基于矩形的碰撞检测**

这里假设碰撞体都是矩形物体。下面创建两个 rect 对象 A 和 B(以下简称 A 和 B)。

检测 A 和 B 是否发生重叠,在讨论是否发生重叠时可以先看一看没有重叠的 4 种状态:

状态 1:B 在 A 的上面,代码如下,如图 5.4 所示。

```
rectB.y + rectB.height < rectA.y
```

状态 2:B 在 A 的右面,代码如下,如图 5.5 所示。

```
rectB.y > rectA.x + rectA.width
```

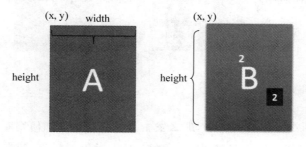

rectB.y > rectA.y + rectA.height

图 5.5  B 在 A 的右面

状态3：B在A的下面，代码如下，如图5.6所示。

```
rectB.y > rectA.y + rectA.height
```

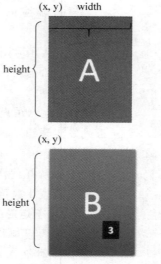

图5.6　B在A的下面

状态4：B在A的左面，代码如下，如图5.7所示。

```
rectB.x + rectB.width < rectA.x
```

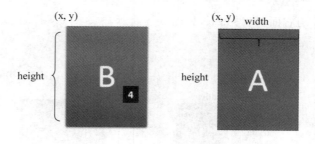

图5.7　B在A的左面

知道了如何判断没有重叠的状态，那么发生重叠的状态该如何判断呢？没错，"取反"即可，创建函数collisionCheck()，该函数传入两个Rect对象参数，当两个Rect对象发生重叠时返回值为true，代码如下：

```
function collisionCheck(rectA,rectB) {
    return !(rectB.y + rectB.height < rectA.y || rectB.y > rectA.x + rectA.width ||
    rectB.y > rectA.y + rectA.height|| rectB.x + rectB.width < rectA.x)
}
```

**2. 基于圆形的碰撞检测**

圆形碰撞同样创建两个 Circle 对象 A 和 B(以下简称 A 和 B)。

圆形间碰撞检测可以简单地通过两圆心间的距离与两圆半径之和的比较进行判断,当两圆心的距离小于两圆半径之和时发生碰撞。

首先需要做的是计算出两圆心间的距离,这里将用到两点间的距离公式,设 A、B 两点的坐标分别为 $(x_1,y_1)$ 和 $(x_2,y_2)$:

$$|AB|=\sqrt{(x_1-x_2)^2+(y_1-y_2)^2} \tag{5.1}$$

当取得两圆心间的距离之后将与两圆半径之和比较,如果距离小于半径之和,则返回值为 true,如图 5.8 所示。

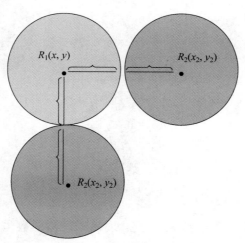

图 5.8 检测圆形碰撞

创建函数 collisionCheckCircle(),该函数传入两个 Rect 对象参数,当两个 Rect 对象发生重叠时返回值为 true,代码如下:

```
function collisionCheckCircle(circleA,circleB) {
    var dx = circleA.x - circleB.x;
    var dy = circleA.y - circleB.y;
    var distance = Math.sqrt(dx * dx + dy * dy);
    return distance < (circleA.radius + circleB.radius);
}
```

**3. 基于矩形与圆形间的碰撞检测**

前面讲解的碰撞检测都是单一形状间的碰撞检测,下面将检测矩形和圆形间的碰撞。

和矩形间的碰撞检测一样,我们先看一看没有发生碰撞的 4 种情况:

(1) Circle.y + Circle.radius < Rect.y。
(2) Circle.x − Circle.radius > Rect.x + Rect.width。
(3) Circle.y − Circle.radius > Rect.y + Rect.height。
(4) Circle.x + Circle.radius < Rect.x。

这 4 种情况如图 5.9 所示。

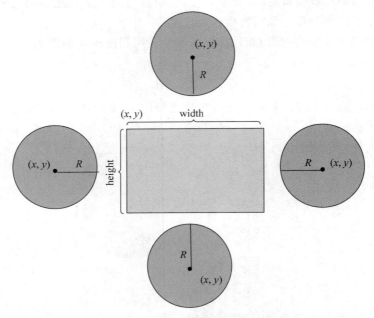

图 5.9　矩形与圆形间的碰撞检测

collisionCheck()函数将没有重叠的状态"取反",向该函数传入 Rect 对象和 Circle 对象,当 Rect 对象与 Circle 对象发生重叠时返回值为 true,代码如下:

```
function collisionCheck(Rect,Circle) {
    return !(Circle.y + Circle.radius < Rect.y ||
    Circle.x - Circle.radius > Rect.x + Rect.width ||
    Circle.y - Circle.radius > Rect.y + Rect.height ||
    Circle.x + Circle.radius < Rect.x)
}
```

## 5.3.2　子弹飞行算法

本游戏中暂时没有拐弯飞行的子弹,所有的子弹都是直线飞行。

### 1. 开花弹算法

开花弹的落点位置是 CircleMatrix() 函数调用后返回的一圈圆形离散坐标,代码如下:

```
function CircleMatrix(num) {              /* 传入多少颗 */
    var len = num || 20;                  /* 默认 20 颗 */
    var arr = [];
    for(var i = 0; i < len; i++){
        arr.push([1291 * Math.sin(2 * Math.PI * i/len),1291 * Math.cos(2 * Math.PI * i/len)]);
    }
    return arr;
}
```

### 2. 从屏幕外 360°随机飞入的自机狙

自机狙指的是朝玩家当前方向射去的子弹,代码如下:

```
function OutLaunch(num) {
    var delay = Math.floor(Math.random() * 500 + 100);
    var rest = num;
    delayLaunch();
    function delayLaunch() {
        var ranX,ranY
        do{
            ranX = Math.floor(Math.random() * 630-15);
            ranY = Math.floor(Math.random() * 630-15);
        } while(ranX > - 10 && ranX < 610 && ranY > - 10 && ranY < 610)
        doLaunch(ranX,ranY,1);
        rest -- ;
        if(rest > 0){
            setTimeout(delayLaunch,delay)
        }
    }
}
```

## 5.4 飞机大战游戏界面实现

本游戏案例的界面分为游戏主界面和游戏控制器手柄界面,游戏主界面是响应式的,可以在 TV 大屏和手机屏之间自由适配。

本游戏案例需要创建两个 JS Ability,即 MainAbility 和 GameControlAbility。将 MainAbility 绑定 default component,用于编写游戏的主界面。将 GameControlAbility 绑定 GControl component,用于编写游戏的控制器界面。

代码的目录结构如图 5.10 所示。

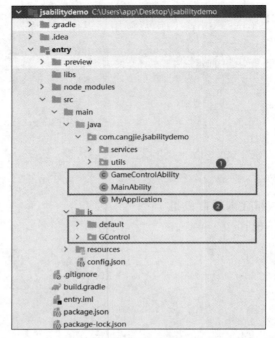

图 5.10　游戏项目的目录结构

### 5.4.1　游戏主界面

游戏主界面包括两个页面：一个游戏引导页和一个游戏主界面。游戏引导页的页面布局如图 5.11 所示。

在 default 目录下添加 nav JS Page，在 config.json 文件中把 nav 页面设置为主页，配置如下：

```
"js": [
  {
"pages": [
"pages/nav/nav",
"pages/index/index",
    ],
"name": "default",
"window": {
"designWidth": 720,
"autoDesignWidth": true
    }
  }
],
```

pages 数组的第 1 个值为组件的首页。

图 5.11　游戏引导页面

游戏引导页面的代码布局如代码示例 5.1 所示。

代码示例 5.1　游戏引导页面　planeGame/js/default/nav.hml

```
< div class = "container">
< button class = "btn" @click = "start(2)">
        开始游戏
</button >
< button class = "btn" @click = "openCtrl()">
        打开远程游戏遥控器
</button >
</div >
```

打开远程游戏遥控器的代码如代码示例 5.2 所示。

代码示例 5.2　打开远程游戏遥控器　planeGame/js/default/nav.js

```
import router from '@system.router';

export default {
    data: {},
    start(flag) {
        let uri = 'pages/nav/nav';
```

```
                switch (flag) {
                    case 2:
                    uri = 'pages/index/index;
                    break;
                }
                router.push({
                    uri: uri,
                })
            },
            openCtrl:async function(){
                let actionData = {
                    uri: 'www.huawei.com'
                };
                let target = {
                    bundleName: "com.cangjie.jsabilitydemo",
                    abilityName: "com.cangjie.jsabilitydemo.GameControlAbility",
                    data: actionData
                };
                let result = await FeatureAbility.startAbility(target);
                let ret = JSON.parse(result);
                if (ret.code == 0) {
                    console.log('success');
                } else {
                    console.log('cannot start browing service, reason: ' + ret.data);
                }
            }
    }
}
```

游戏的主界面其实非常简单,如代码示例 5.3 所示,通过 canvas 绘制,因此 HML 代码非常简单,这里鸿蒙 JavaScript 版 UI 目前还没有 Audio,所以采用 Video 替代 Audio。

**代码示例 5.3　游戏主界面　planeGame/js/default/index.hml**

```
< div class = "container">
< canvas id = "game" class = "game" ontouchstart = "moveDown" ontouchend = "moveUp"></canvas >
< video src = "{{gaemMusic}}" autoplay = "true" style = "display: none;"></video >
</div >
.container {
    display: flex;
    justify-content: center;
    align-items: center;
    width: 100%;
    height: 100%;
}

.game {
    width: 100%;
```

```
    height: 100%;
    background-image: url("/common/imgs/bj.jpg");
    background-repeat: repeat-y;
}
```

在.game 样式中通过 background-image 给 canvas 添加样式背景,后面需要给游戏添加天空的动画背景,所以需要设置 background-repeat:repeat-y。

### 5.4.2 游戏控制手柄界面

游戏手柄界面采用 fixed 定位方式实现,效果如图 5.12 所示。

图 5.12 游戏控制手柄界面

游戏手柄界面布局如代码示例 5.4 所示。

代码示例 5.4 planeGame/js/control/index.hml

```
<div class="container">
<div class="logo">
<image class="logo_btn" id="logo"></image>
</div>
<div class="start">
<image class="start_btn" @click="start"></image>
</div>
<div class="up" @click="runPlane(102)">
<image class="up_btn"></image>
</div>
<div class="down">
<image class="down_btn" @click="runPlane(103)"></image>
</div>
<div class="left">
```

```html
        <image class = "left_btn" @click = "runPlane(104)"></image>
    </div>
    <div class = "right">
        <image class = "right_btn" @click = "runPlane(105)"></image>
    </div>
    <div class = "tv">
        <image class = "tv_btn"></image>
    </div>
</div>
```

样式如代码示例5.5所示。

**代码示例5.5　planeGame/js/control/index.css**

```css
.container {
    flex-direction: column;
    justify-content: center;
    align-items: center;
    background-image: url("/common/bj.jpg");
    background-size: 100%;
    background-repeat: no-repeat;
    object-fit: contain;
}

.logo {
    position: fixed;
    left: 45%;
    top: 25px;
    width: 80px;
    height: 80px;
}

.logo_btn {
    background-image: url("/common/plane_logo.png");
    background-repeat: no-repeat;
    background-size: 100%;
    width: 80px;
    height: 80px;
    border-radius: 40px;
    object-fit: cover;
}

.tv {
    position: fixed;
    right: 20px;
    top: 25px;
    width: 80px;
    height: 80px;
}
```

```css
.tv_btn {
    background-image: url("/common/tv.jpg");
    background-repeat: no-repeat;
    background-size: 100%;
    width: 40px;
    height: 40px;
    object-fit: contain;
}

.start {
    position: fixed;
    left: 150px;
    top: 150px;
    width: 80px;
    height: 80px;
}

.stop_btn {
    background-image: url("/common/stop.png");
    background-repeat: no-repeat;
    background-size: 100%;
    object-fit: cover;
    border-radius: 40px;
}

.stop_btn:active {
    border-width: 10px;
    border-color: red;
}

.start_btn {
    background-image: url("/common/fire_btn.png");
    background-repeat: no-repeat;
    background-size: 100%;
    object-fit: cover;
    border-radius: 40px;
}

.start_btn:active {
    border-width: 20px;
    border-color: red;
}

.up {
    position: fixed;
    right: 200px;
    top: 100px;
    width: 80px;
    height: 80px;
```

```css
}

.up_btn:active {
    border-width: 10px;
    border-color: red;
}

.up_btn {
    background-image: url("/common/up.png");
    background-repeat: no-repeat;
    background-size: 100%;
    object-fit: cover;
    border-radius: 40px;
}

.down {
    position: fixed;
    right: 200px;
    top: 200px;
    width: 80px;
    height: 80px;
}

.down_btn {
    background-image: url("/common/down.png");
    background-repeat: no-repeat;
    background-size: 100%;
    object-fit: cover;
    border-radius: 40px;
}

.down_btn:active {
    border-width: 10px;
    border-color: red;
}

.left {
    position: fixed;
    right: 100px;
    top: 150px;
    width: 80px;
    height: 80px;
}

.left_btn {
    background-image: url("/common/right.png");
    background-repeat: no-repeat;
    background-size: 100%;
    object-fit: cover;
```

```css
    border-radius: 40px;
}

.left_btn:active {
    border-width: 10px;
    border-color: red;
}

.right {
    position: fixed;
    right: 300px;
    top: 150px;
    width: 80px;
    height: 80px;
}

.right_btn {
    background-image: url("/common/left.png");
    background-repeat: no-repeat;
    background-size: 100%;
    object-fit: cover;
    border-radius: 40px;
}

.right_btn:active {
    border-width: 10px;
    border-color: red;
}
```

将游戏控制手柄的界面设置为横屏显示,修改 config.json 文件,在 GameControlAbility 对象中将 orientation 的值设置为 landscape,同时将该 ability 设置为无标题主题,配置如下:

```json
{
"name": "com.cangjie.jsabilitydemo.GameControlAbility",
"icon": "$media:icon",
"description": "$string:gamecontrolability_description",
"label": "鸿蒙飞机大战游戏手柄",
"type": "page",
"orientation": "landscape",
"launchType": "standard",
"metaData": {
"customizeData": [
    {
"name": "hwc-theme",
"value": "androidhwext:style/Theme.Emui.Light.NoTitleBar",
"extra": ""
    }
  ]
 }
}
```

## 5.5 飞机大战核心代码实现——单机篇

本节详细讲解飞机大战游戏的开发细节,通过本节的学习,可以开发一个单机版的飞机大战游戏。

### 5.5.1 加载游戏资源

游戏中涉及很多需要预加载的图片资源,这些游戏资源如果在游戏中进行创建会非常消耗资源,因此需要提前创建出来,这样在游戏运行中需要使用的时候直接引用就可以了。

游戏资源分为几类:主角飞机图片、敌机图片、子弹图片、爆炸图片。

为了方便使用,这里把这几类游戏资源创建成 JS 对象,对游戏资源进行规格设置。所有的图片资源都需要转换成 Image 对象,Image 对象的创建需要耗时,所以需要提前创建,否则会非常影响游戏效果,如代码示例 5.6 所示。

代码示例 5.6 定义游戏资源

```
//图片资源
const resources = [
"/common/imgs/plane/hero/hero01.png",
"/common/imgs/plane/enemy/enemy1.png",
"/common/imgs/plane/bullet/bullet2.png",
"/common/imgs/plane/boom/boom02.png"
]

//主角配置
let playerObj = {
    width: 100,
    height: 91,
    image: {}
}

//炸弹配置
let boomObj = {
    width: 94,
    height: 119,
    image: {}
}

//飞机对象配置
let enemyObj = {
    width: 50,
    height: 50,
    timer: 1000,
    lastTime: 0,
    image: {}
```

```
}
//子弹对象配置
let bulletObj = {
    width: 40,
    height: 50,
    speed: 10,
    lastTime: 0,
    timer: 200,
    image: {}
}
```

在加载资源前,首先需要获取屏幕的宽和高,这里获取的屏幕的宽和高需要重新计算。获取设备信息需要通过鸿蒙 JS API 获取,代码如下:

```
import device from '@system.device';
```

设备信息需要在 onShow()方法中获取,失败信息可以通过 fail 回调方法获取,如代码示例 5.7 所示。

**代码示例 5.7 获取设备信息**

```
initDevice() {
    device.getInfo({
        success: (data) =>{
            //计算屏幕的宽和高(像素)
            this.screen = {
                width: data.windowWidth / data.screenDensity,
                height: data.windowHeight / data.screenDensity
            }
            console.error(JSON.stringify(data))
        }
    });
},
```

**注意**:这里的 data.windowWidth 还不是 px,转换成像素时需要除以 screenDensity 换算出来。

这里封装了一个 loadImage()方法,这种方法把一张图片路径创建成 Image 对象。count 变量可用来计算资源加载情况,创建好一张图片对象后回调计算资源的加载。

当所有图片加载完成后,创建游戏主角,如代码示例 5.8 所示。

**代码示例 5.8 加载游戏资源**

```
loadImage(src) {
    var img = new Image();
    img.src = src;
    img.onload = this.resourceLoadComplete
```

```
        return img;
    },
    loadResourceImages() {
        //加载主角战机
        playerObj.image = this.loadImage(resources[0])

        //加载敌机
        enemyObj.image = this.loadImage(resources[1]);

        //子弹
        bulletObj.image = this.loadImage(resources[2]);

        //爆炸
        boomObj.image = this.loadImage(resources[3]);
    },
    resourceLoadComplete() {
        ++this.count;
        if (this.count >= resources.length) {
            console.error("load complete: count = " + this.count)
            //资源加载完成,创建游戏主角
            this.startGame()
        }
    },
```

### 5.5.2 太空背景动画

太空背景滚动的作用是为了让游戏玩家产生视觉错觉,飞机实际是不动的,是背景在动,让玩家觉得是飞机在飞行。实现要点是让背景图片不断地从下往上运动。

通过 backgroundPosition 属性,设置 backgroundPosition 在 y 轴 repeat。下面的动画实现通过 $element 的 animate()方法调用关键帧动画,如代码示例 5.9 所示。

**代码示例 5.9 太空背景动画**

```
loadGameResource() {
    var game = this.$element("game");
    this.cxt = game.getContext("2d");
    var options = {
        duration: 200000,
        easing: 'linear',
        fill: 'forwards',
        iterations: "Infinity",
    };
    //向上运动
    var frames = [
        {
            backgroundPosition: "0% 0%"
        },
```

```
            {
                backgroundPosition:"0% -1000%"
            },
        ];
        game.animate(frames,options).play();
    },
```

### 5.5.3 游戏动画入口

为了提高整个游戏性能,采用一个 setInterval 来处理所有游戏角色动画。设置游戏循环入口,游戏角色需要不断变化,因此需要把主角飞机、子弹和敌机加入动画循环,如代码示例 5.10 所示。

**代码示例 5.10 游戏动画入口**

```
onShow() {
//循环动画
    this.loopGame();
},

//游戏主入口动画
loopGame() {
//延时时间计算
    var currentTime = 0;
//帧动画
    var looptimer = setInterval(()=>{
//清屏幕
        this.cxt.clearRect(0,0,this.screen.width,this.screen.height);
//屏幕的宽和高通过 device 获取
//绘制主角
        this.player.draw(this.cxt);

    },50)
},
```

这里首先在动画中循环绘制主角,然后在后面的游戏控制中控制游戏主角的移动。

### 5.5.4 绘制游戏主角

定义一个 Player 类,这个类用来创建主角飞机对象,如代码示例 5.11 所示。

**代码示例 5.11 创建主角飞机对象**

```
//主角飞机对象
function Player(screen,img) {
    this.screen = screen;
    this.width = img.width;
```

```
    this.height = img.height;
    //坐标
    this.x = screen.width / 2 - img.width / 2;
    this.y = screen.height - (img.height + 100);
    this.img = img.image;
    //能量
    this.power = 6;
}

//绘制主角
Player.prototype.draw = function (cxt) {

    //判断横坐标是否出了屏幕
    if (this.x <= 0) {
        this.x = 0;
    } else if (this.x >= (this.screen.width - this.width)) {
        this.x = (this.screen.width - this.width)
    }

    //判断纵坐标是否出了屏幕
    if (this.y <= 0) {
        this.y = 0;
    } else if (this.y >= (this.screen.height - this.height)) {
        this.y = (this.screen.height - this.height)
    }
    cxt.drawImage(this.img,this.x,this.y,this.width,this.height);
}
```

绘制主角的 draw()方法,需要判断 x 和 y 坐标在屏幕范围内,如代码示例 5.12 所示。

**代码示例 5.12　控制主角飞机在屏幕范围内移动**

```
//判断横坐标是否出屏幕
  if (this.x <= 0) {
      this.x = 0;
  } else if (this.x >= (this.screen.width - this.width)) {
      this.x = (this.screen.width - this.width)
  }

  //判断纵坐标是否出屏幕
  if (this.y <= 0) {
      this.y = 0;
  } else if (this.y >= (this.screen.height - this.height)) {
      this.y = (this.screen.height - this.height)
  }
```

当所有资源加载成功后,创建主角对象,代码如下:

```
startGame() {
    this.player = new Player(this.screen,playerObj);
},
```

接下来,需要在游戏循环中让游戏主角动起来,代码如下:

```
loopGame() {
//用来控制时间,每次循环累加
    var currentTime = 0;
//帧动画
    var looptimer = setInterval(()=>{
//清屏幕
        this.cxt.clearRect(0,0,this.screen.width,this.screen.height)
//绘制主角
        this.player.draw(this.cxt);

    },50)
},
```

### 5.5.5 绘制游戏敌机

定义敌机类 Enemy,这个类用来绘制敌机,如代码示例 5.13 所示。

**代码示例 5.13  绘制游戏敌机**

```
//飞机对象
function Enemy(x,y,dx,dy,enemyObj,rotation) {
    this.x = x;
    this.y = y;
//动态坐标
    this.dx = dx;
    this.dy = dy;
    this.img = enemyObj.image;
    this.width = enemyObj.width;
    this.height = enemyObj.height;
//左右运动
    this.rotation = rotation;
    this.id = 0;

    if (this.rotation < 0.2) {
        this.dx = - this.dx;
    } else if (this.rotation < 0.7) {
        this.dx = - this.dx;
    } else {
        this.dx = 0;
        this.dy = this.dy;
    }
}
```

```javascript
//绘制敌机
Enemy.prototype.draw = function (cxt) {
    cxt.drawImage(this.img,this.x,this.y,this.width,this.height)
}

//更新敌机运动
Enemy.prototype.update = function (screen, cxt, cb) {

this.y += this.dy;
    this.x += this.dx;

//控制敌机不出屏幕
    if (this.x + this.width >= screen.width) {
        this.dx = -this.dx;
    } else if (this.x <= 0) {
        this.dx = Math.abs(this.dx)
    }

//当敌机在y轴出了屏幕,通过回调cb把出了屏幕的敌机删除
    if (this.y > screen.height + this.height) {
        //删除出了屏幕的敌机
        cb(this.id)
    }

    this.draw(cxt)
}

//创建敌机
function createEnemy(screen) {
    var x = Math.random() * (screen.width - enemyObj.width);
    var y = -enemyObj.height;
    var dx = 3;
    var dy = 3;
    var rotation = Math.random();
    return new Enemy(x,y,dx,dy,enemyObj,rotation);
}
```

初始化敌机,设置敌机编号,并把创建的敌机放到敌机数组中,如代码示例5.14所示。

**代码示例5.14  初始化敌机,设置敌机编号**

```javascript
initEnemy() {
//循环统计敌机编号
    ++this.enemyIndex;
//创建敌机对象
    var enemy = createEnemy(this.screen);
//设置敌机编号
    enemy.id = this.enemyIndex;
//把敌机按编号放到敌机数组中
```

```
        this.enemyArray[this.enemyIndex] = enemy;
    //绘制敌机
        enemy.draw(this.cxt);
    },
```

在主动画函数中循环控制敌机下落,设置每秒创建一架敌机,这里需要判断时间间隔,通过变量 currentTime 与敌机对象中的 timer 比较时间,这里的 timer 是延迟的时间,如代码示例 5.15 所示。

**代码示例 5.15  敌机动画**

```
loopGame() {
//计时器
    var currentTime = 0;
    var looptimer = setInterval(() => {

        this.cxt.clearRect(0,0,this.screen.width,this.screen.height)
        this.player.draw(this.cxt);

        //每 1s 创建一架敌机, enemyObj.timer 用于控制时间间隔
        if (currentTime >= enemyObj.lastTime + enemyObj.timer) {
//更新间隔时间
            enemyObj.lastTime = currentTime
            console.error("------ 每 1s 创建一架敌机 ------- ")
            this.initEnemy()
        }

//敌机运动,对已经不在屏幕内的敌机,从数组删除
        this.enemyArray.forEach(enemy => {
            enemy.update(this.screen,this.cxt,(id) => {
                console.error("------ 删除敌机 ------- ")
                delete this.enemyArray[id]; //删除 id 的敌机
            })
        })

        currentTime += 50;
    },50)
},
```

敌机下落,这里涉及碰撞检测,如代码示例 5.16 所示。

碰撞检测函数传入两个矩形对象,如果返回值为 true,则表示两个矩形相交了;否则,返回值为 false。

**代码示例 5.16  碰撞检测**

```
collides(a, b) {
    return a.x < b.x + b.width &&
    a.x + a.width > b.x &&
```

```
    a.y < b.y + b.height &&
    a.y + a.height > b.y;
},
```

### 5.5.6 绘制子弹对象

定义子弹类,子弹需要根据主角的位置进行移动,需要将所有 $x$ 和 $y$ 参数设置为主角飞机的位置并计算出来,如代码示例 5.17 所示,子弹发射效果如图 5.13 所示。

图 5.13 子弹随飞机发射

代码示例 5.17 定义子弹类

```
function Bullet(x,y,obj) {
    this.x = x;
    this.y = y;
    this.height = obj.height;
    this.width = obj.width;
    this.speed = obj.speed;
    this.id = 0;
    this.img = obj.image
}

//更新子弹位置
Bullet.prototype.update = function (cxt, cb) {
    this.y += - this.speed;
    if (this.y < - this.height) {
        cb(this.id)
```

```
        }
        this.draw(cxt)
}

//绘制子弹
Bullet.prototype.draw = function (cxt) {
    cxt.drawImage(this.img,this.x,this.y,this.width,this.height);
}

exports default Bullet;
```

根据主角生成子弹,这里子弹的坐标需要计算,并且需要对每个子弹进行编号,然后加入子弹数组中,如代码示例5.18所示。

代码示例5.18 根据主角生成子弹

```
//创建子弹对象
fireBullet() {
//每次创建累加索引
    ++this.bulleteIndex;
    var x = (this.player.x + playerObj.width / 2) - bulletObj.width / 2;
    var y = this.player.y - playerObj.height / 2;
//创建子弹对象
    var bullet = new Bullet(x,y,bulletObj);
    bullet.id = this.bulleteIndex;
//把当前索引的子弹放到子弹数组中
    this.bulletsArray[this.bulleteIndex] = bullet;
},
```

子弹的发射需要通过主循环入口循环子弹数组,以此创建子弹,生成的子弹跟随主角Player运动,这里设置200ms创建一颗子弹,当子弹出了屏幕后删除子弹,如代码示例5.19所示。

代码示例5.19 循环子弹数组创建子弹

```
loopGame() {
    var currentTime = 0;
    var looptimer = setInterval(()=>{
        this.cxt.clearRect(0,0,this.screen.width,this.screen.height)
        this.player.draw(this.cxt);

        //每1s创建一架敌机
        if (currentTime >= enemyObj.lastTime + enemyObj.timer) {
            enemyObj.lastTime = currentTime
            console.error("------每1s创建一架敌机-------")
            this.initEnemy()
        }

        this.enemyArray.forEach(enemy =>{
```

```
            enemy.update(this.screen,this.cxt,(id) =>{
                console.error("------ 删除敌机 -------")
                delete this.enemyArray[id];
            })
        })

        //每 200ms 生成一颗子弹
        if (currentTime >= bulletObj.lastTime + bulletObj.timer) {
            bulletObj.lastTime = currentTime
            this.fireBullet()
        }

//子弹运动
        this.bulletsArray.forEach(bullet =>{
            bullet.update(this.cxt,id =>{
                console.error("------ 删除子弹 -------")
                delete this.bulletsArray[id];
            })
        })

        //碰撞检测
        this.handleCollisions()

//主角被碰撞后,减主角的生命值,每碰撞一次减两个生命值
//当 power 为 0 时,游戏结束
        if (this.player.power <= 0) {
            //清除定时器
            clearInterval(looptimer);
            looptimer = null;
        }

        currentTime += 50;
    },50)
},
```

这里需要处理子弹与敌机的碰撞检测,以及敌机与主角的碰撞检测,如代码示例 5.20 所示。

**代码示例 5.20　敌机与主角的碰撞检测**

```
handleCollisions() {
//子弹是否和敌机碰撞
    this.bulletsArray.forEach(bullet =>{
        this.enemyArray.forEach(enemy =>{
            if (this.collides(bullet,enemy)) {
                this.initBoom(enemy)
                delete this.bulletsArray[bullet.id]
                delete this.enemyArray[enemy.id]
                prompt.showToast({
```

```
                message: "消灭一架敌机:" + enemy.id
            })
        }
    })
})
//敌机是否与主角碰撞
this.enemyArray.forEach(enemy =>{
    if (this.collides(this.player,enemy)) {
//主角碰撞,生命值减 2
        this.player.power += -2;
//绘制碰撞的爆炸效果
        this.initBoom(enemy)
        delete this.enemyArray[enemy.id]
        prompt.showToast({
            message: "还有" + this.player.power / 2 + "条生命!"
        })
    }
})
},
```

### 5.5.7 绘制爆炸效果

上面的敌机与主角,或者子弹与敌机的碰撞,都会有碰撞爆炸效果,如图 5.14 所示。

图 5.14 碰撞检测

绘制碰撞的爆炸效果,如代码示例 5.21 所示。

**代码示例 5.21 爆炸效果绘制**

```
initBoom(enemy) {
this.cxt.drawImage(boomObj.image,enemy.x,enemy.y,boomObj.width,boomObj.height)
},
```

### 5.5.8 操作主角飞机

这里是单机版的飞机大战,可以通过单击屏幕的不同位置操作主角飞机,这里需要处理 touch 事件,如代码示例 5.22 所示。

代码示例 5.22 操作主角飞机

```
moveDown(e) {
    var x = e.touches[0].localX;
    var y = e.touches[0].localY;
    //比较坐标
    if (this.player.y > y) {
        this.player.y -= 10
    } else {
        this.player.y += 10
    }

    if (this.player.x > x) {
        this.player.x -= 10
    } else {
        this.player.x += 10
    }
},
```

## 5.6 飞机大战核心代码实现——鸿蒙篇

上面通过 8 个小节,我们完成了一个单机版的飞机大战游戏,下面介绍如何给游戏添加鸿蒙的分布式能力。

### 5.6.1 多设备间游戏流转

通过鸿蒙分布式 UI 能力,可以对同一个应用,通过多个设备来控制,例如使用 TV 显示游戏场景,通过手机游戏手柄来控制游戏,如图 5.15 所示,通过蓝牙音箱来播放游戏声音。

注意:本书中涉及的手机之间的应用流转,需要开发者注册华为开发者账号,同时在手机设备上进行登录,否则无法流转。

下面开始介绍如何实现游戏的遥控器,如图 5.15 所示,通过手机控制 TV 的游戏角色,这里实现两个重要的功能:

(1) 当单击 TV 连接远程 TV 设备后,启动游戏。

(2) 单击控制器的上移、下移和左右移动按键操作游戏角色。

下面介绍如何实现多设备间的流转,当单击 TV 连接图标后,可通过手机端遥控器在智慧屏上启动飞机大战游戏。

图 5.15 游戏手柄控制

### 1. 申请分布式访问权限

打开 Config.json 文件，添加权限申请配置，如代码示例 5.23 所示。

代码示例 5.23　修改 Config.json

```
"reqPermissions": [
  {
"name": "ohos.permission.DISTRIBUTED_DATASYNC"
  },
  {
"name": "ohos.permission.servicebus.ACCESS_SERVICE"
  },
  {
"name": "ohos.permission.servicebus.BIND_SERVICE"
  },
  {
"name": "ohos.permission.DISTRIBUTED_DEVICE_STATE_CHANGE"
  },
  {
"name": "ohos.permission.GET_DISTRIBUTED_DEVICE_INFO"
  },
  {
"name": "ohos.permission.GET_BUNDLE_INFO"
  }
]
```

在 MainAbility 中获取权限，如代码示例 5.24 所示。

代码示例 5.24　获取权限

```
public class MainAbility extends AceAbility {
    @Override
```

```java
    public void onStart(Intent intent) {
        setInstanceName("default");
        super.onStart(intent);
        requestPermission();
    }

    //获取权限
    private void requestPermission() {
        String[] permission = {
"ohos.permission.READ_USER_STORAGE",
"ohos.permission.WRITE_USER_STORAGE",
"ohos.permission.DISTRIBUTED_DATASYNC",
"ohos.permission.MICROPHONE",
"ohos.permission.GET_DISTRIBUTED_DEVICE_INFO",
"ohos.permission.KEEP_BACKGROUND_RUNNING",
"ohos.permission.NFC_TAG"};
        List<String> applyPermissions = new ArrayList<>();
        for (String element : permission) {
            if (verifySelfPermission(element) != 0) {
                if (canRequestPermission(element)) {
                    applyPermissions.add(element);
                }
            }
        }
        requestPermissionsFromUser(applyPermissions.toArray(new String[0]), 0);
    }

    @Override
    public void onStop() {
        super.onStop();
    }
}
```

### 2. 实现多设备间游戏流转

当单击 TV 图标后，在弹出的可以连接的在线设备中选择目标设备进行游戏启动，这里如果项目同时配置了适配多种设备，就可在对应的设备中启动该游戏，配置如图 5.16 所示。

```
"module": {
  "package": "com.cangjie.jsabilitydemo",
  "name": ".MyApplication",
  "deviceType": [
    "phone",
    "tv"
  ],
```

图 5.16　设置可以运行的鸿蒙设备

单击 TV 图标,远程启动不同设备,如代码示例 5.25 所示。

代码示例 5.25  远程启动 TV 设备

```
openTV:async function(){
    let actionData = {
        uri: 'www.huawei.com'
    };
    let target = {
        bundleName: "com.cangjie.jsabilitydemo",
        abilityName: "com.cangjie.jsabilitydemo.GameAbility",
        data: actionData
    };

    let result = await FeatureAbility.startAbility(target);
    let ret = JSON.parse(result);
    if (ret.code == 0) {
        console.log('success');
    } else {
        console.log('cannot start browing service, reason: ' + ret.data);
    }
}
```

## 5.6.2  实现游戏远程控制

游戏远程控制,可以通过 PA 订阅的模式实现,通过分布式数据库保存需要共享的数据,如图 5.17 所示。

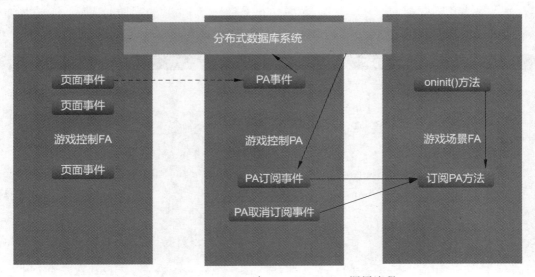

图 5.17  Page Ability 与 Service Ability 调用流程

### 1. 游戏控制器端发送控制事件

游戏控制界面通过 ontouchstart 事件发送飞机的飞行方向代码,如代码示例 5.26 所示。n 表示方向的值:1 表示向上,2 表示向左,3 表示向右,4 表示向下。

代码示例5.26 游戏控制界面

```
<div class = "container">
<div class = "logo">
<image class = "logo_btn" id = "logo"></image>
</div>
<div class = "start">
<image class = "start_btn" @click = "shoot"></image>
</div>
<div class = "up" @ontouchstart = "runPlane(1)">
<image class = "up_btn"></image>
</div>
<div class = "down" @ontouchstart = "runPlane(2)">
<image class = "down_btn"></image>
</div>
<div class = "left" @ontouchstart = "runPlane(3)">
<image class = "left_btn"></image>
</div>
<div class = "right" @ontouchstart = "runPlane(4)">
<image class = "right_btn"></image>
</div>
<div class = "tv" @click = "openTV">
<image class = "tv_btn"></image>
</div>
</div>
```

将发送控制代码的方法封装到下面的工具模块中,如代码示例 5.27 所示。

代码示例5.27 /common/game/utils.js

```
//abilityType: 0 - Ability; 1 - Internal Ability
const ABILITY_TYPE_EXTERNAL = 0;
const ABILITY_TYPE_INTERNAL = 1;
//syncOption(Optional, default sync): 0 - Sync; 1 - Async
const ACTION_SYNC = 0;
const ACTION_ASYNC = 1;
const ACTION_MESSAGE_CODE_PLUS = 1001;


export const gameAbility = {
    callAbility: async function(data){
        let action = {};
        action.bundleName = 'com.cangjie.jsabilitydemo';
        action.abilityName = 'com.cangjie.jsabilitydemo.services.GamePlaneAbility';
        action.messageCode = data.code;
        action.data = data;
```

```
            action.abilityType = ABILITY_TYPE_EXTERNAL;
            action.syncOption = ACTION_SYNC;
            let result = await FeatureAbility.callAbility(action);
            return JSON.parse(result);
    },
    //根据 messeage_code 订阅不同事件
    subAbility: async function(messeage_code,callBack){
            let action = {};
            action.bundleName = 'com.cangjie.jsabilitydemo';
            action.abilityName = 'com.cangjie.jsabilitydemo.services.GamePlaneAbility';
            action.messageCode = messeage_code;
            action.abilityType = ABILITY_TYPE_EXTERNAL;
            action.syncOption = ACTION_SYNC;
            let result = await FeatureAbility.subscribeAbilityEvent(action, function(callbackData) {
                var callbackJson = JSON.parse(callbackData);
                this.eventData = JSON.stringify(callbackJson.data);
                callBack && callBack(callbackJson);
            });
            return JSON.parse(result);
    },
    //根据 message_code 取消订阅不同事件
    unSubAbility : async function(messeage_code){
        let action = {};
        action.bundleName = 'com.cangjie.jsabilitydemo';
        action.abilityName = 'com.cangjie.jsabilitydemo.services.GamePlaneAbility';
        action.messageCode = messeage_code;
        action.abilityType = ABILITY_TYPE_EXTERNAL;
        action.syncOption = ACTION_SYNC;
        let result = await FeatureAbility.unsubscribeAbilityEvent(action);
        return JSON.parse(result);
    }
}
```

单击 runPlane()方法调用 callAbility()方法,发送命令代码给远程 Service Ability,这里同时发送一个表示方向的值 direction,代码如下:

```
import {gameAbility} from "../../common/game/utils.js"

runPlane(n) {
    gameAbility.callAbility({
        code: 1003, //调用编号,自定义的
        direction: n
    }).then(result =>{
    })
}
```

游戏控制端将命令发送给 Service Ability,需要在 action 中指定需要调用的 Service

Ability 的 bundleName 和 abilityName,abilityName 是我们自己创建的 Service Ability,可以在 config.json 文件中查看。

### 2. 创建控制游戏的 Service Ability

在 java 目录中,创建名为 GamePlaneAbility 的 Service Ability。

在 GamePlaneAbility 类的内部创建一个用来接受客户端调用的远程代理类 RemoteObject,作为它的内部类,如代码示例 5.28 所示。

为了临时存储从客户端发送过来的命令,可以使用分布式 KV 数据库临时存储命令信息,发送给订阅端后,删除数据,使用这种方法的目的是使用鸿蒙分布式数据库系统。

**代码示例 5.28    java/service/GamePlaneAbility**

```java
public class GamePlaneAbility extends Ability {

    private static final String TAG = "GamePlaneAbility";
    private MyRemote remote = new MyRemote();
    private KvManagerConfig config;
    private KvManager kvManager;
    private String storeID = "test";
    private SingleKvStore singleKvStore;

    @Override
    protected void onStart(Intent intent) {
        super.onStart(intent);
        //初始化 manager
        config = new KvManagerConfig(this);
        kvManager = KvManagerFactory.getInstance().createKvManager(config);

        //创建数据库
        Options options = new Options();
        options.setCreateIfMissing(true).setEncrypt(false).setKvStoreType(KvStoreType.SINGLE_VERSION);
        singleKvStore = kvManager.getKvStore(options, storeID);

    }

    @Override
    protected IRemoteObject onConnect(Intent intent) {
        super.onConnect(intent);
        return remote.asObject();
    }

    class MyRemote extends RemoteObject implements IRemoteBroker {

        private static final int ERROR = -1;
        private static final int SUCCESS = 0;
        private static final int SUBSCRIBE = 1005;
        private static final int UNSUBSCRIBE = 1006;
        private static final int GHE_DEVICE = 1001;
```

```java
            private static final int OPEN_DEVICE = 1002;
            private static final int OPEDRECTION = 1003;

            private Set< IRemoteObject > remoteObjectHandlers = new HashSet< IRemoteObject >();

        MyRemote() {
            super("飞机控制游戏代理");
        }

        @Override
        public boolean onRemoteRequest( int code, MessageParcel data, MessageParcel reply,
MessageOption option) throws RemoteException {

            switch (code) {
                case OPEDRECTION:{
                    String zsonStr = data.readString();
                    ReqParam param = ZSONObject.stringToClass(zsonStr,ReqParam.class);
                    Map< String, Object > zsonResult = new HashMap< String, Object >();
                    System.out.println("-------- OPEDRECTION --------");
                    zsonResult.put("data", param.getDirection());
                    try {
                        singleKvStore.putInt("hello", param.getDirection());
                    } catch (KvStoreException e) {
                        e.printStackTrace();
                    }
                    zsonResult.put("code", SUCCESS);
                    reply.writeString(ZSONObject.toZSONString(zsonResult));
                    break;
                }
                //开启订阅,保存对端的remoteHandler,用于上报数据
                case SUBSCRIBE: {
                    remoteObjectHandlers.add(data.readRemoteObject());
                    startNotify();
                    //return result, the key field should be negotiated with JS side.
                    Map< String, Object > zsonResult = new HashMap< String, Object >();
                    zsonResult.put("code", SUCCESS);
                    reply.writeString(ZSONObject.toZSONString(zsonResult));
                    break;
                }
                //取消订阅,置空对端的remoteHandler
                case UNSUBSCRIBE: {
                    remoteObjectHandlers.remove(data.readRemoteObject());
                    //return result, the key field should be negotiated with JS side.
                    Map< String, Object > zsonResult = new HashMap< String, Object >();
                    zsonResult.put("code", SUCCESS);
                    reply.writeString(ZSONObject.toZSONString(zsonResult));
                    break;
                }
                //获取设备
```

```java
                    case GHE_DEVICE:{
                        Map<String,Object> zsonResult = new HashMap<String,Object>();
                        List deviceList = DeviceUtils.getDevice();
                        zsonResult.put("data",deviceList);

                        zsonResult.put("code", SUCCESS);
                        reply.writeString(ZSONObject.toZSONString(zsonResult));
                        System.out.println("++++++ +" + deviceList + "++++++++");
                        break;
                    }

                    case OPEN_DEVICE:{
                        Map<String,Object> zsonResult = new HashMap<String,Object>();
                        String deviceId = DeviceUtils.getDeviceId();
                        Intent intent = DeviceUtils.getAbilityIntent(deviceId," com.cangjie.jsabilitydemo.GameAbility");
                        startAbility(intent);
                        zsonResult.put("code", SUCCESS);
                        zsonResult.put("data",intent);

                        reply.writeString(ZSONObject.toZSONString(zsonResult));
                        break;
                    }

                    default: {
                        reply.writeString("service not defined");
                        return false;
                    }
                }
                return true;
            }

            public void startNotify() {
                new Thread(() -> {
                    while (true) {
                        try {
                            Thread.sleep(5);
                            testReportEvent();
                        } catch (RemoteException | InterruptedException e) {
                            break;
                        }
                    }
                }).start();
            }

            private void testReportEvent() throws RemoteException {
                int num = 0;
                try {
                    num = singleKvStore.getInt("hello");
```

```
            singleKvStore.delete("hello");
        } catch (KvStoreException e) {
            e.printStackTrace();
        }
        MessageParcel data = MessageParcel.obtain();
        MessageParcel reply = MessageParcel.obtain();
        MessageOption option = new MessageOption();
        Map<String, Object> zsonEvent = new HashMap<String, Object>();

        zsonEvent.put("direction", num);
        data.writeString(ZSONObject.toZSONString(zsonEvent));

        for (IRemoteObject item : remoteObjectHandlers) {
            item.sendRequest(100, data, reply, option);
        }

        reply.reclaim();
        data.reclaim();
    }

    @Override
    public IRemoteObject asObject() {
        return this;
    }
}
```

上面的代码在接收到客户端发送过来的字符串后,通过 ZSONObject.stringToClass 把字符串转换成 Java 类对象,这里使用了一个预先定义好的类 ReqParam,如代码示例 5.29 所示。

**代码示例 5.29　用来转换请求的对象**

```
public class ReqParam {
    private int code;
    private int direction;

    public int getCode() {
        return code;
    }

    public void setCode(int code) {
        this.code = code;
    }

    public int getDirection() {
        return direction;
    }
    public void setDirection(int direction) {
        this.direction = direction;
    }
}
```

### 3. 在游戏主界面 FA 中订阅 PA 中的广播方法

在游戏主界面的 onInit() 方法中订阅 Service Ability 推送过来的控制代码,在游戏端调用 move() 方法实现游戏角色控制,如代码示例 5.30 所示。

代码示例 5.30 订阅 PA 中的广播方法

```
//控制手柄控制飞机和子弹运动
move(direction) {
if (direction == 1) {
        this.player.y -= 30
    } else if (direction == 2) {
        this.player.x -= 30
    } else if (direction == 3) {
        this.player.x += 30
    } else {
        this.player.y += 30;
    }
},

subRemoteService(){
    gameAbility.subAbility(1005,(data) =>{
        console.error(JSON.stringify(data));
        //通过订阅返回的 data,判断方向
        let direction = data.data.direction
        if(direction > 0) {
            this.move(direction)
        }
        prompt.showToast({
            message:data.data.direction
        })

    }).then(result =>{
        //订阅状态返回
        if(result.code!= 0)
        return console.warn("订阅失败");
        console.log("订阅成功");
    });
},
```

通过以上步骤就可以完成手机控制 TV 游戏的场景。通过鸿蒙的分布式能力,实现游戏远程控制,相对来讲还是非常简单的。

## 5.7 本章小结

本章为鸿蒙"飞机大战"游戏的开发介绍,读者可以基于本章的代码实现自己的飞机大战游戏。

# 第 6 章 原子化服务和服务卡片开发

在万物互联时代，人均持有设备的数量不断攀升，设备和场景的多样性使应用开发变得更加复杂、应用入口更加丰富。在此背景下，应用提供方和用户迫切需要一种新的服务提供方式，使应用开发更简单、使服务（如听音乐、打车等）的获取和使用更便捷。为此，HarmonyOS除支持传统的需要安装应用的方式外，还支持提供特定功能的免安装的应用（原子化服务）。

## 6.1 什么是原子化服务

原子化服务是 HarmonyOS 提供的一种面向未来的服务提供方式，此服务提供方式有独立入口（用户可通过单击的方式直接触发）、免安装（无须显式安装，由系统程序框架后台安装后即可使用）、可为用户提供一个或多个便捷服务的用户应用程序形态。例如：某传统方式需要安装购物应用 A，在按照原子化服务理念调整设计后，成为由"商品浏览""购物车""支付"等多个便捷服务组成的、可以免安装的购物原子化服务 A ＊，如图 6.1 所示。

图 6.1 传统应用到原子化服务

原子化服务基于 HarmonyOS API 开发，支持运行在 1+8+N 设备上，供用户在合适的场景、合适的设备上便捷使用。原子化服务相对于传统方式的需要安装的应用形态更加轻量，同时提供更丰富的入口、更精准的分发。

### 6.1.1 原子化服务特征

原子化服务特征：随处可及、服务直达、跨设备、一次开发多端部署。

**1. 随处可及**

服务发现：原子化服务可在服务中心发现并使用。

智能推荐：原子化服务可以基于合适场景被主动推荐给用户使用，用户可在服务中心和小艺建议中发现系统推荐的服务。

**2．服务直达**

原子化服务支持免安装使用。

服务卡片：支持用户在不打开原子化服务的情况下获取服务内重要信息的展示和动态变化，如天气、关键事务备忘、热点新闻列表等。

**3．跨设备**

原子化服务支持运行在1+8+N设备上，如手机、平板等设备。

支持跨设备分享：例如接入华为分享后，用户可将原子化服务分享给好友，好友确认后可打开分享的服务。

支持跨端迁移：例如手机上未完成的邮件，迁移到平板上继续编辑。

支持多端协同：例如手机用作文档翻页和批注，配合智慧屏显示完成分布式办公；手机作为手柄与智慧屏配合玩游戏。

**4．一次开发多端部署**

对于开发者而言，原子化服务只需开发一次，便可以部署在各种HarmonyOS终端上，大大降低了开发成本。

对于传统的App软件开发者来讲，一个绕不开的问题就是同一个App需要分别针对不同的设备进行适配。例如程序员在手机上开发了一款应用，针对手表则需要重新适配、发布到手表的应用市场。针对大屏适配后，再发布到大屏的应用市场，严重影响了应用的开发效率和变现能力。

HarmonyOS在架构设计之初，就提出了一次构建支持多端部署的架构设计原则。HarmonyOS通过提供用户程序框架、Ability框架及UI框架，能够保证开发的应用在多终端运行时保证一致性。多终端软件平台API具备一致性，确保用户程序的运行兼容性。如此一来，开发者仅需为不同形态的设备配置不同参数，IDE就能够自动生成支持多设备分发的App包。App包上架应用市场后，应用市场会自动按照设备类型进行HAP包的拆分、组装和分发，进而端到端实现了一次开发，支持多端部署的设计。

## 6.1.2 原子化服务与传统应用的区别

原子化服务由1个或多个HAP包组成，1个HAP包对应1个FA或1个PA。每个FA或PA均可独立运行，完成1个特定功能；1个或多个功能（对应FA或PA）完成1个特定的便捷服务。

原子化服务与传统应用的区别如表6.1所示。

表6.1 原子化服务与传统应用的区别

| 项目 | 原子化服务 | 传统应用 |
| --- | --- | --- |
| 软件包形态 | App Pack(.app) | App Pack(.app) |
| 分发平台 | 由原子化服务平台(Huawei Ability Gallery)管理和分发 | 由应用市场(AppGallery)管理和分发 |

续表

| 项　　目 | 原子化服务 | 传统应用 |
|---|---|---|
| 安装后有无桌面icon | 无桌面icon,但可手动添加到桌面,显示形式为服务卡片 | 有桌面icon |
| HAP包免安装要求 | 所有HAP包(包括Entry HAP和FeatureHAP)均需满足免安装要求 | 所有HAP包(包括Entry HAP和Feature HAP)均为非免安装的 |

### 6.1.3　原子化服务上架流程

原子化服务上架流程,如图6.2所示,具体步骤如下:

(1) 开发阶段,IDE基于包格式编译打包,生成支持多设备的应用包。

(2) 上架到华为应用市场。

(3) 在云侧对App进行拆包,部署到CDN(Content Delivery Network,内容分发网络),包信息同步到服务分发中心。

(4) 在端侧运行过程中,根据自身设备类型获取相应的HAP及整体摘要信息。

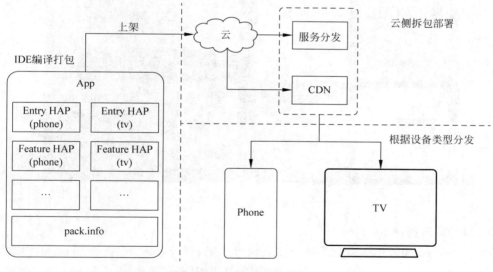

图6.2　一次开发、多端部署

### 6.1.4　原子化服务开发要求

原子化服务相对于传统方式的需要安装的应用更加轻量,同时提供更丰富的入口、更精准的分发,但需要满足一些开发规则要求。

**1. 包大小限制**

原子化服务内所有HAP包(包括Entry HAP和Feature HAP)均需满足免安装要求。(说明:原子化服务由一个或多个HAP包组成,1个HAP包对应1个FA或1个PA。)

免安装的 HAP 包不能超过 10MB,以提供秒开体验。超过此大小的 HAP 包不符合免安装要求,也无法在服务中心展示。

### 2. 在服务中心配置原子化服务

如果需要在服务中心展示已开发的原子化服务程序,如图 6.3 所示,则该服务对应 HAP 包必须包含 FA,并且 FA 中必须指定一个唯一的 mainAbility(定位为用户操作入口),mainAbility 必须为 Page Ability。同时,mainAbility 中至少配置 2×2(小尺寸)规格的默认服务卡片(也可以同时提供其他规格的卡片)及该便捷服务对应的基础信息(包括图标、名称、描述、快照)。

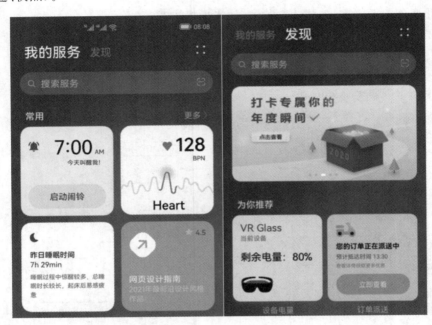

图 6.3　服务和发现中心

### 3. 项目类型选择 Service

通过 DevEco Studio 工程向导创建工程时,Project Type 字段应选择 Service,同时勾选 Show in Service Center,如图 6.4 所示。此时在工程中将自动指定 mainAbility,并添加默认服务卡片信息,开发者根据实际业务设计继续开发即可。

## 6.1.5　原子化服务开发流程

原子化服务项目开发与普通的 App 项目开发除 6.1.4 节提到的差异外,并无其他差异,下面详细介绍原子化服务开发的步骤。

### 1. 创建原子化服务项目

原子化服务项目的工程目录结构与 App 安装项目是完全一样的,区别只是在配置设置方面的差异,如图 6.5 所示。

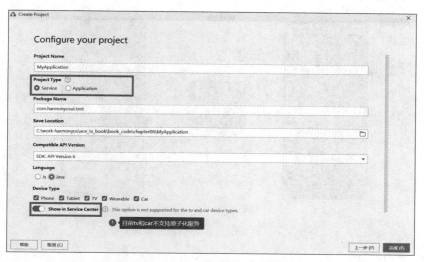

图 6.4 设置为原子化服务项目

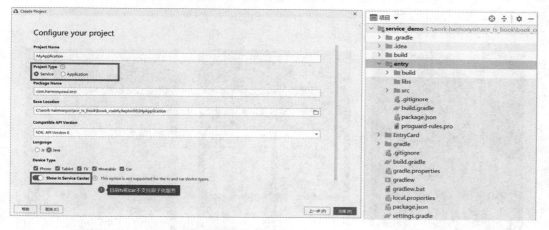

图 6.5 原子化服务项目工程目录

一个原子化服务项目,默认有一个服务卡片项目,服务卡片的作用类似于 Application 项目的桌面图标的作用,可以添加到桌面,如图 6.6 所示,服务卡片功能比普通的桌面图标强大很多。

开发好项目后,单击编译后便可运行到签名好的真机上,如图 6.7 所示,运行成功后,既可以在手机原子化服务中心查看(原子化服务不会在桌面上生成图标),也可以把原子化项目的服务卡片添加到手机桌面上。

**2. 进入手机服务中心**

进入手机服务中心的方式,可以通过快捷方式进入"服务中心",如从屏幕左或右下角向斜上方滑动,如图 6.8 所示。

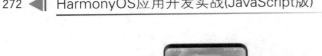

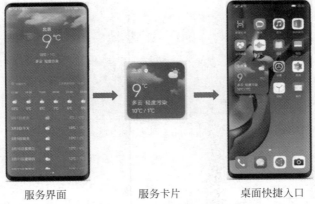

　　服务界面　　　　　服务卡片　　　　桌面快捷入口

图 6.6　服务卡片作为桌面快捷入口

图 6.7　DevEco Studio 编译

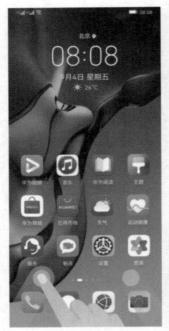
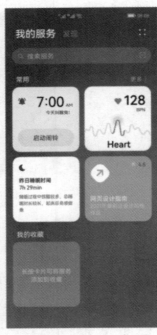

从屏幕左下角或右下角向斜上方　　"我的服务"版块：展示常用服　　"发现"版块：全量的服务供用
滑动，即可进入"服务中心"　　　　务和用户主动收藏的服务　　　　户进行管理和使用

图 6.8　进入手机服务中心

3. 编译运行原子化服务

编译后已安装的原子化服务项目可以在我的服务中心查找,常用栏中的第一栏就是编译好的项目的入口,如图6.9所示。

图6.9 在原子化服务中心查看原子服务项目

## 6.2 什么是服务卡片(Service Widget)

微软的Windows Phone系统是移动设备卡片设计系统的鼻祖,但是由于当年受众面很窄,Windows Phone系统虽然很流畅、用了很多磁吸卡片设计,但是问题就在于应用太少,单纯地把图标切换成卡片,而非做成零层交互逻辑。

零层交互逻辑是让用户仅在桌面就能使用App中的部分功能,这种性质的操作没有"打开开程序→打开下级程序(新页面)→打开下下级程序(新页面)……"一系列复杂路径,这就是零层级交互。

从生活习惯来讲,很多用户有查看天气的习惯,有时候需要了解未来一周的天气,但更多时候只想知道当天的24h的天气情况,看了今天的天气预报来决定出门需要穿什么样的衣服,以及需不需要带雨伞。

如果采用传统方式查看天气预报信息,则需要打开相应的应用或者相应的桌面插件。如果是服务卡片,则只需将"天气"这个原子组件固定在桌面,每次查看时拖动时间轴就可以查看了,不需要二次打开,这也是零层级交互。

目前各种移动手机操作系统都有服务卡片设计,如图6.10所示,但是HarmonyOS除了强调零层操作之外,更把生态互联作为主打方向,实现多终端的应用同步,在鸿蒙的生态下,可以做到更好的卡片呈现效果。

图 6.10　不同系统的 Widget 设计

## 6.2.1　服务卡片定义

HarmonyOS 服务卡片(以下简称"卡片")是 FA 的一种界面展示形式,如图 6.11 所示,将 FA 的重要信息或操作直接放置到卡片中,用户通过直接操作卡片就可以达到应用的使用体验,这样大大减少了应用的使用层级。卡片常常嵌入其他应用中作为其界面的一部分显示(也可以使用原子化服务将应用保存到服务中心中,这种方式不需要安装应用),并支持拉起页面,以及发送消息等基础的交互功能。

卡片的桌面效果如图 6.11 所示,桌面包含普通的图标和卡片,如天气服务卡片展示了多时段天气,运动健康的服务卡片展示了运动步数等重要信息。在应用图标下方有提示条,表示该应用嵌有服务卡片,可以通过手指在该图标上滑出服务卡片,如天气、图库、运动健康、备忘录等均可以上滑图标呼出服务卡片。

## 6.2.2　服务卡片的三大特征

服务卡片的三大特征包括易用可见、智能可选和多端可变。

(1) 易用可见:凸显服务信息的内容外露,减少层级跳转。

(2) 智能可选:全天可变的数据信息,支持自定义类型的服务卡片设计。

(3) 多端可变:适配多端设备的自适应属性。

图 6.11　服务卡片

## 6.2.3　服务卡片的设计规范

**1. 服务卡片的数量要求**

每个应用均能配置多个服务卡片，但是每个 Page Ability 配置的服务卡片的总数不能超过 16 个。以天气应用为例，一个天气应用可以制作当地天气、多时段天气、多天天气、穿衣提醒等不同内容的卡片信息，但如果将这些卡片配置在同一个 Page Ability 上，则总数量不可以超过 16 个。

系统为每个应用提供了 4 种尺寸规格以供选择，小尺寸的卡片尺寸为必选项。

每个服务卡片被用户使用时可以创建多个实例，如用户手动添加了当地天气的服务卡片，可以通过对单个服务卡片的重复添加实现多个实例。

**2. 服务卡片的尺寸选择**

每个应用在选择服务卡片时，应按需选择对应尺寸，以此保证内容展示的最大化。

服务卡片的尺寸分为微（1×2）、小（2×2）、中（2×4）、大（4×4）共 4 种尺寸，其中小尺寸为必选尺寸。

同一种尺寸还支持多个不同内容布局的卡片，可以通过服务卡片管理界面进行选择，应用方可以指定某一个服务卡片作为默认卡片展示。

**1）微尺寸**

微尺寸卡片能提供简单的数据信息、快捷指令和快捷入口。例如运动数据卡片仅展示步数数据，行程信息卡片仅提供航班号和登机时间等，如图 6.12(a) 所示。

图 6.12 小尺寸服务卡片

2) 小尺寸

小尺寸卡片能提供精简的服务信息内容、丰富的展示特性信息、即时信息或操作提示等。例如行程信息卡片除了提供航班信息,还提供了查看详情的操作,如图 6.12(b)所示。

3) 中尺寸和大尺寸

中尺寸和大尺寸卡片能提供两种以上维度的服务信息展示和多个可交互的热区展示,展示更加沉浸丰富的视觉图片及采用多个宫格和列表内容展示。例如运动数据采用中尺寸卡片展示了步数和运动强度两个维度的信息。玩机技巧卡片采用宫格方式推荐了多个应用入口,如图 6.13 所示。

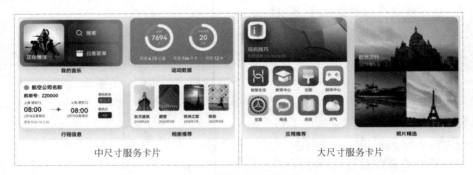

图 6.13 中、大尺寸服务卡片

**3. 服务卡片的内容构成**

服务卡片由多种设计元素组合而成,以下 7 种常见信息元素可以作为内容构成服务卡片:图标、数据、文本、按钮、图片、宫格、列表。

在组织服务卡片内容的时候可以按照尺寸的大小来判断应当选择哪种元素,包括元素组合的数量,如图 6.14 所示。

对于微尺寸卡片,建议卡片上的内容元素不超过两种,可以在"图标、数据、文本、图片"中最多选择两种元素进行组合,以达到内容收益尽可能多且不显得繁杂的效果。对于小尺寸卡片,建议卡片上的内容不超过 3 种,可以在"图标、数据、文本、按钮、图片"中最多选择 3 种元素进行组合。中/大尺寸卡片可根据业务需要自由选择。

图 6.14　卡片尺寸和元素对应关系图

## 6.2.4　服务卡片的整体架构

服务卡片的整体框架主要包含三部分：卡片使用方、卡片管理服务和卡片提供方。

**1. 卡片使用方**

显示卡片内容的宿主应用，控制卡片在宿主中展示的位置，如桌面、服务中心、搜索等。

**2. 卡片管理服务**

用于管理系统中所添加卡片的常驻代理服务，包括卡片对象的管理与使用，以及卡片周期性刷新等。

**3. 卡片提供方**

提供卡片显示内容的 HarmonyOS 应用或原子化服务，控制卡片的显示内容、控件布局及控件单击事件。

服务卡片的整体架构如图 6.15 所示。

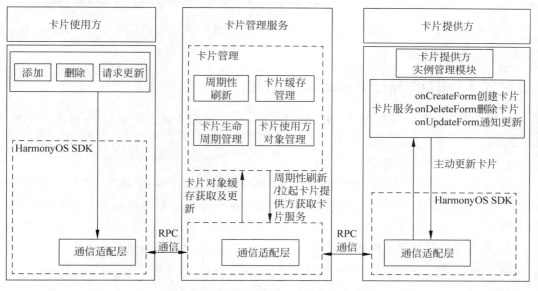

图 6.15　服务卡片的整体架构

## 6.3 服务卡片开发详细讲解

目前开发服务卡片有两种语言可以选择：Java 和 JavaScript。Java 卡片与 JS 卡片存在一些功能差异，如表 6.2 所示。

表 6.2 Java 卡片与 JS 卡片的功能差异

| 场景 | Java 卡片 | JS 卡片 | 支持的版本 |
| --- | --- | --- | --- |
| 实时刷新(类似时钟) | Java 使用 ComponentProvider 做实时刷新代价比较大 | JS 可以做到端侧刷新，但是需要定制化组件 | HarmonyOS 2.0 及以上 |
| 开发方式 | Java UI 在卡片提供方需要同时对数据和组件进行处理。生成 ComponentProvider 远端渲染 | JS 卡片在使用方加载渲染，提供方只要处理数据、组件和逻辑分离 | HarmonyOS 2.0 及以上 |
| 组件支持 | Text、Image.DirectionalLayout.PositionLayout、DependentLayout | div、list.list-item.swiper.stack、image.text.span.progress.button（定制：chart、clock、calendar） | HarmonyOS 2.0 及以上 |
| 卡片内动效 | 不支持 | 暂不开放 | HarmonyOS 2.0 及以上 |
| 阴影模糊 | 不支持 | 支持 | HarmonyOS 2.0 及以上 |
| 动态适应布局 | 不支持 | 支持 | HarmonyOS 2.0 及以上 |
| 自定义卡片跳转页面 | 不支持 | 支持 | HarmonyOS 2.0 及以上 |

目前官方推荐使用 JavaScript 开发服务卡片，JS 卡片使用 HML＋CSS＋JSON 开发，相对来讲更加便捷和简单。

### 6.3.1 创建 JavaScript 服务卡片

应用(Application)项目和原子化服务(Service)项目都可创建服务卡片，下面介绍创建 JavaScript 服务卡片的具体步骤。

1. 根据模板创建服务卡片

DevEco Studio 为开发者提供了 15 个服务卡片模板，如图 6.16 所示，开发者可以选择合适的模板来创建服务卡片，也可以在这些卡片模板的基础上根据自己的设计要求进行二次开发。

2. 服务卡片配置文件详细讲解

创建成功后，在 config.json 文件的 module 中会生成 js 模块，用于对应卡片的 js 相关资源，配置如代码示例 6.1 所示。

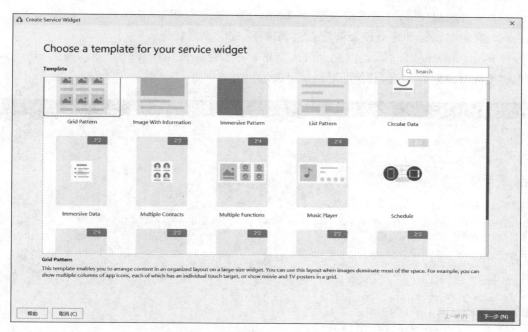

图 6.16　服务卡片模板

代码示例 6.1

```
"js": [
  {
"name": "card",
"pages": [
"pages/index/index"
    ],
"window": {
"designWidth": 720,
"autoDesignWidth": true
    },
"type": "form"
  }
]
```

config.json 文件中 abilities 配置 forms 模块的细节如代码示例 6.2 所示,各属性的详情可见表 6.3。

配置文件中,应注意如下配置:

(1) js 模块中的 name 字段要与 forms 模块中的 js ComponentName 字段的值一致,为 js 资源的实例名。

(2) forms 模块中的 name 为卡片名,即在 onCreateForm 中根据 AbilitySlice.PARAM _FORM_NAME_KEY 可取到的值。

(3) 卡片的 Ability 中还需要配置 visible：true 和 formsEnabled：true。

(4) 在定时刷新和定点刷新都配置的情况下，定时刷新优先。

(5) defaultDimension 需要设置为默认规格，2×2 为默认，必须设置。supportDimensions 中也必须同时设置为 2×2，其他的规格是可选的。

代码示例 6.2

```
"forms": [
    {
"name": "Form_Js",
"description": "form_description",
"type": "JS",
"jsComponentName": "card",
"formConfigAbility": "ability://com.huawei.demo.SecondFormAbility",
"colorMode": "auto",
"isDefault": true,
"updateEnabled": true,
"scheduledUpdateTime": "10:30",
"updateDuration": 1,
"defaultDimension": "2 * 2",
"supportDimensions": [
"2 * 2",
"2 * 4",
"4 * 4"
    ],
"metaData": {
"customizeData": [
        {
"name": "originWidgetName",
"value": "com.huawei.weather.testWidget"
        }
      ]
    }
  }
]
```

表 6.3 服务卡片配置属性

| 属性名称 | 子属性名称 | 含义 | 数据类型 | 是否可缺省 |
| --- | --- | --- | --- | --- |
| name | - | 表示卡片的类名。字符串的最大长度为 127 字节 | 字符串 | 否 |
| description | - | 表示卡片的描述。取值可以是描述性内容，也可以是对描述性内容的资源索引，以支持多语言。字符串的最大长度为 255 字节 | 字符串 | 可缺省，缺省为空 |

续表

| 属性名称 | 子属性名称 | 含义 | 数据类型 | 是否可缺省 |
| --- | --- | --- | --- | --- |
| isDefault | - | 表示该卡片是否为默认卡片，每个 Ability 有且只有一个默认卡片<br>true：默认卡片<br>false：非默认卡片 | 布尔值 | 否 |
| type | - | 表示卡片的类型，取值范围如下。<br>Java：Java 卡片。<br>JS：JS 卡片 | 字符串 | 否 |
| colorMode | - | 表示卡片的主题样式，取值范围如下。<br>auto：自适应。<br>dark：深色主题。<br>light：浅色主题 | 字符串 | 可缺省，缺省值为 auto |
| supportDimensions | - | 表示卡片支持的外观规格，取值范围如下。<br>1×2：表示 1 行 2 列的二宫格。<br>2×2：表示 2 行 2 列的四宫格。<br>2×4：表示 2 行 4 列的八宫格。<br>4×4：表示 4 行 4 列的十六宫格 | 字符串数组 | 否 |
| defaultDimension | - | 表示卡片的默认外观规格，取值必须在该卡片 supportDimensions 配置的列表中 | 字符串 | 否 |
| landscapeLayouts | - | 表示卡片外观规格对应的横向布局文件，与 supportDimensions 中的规格一一对应。<br>仅当卡片类型为 Java 卡片时，需要配置该标签。<br>字符串数组 | 字符串数组 | 否 |
| portraitLayouts | - | 表示卡片外观规格对应的竖向布局文件，与 supportDimensions 中的规格一一对应。<br>仅当卡片类型为 Java 卡片时，需要配置该标签 | 字符串数组 | 否 |
| updateEnabled | - | 表示卡片是否支持周期性刷新，取值范围：<br>true：表示支持周期性刷新，可以在定时刷新（updateDuration）和定点刷新（scheduledUpdateTime）两种方式中任选其一，优先选定定时刷新。<br>false：表示不支持周期性刷新 | 布尔类型 | 否 |

续表

| 属性名称 | 子属性名称 | 含义 | 数据类型 | 是否可缺省 |
|---|---|---|---|---|
| scheduledUpdateTime | - | 表示卡片的定点刷新的时刻,采用24h制,精确到分钟 | 字符串 | 可缺省,缺省值为0:0 |
| updateDuration | - | 表示卡片定时刷新的更新周期,单位为30min,取值为自然数。当取值为0时,表示该参数不生效。当取值为正整数N时,表示刷新周期为30×N分钟 | 数值 | 可缺省,缺省值为0 |
| formConfigAbility | - | 表示卡片的配置跳转链接,采用URI格式 | 字符串 | 可缺省,缺省值为空 |
| jsComponentName | - | 表示JS卡片的Component名称。字符串的最大长度为127字节。仅当卡片类型为JS卡片时,需要配置该标签 | 字符串 | 否 |
| metaData | - | 表示卡片的自定义信息,包含customizeData数组标签 | 对象 | 可缺省,缺省值为空 |
| customizeData |  | 表示自定义的卡片信息 | 对象数组 | 可缺省,缺省值为空 |
| | name | 表示数据项的键名称。字符串的最大长度为255字节 | 字符串 | 可缺省,缺省值为空 |
| | value | 表示数据项的值。字符串的最大长度为255字节 | 字符串 | 可缺省,缺省值为空 |

### 3. 服务卡片模板详细讲解

DevEco Studio 提供了15个不同场景下的卡片模板,下面详细讲解几个典型场景下的模板的用法。

图表数据卡片模板(Circular Data)是一个小尺寸(2×2)数据卡片,如图 6.17 所示,该模板用于以环形图和文本格式显示用户定义的内容,突出关键数据的比例。

创建好 Circular Data 卡片后,卡片的默认名为 widget,这里使用默认配置,默认的卡片配置如代码示例6.3所示。

代码示例 6.3

```
"forms": [
  {
    "jsComponentName": "widget",
    "isDefault": true,
    "scheduledUpdateTime": "10:30",
    "defaultDimension": "2 * 2",
    "name": "widget",
    "description": "This is a service widget",
```

```
"colorMode": "auto",
"type": "JS",
"supportDimensions": [
"2 * 2"
    ],
"updateEnabled": true,
"updateDuration": 1
    }
]
```

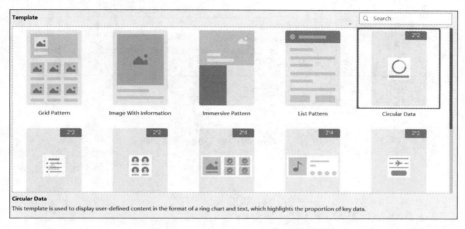

图 6.17　Circular Data 卡片模板

创建的工程目录结构如图 6.18 所示,添加卡片后,IDE 会修改 MainAbility.java 的代码,增加与卡片相关的回调方法和添加一个 widget 的目录,在该目录中增加卡片控制器和与卡片绑定相关的代码类。

在 js 目录中 ide 为卡片添加了一个 js component (widget),这个 widget 和默认的 default 目录在同一个目录下,这两个 js component 的区别是,default 的类型是 page(页面),widget 的类型是 form(卡片)。

打开卡片的 index.html 文件,如图 6.19 所示,这里使用了 chart 组件,chart 组件可以动态地显示 MainAbility 绑定的数据,当单击卡片中间的 chart 图时,可以重新绑定新计算的比例。

卡片页面的效果如图 6.20 所示,100%(数字)显示在圆形进度条的中间,这里结合 stack 组件和 chart 组件,采用堆叠的方式,如代码示例 6.4 所示。

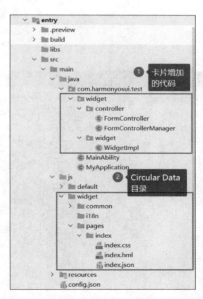

图 6.18　Circular Data 卡片模板代码目录

```
  2
  3  <div class="list_pattern_layout">
  4      <stack class="stack_container">
  5          <div class="title_container">
  6              <text class="title_text">{{ titleContent }}</text>
  7          </div>
  8          <div class="div_chart_contain">
  9              <chart class="chart_rainbow"
 10                  type="rainbow"
 11                  segments="{{ segments }}"
 12                  effects="true"
 13                  onclick="freeUpEvent"/>
 14          </div>
 15      </stack>
 16      <text class="text_title">{{ textContent }}</text>
 17  </div>
```

图 6.19 卡片页面

图 6.20 卡片预览效果图

**代码示例 6.4**

```
< div class = "list_pattern_layout">
< stack class = "stack_container">
< div class = "title_container">
< text class = "title_text">{{ titleContent }}</text>
</div>
< div class = "div_chart_contain">
< chart class = "chart_rainbow"
            type = "rainbow"
            segments = "{{ segments }}"
            effects = "true"
            onclick = "freeUpEvent"/>
</div>
</stack>
< text class = "text_title">{{ textContent }}</text>
</div>
```

预览效果只能静态展示，动态地绑定数据需要通过模拟器或者通过真机测试查看。
index.hml 文件中绑定的数据通过 index.json 文件定义的数据结构绑定，如图 6.21 所示。

```html
<div class="list_pattern_layout">
    <stack class="stack_container">
        <div class="title_container">
            <text class="title_text">{{ titleContent }}</text>
        </div>
        <div class="div_chart_contain">
            <chart class="chart_rainbow"
                type="rainbow"
                segments="{{ segments }}"
                effects="true"
                onclick="freeUpEvent"/>
        </div>
    </stack>
    <text class="text_title">{{ textContent }}</text>
</div>
```
index.hml

```json
{
  "data": {
    "titleContent": "100%",
    "textContent": "",
    "segments": []
  },
  "actions": {
    "freeUpEvent": {
      "action": "message",
      "params": {}
    }
  }
}
```
index.json

图 6.21　卡片数据的绑定关系

为了提升卡片的性能，卡片中并没有使用 JS 框架，所以并没有 JS 文件，只有 JSON 文件，JSON 文件的作用只用于定义 HML 的数据结构，卡片在创建的时候，MainAbility.java 可以把数据动态地绑定和解析到 HML 界面，如代码示例 6.5 所示。

代码示例 6.5　MainAbility.java

```java
@Override
protected ProviderFormInfo onCreateForm(Intent intent) {
    HiLog.info(TAG, "onCreateForm");
    long formId = intent.getLongParam(AbilitySlice.PARAM_FORM_IDENTITY_KEY, INVALID_FORM_ID);
    String formName = intent.getStringParam(AbilitySlice.PARAM_FORM_NAME_KEY);
    int dimension = intent.getIntParam(AbilitySlice.PARAM_FORM_DIMENSION_KEY, DEFAULT_DIMENSION_2X2);
    HiLog.info(TAG, "onCreateForm: formId = " + formId + ",formName = " + formName);
    FormControllerManager formControllerManager = FormControllerManager.getInstance(this);
    FormController formController = formControllerManager.getController(formId);
    formController = (formController == null) ? formControllerManager.createFormController(formId,
            formName, dimension) : formController;
    if (formController == null) {
        HiLog.error(TAG, "Get null controller. formId: " + formId + ", formName: " + formName);
        return null;
    }
    return formController.bindFormData(formId);
}
```

在 MainAbility 中通过 onCreateForm( )方法调用 WidgetImpl.java 文件中的 bindFormData( )方法,以此进行数据绑定,如代码示例6.6所示。

**代码示例6.6　WidgetImpl.java**

```java
@Override
public ProviderFormInfo bindFormData(long formId) {
    HiLog.info(TAG, "bind form data for a new service widget.");
    ProviderFormInfo providerFormInfo = new ProviderFormInfo();
    if (dimension == DEFAULT_DIMENSION_2X2) {
        ZSONObject formData = calculateCurrentUsedStorage(storageUsedPercent);
        providerFormInfo.setJsBindingData(new FormBindingData(formData));
    }
    return providerFormInfo;
}
```

calculateCurrentUsedStorage( )方法每次调用时都会动态计算并返回一个 ZSONObject 对象,如代码示例6.7所示。

**代码示例6.7　WidgetImpl.java**

```java
private ZSONObject calculateCurrentUsedStorage(int usedPercent) {
    int currentUsedStorage = TOTAL_STORAGE_CAPACITY * usedPercent / HUNDRED_PERCENT;
    ZSONArray zsonArray = new ZSONArray();
    ZSONObject zsonObject = new ZSONObject();
    zsonObject.put("name", "segment");
    zsonObject.put("value", usedPercent);
    zsonArray.add(zsonObject);

    ZSONObject newFormData = new ZSONObject();
    newFormData.put(TITLE_CONTENT, usedPercent + "%");
    newFormData.put(TEXT_CONTENT, currentUsedStorage + "GB" + "/" + TOTAL_STORAGE_CAPACITY + "GB");
    newFormData.put(SEGMENTS, zsonArray);

    return newFormData;
}
```

每单击一次卡片都会触发 MainAbility.java 文件中的 onTriggerFormEvent( )方法,在界面发送的动作类型为 message 时触发,如代码示例6.8所示。

**代码示例6.8　WidgetImpl.java**

```java
@Override
public void onTriggerFormEvent(long formId, String message) {
    HiLog.info(TAG, "handle trigger form event.");
    if (!(context instanceof Ability)) {
        return;
    }
```

```
    Ability ability = (Ability) context;
    HiLog.info(TAG, "onTriggerFormEvent() ability:" + ability + ", formId: " + formId + ", 
message: " + message);
    //Do something specific to this kind of form.
    storageUsedPercent = (--storageUsedPercent) < 0 ? HUNDRED_PERCENT : storageUsedPercent;
    ZSONObject formData = calculateCurrentUsedStorage(storageUsedPercent);
    try {
        ability.updateForm(formId, new FormBindingData(formData));
    } catch (FormException e) {
        HiLog.error(TAG, e.getMessage());
    }
}
```

### 6.3.2 服务卡片界面实现

JavaScript 卡片的界面开发提供了一些基础组件，但目前不支持 canvas 组件，这里详细介绍部分常用的卡片组件的使用。

**1. 基础组件（div）**

div 是开发服务卡片的基础容器，用作页面结构的根节点或将内容进行分组，如图 6.22 所示。

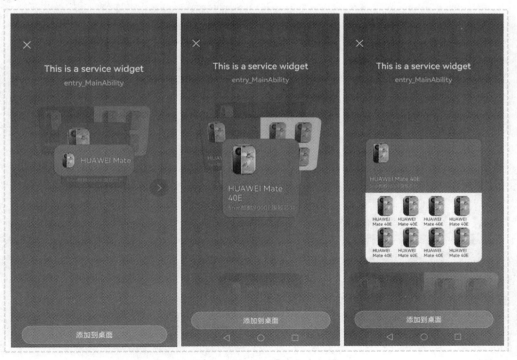

图 6.22　div 组件用来划分区域

使用div组件布局，如代码示例6.9所示，代码中通过变量对不同尺寸的卡片进行显示或者隐藏控制，如{{ mini }}表示在微尺寸上渲染，所以会显示，{{ dim2X4 }}表示在2×4尺寸卡片上渲染，所以会显示，mini和dim2X4变量的值是在卡片创建的时候通过MainAbility.java文件绑定的。

代码示例6.9　chapter06\jscard\entry\src\main\js\widget\pages\index.hml

```html
<div class = "grid_pattern_layout">
    <div if = "{{ mini }}" class = "mini_container">
        <image src = "{{img}}" class = "mini_image"></image>
        <text class = "mini_text">{{ miniTitle }}</text>
    </div>
    <div class = "normal_container">
        <div class = "title_container">
            <div class = "pic_title_container">
                <image src = "{{img}}" class = "title_img"></image>
                <div style = "flex-direction : column;">
                    <text class = "title">{{ title }}</text>
                    <text class = "content">{{ content }}</text>
                </div>
            </div>
        </div>
        <div if = "{{ dim2X4 }}" class = "preview_container">
            <div class = "preview_sub_container" style = "align-items : flex-start;">
                <image src = "{{img}}" class = "preview_img"></image>
                <image src = "{{img}}" class = "preview_img"></image>
            </div>
            <div class = "preview_sub_container" style = "align-items : flex-end;">
                <image src = "{{img}}" class = "preview_img"></image>
                <image src = "{{img}}" class = "preview_img"></image>
            </div>
        </div>
    </div>
    <div class = "detail_container">
        <div class = "sub_detail_container">
            <div class = "detail_unit">
                <image src = "{{img}}" class = "detail_image"></image>
                <text class = "detail_title">{{ detailTitle }}</text>
            </div>
            <div class = "detail_unit">
                <image src = "{{img}}" class = "detail_image"></image>
                <text class = "detail_title">{{ detailTitle }}</text>
            </div>
            <div class = "detail_unit">
                <image src = "{{img}}" class = "detail_image"></image>
                <text class = "detail_title">{{ detailTitle }}</text>
            </div>
            <div class = "detail_unit">
                <image src = "{{img}}" class = "detail_image"></image>
                <text class = "detail_title">{{ detailTitle }}</text>
```

```html
    </div>
  </div>
  <div class = "sub_detail_container">
    <div class = "detail_unit">
      <image src = "{{img}}" class = "detail_image"></image>
      <text class = "detail_title">{{ detailTitle }}</text>
    </div>
    <div class = "detail_unit">
      <image src = "{{img}}" class = "detail_image"></image>
      <text class = "detail_title">{{ detailTitle }}</text>
    </div>
    <div class = "detail_unit">
      <image src = "{{img}}" class = "detail_image"></image>
      <text class = "detail_title">{{ detailTitle }}</text>
    </div>
    <div class = "detail_unit">
      <image src = "{{img}}" class = "detail_image"></image>
      <text class = "detail_title">{{ detailTitle }}</text>
    </div>
  </div>
</div>
</div>
```

样式如代码示例 6.10 所示。

**代码示例 6.10** chapter06\jscard\entry\src\main\js\widget\pages\index.css

```css
.grid_pattern_layout {
    flex-direction: column;
    align-items: flex-start;
    width: 100%;
    height: 100%;
}

.mini_container {/* show in 1X2 */
    display-index: 2;
    align-items: center;
    width: 100%;
    height: 100%;
    background-color: #f00;
}

.mini_image {
    width: 30px;
    height: 30px;
    border-radius: 10px;
    margin-start: 12px;
    margin-end: 8px;
```

```css
    object-fit: contain;
}

.mini_text {
    font-size: 14px;
    color: #e5ffffff;
}

.normal_container {
    display-index: 1;
    flex-direction: column;
}

.title_container {/* show in 2X2,2X4,4X4 */
    display-index: 2;
    align-items: center;
    flex-weight: 150;
    min-height: 120px;
}

.pic_title_container {
    background-color: #f00;
    flex-direction: column;
    height: 100%;
    padding-top: 12px;
    padding-start: 12px;
    padding-bottom: 12px;
    justify-content: space-between;
}

.title_img {
    height: 45%;
    aspect-ratio: 1;
    border-radius: 14px;
    object-fit: contain;
}

.title {
    font-size: 16px;
    color: #e5ffffff;
    margin-bottom: 2px;
}

.content {
    font-size: 12px;
    color: #99ffffff;
    text-overflow: ellipsis;
    max-lines: 1;
}
```

```css
.preview_container {
    position: relative;
    width: 100%;
    height: 100%;
    background-color: #fdfefe;
    padding: 8% 10%;
    justify-content: space-between;
}

.preview_sub_container {
    flex-direction: column;
    justify-content: space-between;
}

.preview_img {
    height: 45%;
    aspect-ratio: 1;
    border-radius: 14px;
    object-fit: contain;
}

.detail_container {/* show in 4X4 */
    display-index: 1;
    flex-direction: column;
    padding: 4% 6%;
    flex-weight: 194;
    min-height: 155.2px;
}

.sub_detail_container {
    height: 49%;
    justify-content: space-between;
    align-items: center;
}

.detail_unit {
    flex-direction: column;
    align-items: center;
    justify-content: center;
}

.detail_image {
    height: 64%;
    aspect-ratio: 1;
    border-radius: 15px;
    object-fit: contain;
}

.detail_title {
```

```css
    margin-top: 2px;
    font-size: 12px;
    color: #e5000000;
}
```

数据绑定如代码示例6.11所示。

**代码示例6.11    chapter06\jscard\entry\src\main\js\widget\pages\index.json**

```json
{
"data": {
"mini": false,
"dim2X4": false,
"miniTitle": "HUAWEI Mate",
"title": "HUAWEI Mate 40E",
"content": "5nm 麒麟 9000E 旗舰芯片",
"detailTitle": "HUAWEI Mate 40E",
"img": "/common/images/img1.png"
    }
}
```

通过MainAbility.java文件中的bindFormData()方法绑定数据,如代码示例6.12所示。

providerFormInfo.setJsBindingData(new FormBindingData(zsonObject))方法绑定JSON(zsonObject)格式化数据给卡片模板的数据结构。

**代码示例6.12    chapter06\jscard\entry\src\main\java\com\harmonyosui\test\widget\widget**

```java
@Override
public ProviderFormInfo bindFormData(long formId) {
    HiLog.info(TAG, "bind form data");
    ZSONObject zsonObject = new ZSONObject();
    ProviderFormInfo providerFormInfo = new ProviderFormInfo();
    if (dimension == DIMENSION_1X2) {
        zsonObject.put("mini", true);
    }
    if (dimension == DIMENSION_2X4) {
        zsonObject.put("dim2X4", true);
    }
    providerFormInfo.setJsBindingData(new FormBindingData(zsonObject));
    return providerFormInfo;
}
```

### 2. 列表组件(list)

对于连续相同的部分可以使用list组件,list组件不但可以显示更多的内容,而且代码更少,如图6.23所示。

图6.23的代码实现如代码示例6.13所示。

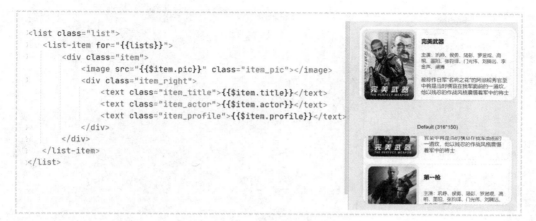

图 6.23 list 组件

代码示例 6.13　chapter06\jscard\entry\src\main\js\listcard\pages\index.hml

```html
<list class="list">
<list-item for="{{lists}}">
<div class="item">
<image src="{{$item.pic}}" class="item_pic"></image>
<div class="item_right">
<text class="item_title">{{$item.title}}</text>
<text class="item_actor">{{$item.actor}}</text>
<text class="item_profile">{{$item.profile}}</text>
</div>
</div>
</list-item>
</list>
```

样式如代码示例 6.14 所示。

代码示例 6.14　chapter06\jscard\entry\src\main\js\listcard\pages\index.css

```css
.item {
    flex-direction: row;
    margin: 5px;
    border: 1px solid #f1f3f5;
    border-radius: 10px;
}

.item_pic {
    width: 180px;
    height: 100%;
    padding: 5px;
    border-radius: 10px;
    background-color: #f1f3f5;
}
```

```css
.item_right {
    width: 100%;
    height: 100%;
    padding: 10px;
    flex-direction: column;
    justify-content: space-around;
}
.item_title {
    font-size: 12px;
    font-weight: 600;
}
.item_actor {
    font-size: 10px;
    font-weight: 300;
}
.item_profile {
    font-size: 11px;
    font-weight: 100;
}
```

**3. 堆叠组件(stack)**

stack 堆叠容器中的子组件按照顺序依次入栈,后一个子组件覆盖前一个子组件,如图 6.24 所示。

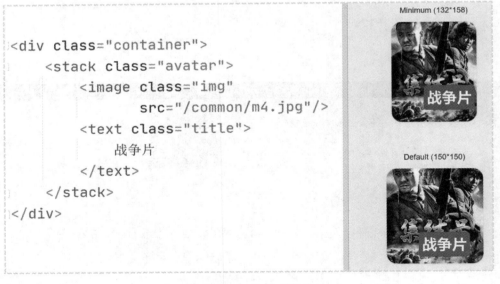

图 6.24　stack 滑动组件

stack 堆叠效果如代码示例 6.15 所示。

代码示例 6.15　chapter06\jscard\entry\src\main\js\stackdemo\pages\index.hml

```
<div class = "container">
<stack class = "avatar">
<image class = "img"
            src = "/common/m4.jpg"/>
<text class = "title">
        战争片
</text>
</stack>
</div>
```

**4. 滑动组件（swiper）**

swiper 组件是放置在桌面上的服务卡片，在左右滑动操作的时候，会使系统分不清楚用户是要左右滑动屏幕，还是要左右滑动卡片，所以目前服务卡片的 swiper 容器不支持手势滑动切换子组件。只有通过单击图片侧面的控制条才可以实现上下滑动。

swiper 组件卡片如图 6.25 所示。

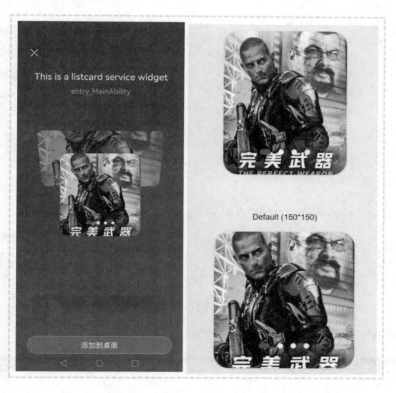

图 6.25　swiper 滑动组件

swiper 滑动卡片的代码实现如图 6.26 所示。

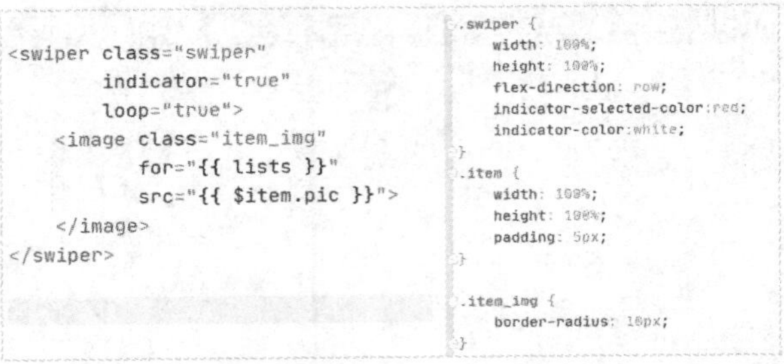

```
<swiper class="swiper"
        indicator="true"
        loop="true">
    <image class="item_img"
           for="{{ lists }}"
           src="{{ $item.pic }}">
    </image>
</swiper>
```

```
.swiper {
    width: 100%;
    height: 100%;
    flex-direction: row;
    indicator-selected-color:red;
    indicator-color:white;
}
.item {
    width: 100%;
    height: 100%;
    padding: 5px;
}
.item_img {
    border-radius: 18px;
}
```

图 6.26　swiper 卡片代码实现

### 5. 图表组件（chart）

chart 图表组件用于在卡片中呈现线图、柱状图、量规图界面，如图 6.27 所示。可以通过 type 属性设置图表类型（不支持动态修改），可选项有 bar（柱状图）、line（线图）、gauge（量规图）、progress（进度类圆形图表）、loading（加载类圆形图表）、rainbow（占比类圆形图表）。

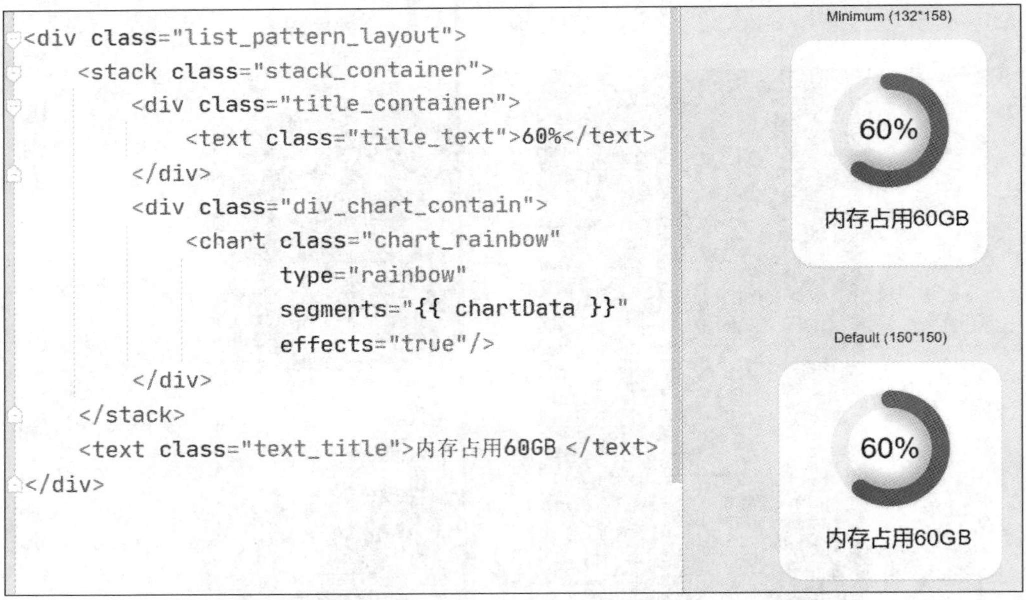

```
<div class="list_pattern_layout">
    <stack class="stack_container">
        <div class="title_container">
            <text class="title_text">60%</text>
        </div>
        <div class="div_chart_contain">
            <chart class="chart_rainbow"
                   type="rainbow"
                   segments="{{ chartData }}"
                   effects="true"/>
        </div>
    </stack>
    <text class="text_title">内存占用60GB </text>
</div>
```

图 6.27　图表组件卡片中用法

图 6.27 的代码实现如代码示例 6.16 所示。

**代码示例 6.16**　chapter06\jscard\entry\src\main\js\chart\pages\index.hml

```
< div class = "list_pattern_layout">
< stack class = "stack_container">
< div class = "title_container">
```

```
<text class = "title_text">60％</text>
</div>
<div class = "div_chart_contain">
<chart class = "chart_rainbow"
                type = "rainbow"
                segments = "{{ chartData }}"
                effects = "true"/>
</div>
</stack>
<text class = "text_title">内存占用 60GB</text>
</div>
```

chartData 数据如代码示例 6.17 所示。

**代码示例 6.17　chapter06\jscard\entry\src\main\js\chart\pages\index.json**

```
{
"data": {
"chartData": [
    {"name": "segment"},
    {"value": 60}
  ]
 }
}
```

### 6.3.3　服务卡片数据绑定

卡片的数据需要通过 Page Ability 提供,下面通过开发者精选卡片案例来详细介绍如何通过网络读取数据,并绑定到卡片上,如图 6.28 所示。

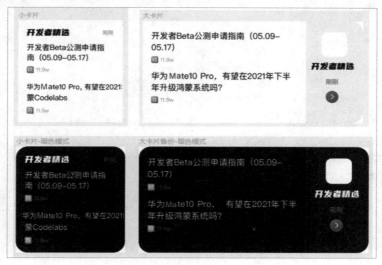

图 6.28　开发者精选卡片

### 1. 添加网络权限

要在 config.json 配置文件的 module 中添加 reqPermissions: [{name:ohos.permission.INTERNET}],如代码示例 6.18 所示。

代码示例 6.18

```json
{
  ......
  "module": {
    ......
    "reqPermissions": [{"name":"ohos.permission.INTERNET"}]
  }
}
```

### 2. 通过网络请求数据

使用 NetManager 封装一个网络请求方法,如代码示例 6.19 所示。

代码示例 6.19 chapter06\jscard\entry\src\main\java\com\harmonyosui\test\http\netTools.java

```java
public class NetTools {
    public interface Callback {
        void getData(String json);
    }
    public static void getApiData(String urlStr,Callback callback) {
        //实例化 NetManager 的对象
        NetManager netManager = NetManager.getInstance(null);
        NetHandle defaultNet = netManager.getDefaultNet();
        HttpURLConnection connection = null;
        InputStream inputStream = null;
        BufferedReader reader = null;
        try {
            URLConnection urlConnection = defaultNet.openConnection(new URL(urlStr), Proxy.NO_PROXY);
            if (urlConnection instanceof HttpURLConnection) {
                connection = (HttpURLConnection) urlConnection;
            }
            connection.setRequestMethod("GET");
            connection.connect();
            //连接的结果
            if (connection.getResponseCode() == 200) {

                //STEP1:获取 URL stream
                inputStream = connection.getInputStream();
                reader = new BufferedReader(new InputStreamReader(inputStream));
                String line;
                String jsonStr = "";
                while ((line = reader.readLine()) != null) {
                    jsonStr = jsonStr + line;
                }
```

```
                System.out.println(jsonStr);
                callback.getData(jsonStr);

            }
        } catch (IOException e) {
            e.printStackTrace();
        } finally {
            if (connection != null) {
                connection.disconnect();
            }
            if (reader != null) {
                try {
                    reader.close();
                } catch (IOException e) {
                    e.printStackTrace();
                }
            }
        }
    }
}
```

### 3. 解析 JSON 数据

获取网络接口数据后,需要把字符串解析成 ZSONObject 对象,如代码示例 6.20 所示。

**代码示例 6.20**

```
private void initData() {
    EventHandler eventHandler = new EventHandler(EventRunner.create(true));
    eventHandler.postTask(new Runnable() {
        @Override
        public void run() {
            NetTools.getApiData("http://apiurl:3000/", json -> {
                //把 jsonstr 转换成 object
                ZSONObject zsonObject = ZSONObject.stringToZSON(json);
                ZSONArray products = zsonObject.getZSONArray("products");
                ...
            });
        }
    });
}
```

## 6.3.4 服务卡片数据更新

通过 IDE 创建服务卡片后,IDE 会自动在 MainAbility 中添加服务卡片对应的回调代码,以及卡片绑定相关类的实现,如代码示例 6.21 所示。

**代码示例 6.21**

```java
public class MainAbility extends Ability {
    //在服务卡片上右击(或上滑)时,通知接口
    protected ProviderFormInfo onCreateForm(Intent intent) {...}
    //在服务卡片请求更新或定时更新时,通知接口
    protected void onUpdateForm(long formId) {...}
    //在服务卡片被删除时,通知接口
    protected void onDeleteForm(long formId) {..}
    //JS 服务卡片单击时,通知接口
    protected void onTriggerFormEvent(long formId, String message) {...}
}
```

**1. 通过配置定时更新卡片**

定时更新只需要在 config.json 文件中开启服务卡片的周期性更新,在 onUpdateForm(long formId)方法下执行数据获取更新。

config.json 文件中 abilities 的 forms 模块用于配置细节,如代码示例 6.22 所示。

**代码示例 6.22**

```
"forms": [
  {
"jsComponentName": "widget2",
"isDefault": true,
//定点刷新的时刻,采用 24h 制,精确到分钟.当满足"updateDuration": 0 条件时,才会生效
"scheduledUpdateTime": "10:30",
"defaultDimension": "1 * 2",
"name": "widget2",
"description": "This is a service widget",
"colorMode": "auto",
"type": "JS",
"supportDimensions": [
"1 * 2"
    ],
"updateEnabled": true,     //表示卡片是否支持周期性刷新
"updateDuration": 1        //卡片定时刷新的更新周期,1 为 30 分钟,2 为 60 分钟,N 为 30×N 分钟
  }
]
```

开启服务卡片的周期性更新后就可以在 updateFormData()方法实现服务卡片的数据更新了。updateFormData()方法在配置文件设置的 updateDuration 为 30 分钟到达后会被调用,这里更新了 dim2X2 和 dim2X4 属性,如代码示例 6.23 所示。

**代码示例 6.23** chapter06\jscard\entry\src\main\java\...\widget\newscard\NewsCardImpl.java

```java
public void onUpdateFormData(long formId) {
    updateFormData(formId);
}
```

```java
@Override
public void updateFormData(long formId, Object... vars) {
    ZSONObject zsonObject = new ZSONObject();
    if (dimension == DIMENSION_2X2) {
        zsonObject.put("dim2X2", true);
    }
    if (dimension == DIMENSION_2X4) {
        zsonObject.put("dim2X4", false);
    }

    FormBindingData formBindingData = new FormBindingData(zsonObject);
    try {
        ((MainAbility)context).updateForm(formId,formBindingData);
    } catch (FormException e) {
        e.printStackTrace();
    }
}
```

dim2X2 和 dim2X4 属性定义在 index.json 文件中,如代码示例 6.24 所示。

**代码示例 6.24　chapter06\jscard\entry\src\main\js\NewsCard\pages\index\index.json**

```json
{
"data": {
"dim2X2": false,
"dim2X4": false
  }
}
```

界面效果实现如代码示例 6.25 所示。

**代码示例 6.25　chapter06\jscard\entry\src\main\js\NewsCard\pages\index\index.hml**

```html
< div class = "card">
< div class = "dim2X2" if = "{{dim2X2}}">
< image class = "mini_img" src = "/common/开发者精选.png"/>
< text class = "mini_time">刚刚</text>
</div>
< div class = "dim2X4">
< div class = "left">
< div class = "item">
< text class = "item_title">
                    华为 Mate10 Pro,有望在 2021 年下半年升级鸿蒙操作系统吗?
</text>
< div class = "item_footer">
< image class = "rec_img" src = "/common/推荐.png"/>
< text class = "num"> 30w </text>
</div>
</div>
< div class = "item">
```

```
<text class = "item_title">
            华为 Mate10 Pro,有望在 2021 年下半年升级鸿蒙操作系统吗?
</text>
<div class = "item_footer">
<image class = "rec_img" src = "/common/推荐.png"/>
<text class = "num">30w</text>
</div>
</div>
</div>
<div class = "right" if = "{{dim2X4}}">
<div class = "bar">
<image class = "bar_logo" src = "/common/icon.png" />
<image class = "bar_tl"/>
<text class = "bar_time">刚刚</text>
<image class = "bar_img" src = "/common/arrow.png" />
</div>
</div>
</div>
</div>
```

### 2. 通过编程控制更新卡片

在配置文件中定义的更新频率是按照 30 分钟或者按照每天指定的时间调用更新函数的,如果需要自定义更新时间,则可以在 onCreateForm()方法中使用定时器 Timer,如代码示例 6.26 所示。

**代码示例 6.26 onCreateForm**

```
if("jscard".equals(formName)){
    Timer timer = new Timer();
    timer.schedule(new TimerTask() {
        @Override
        public void run() {
            //JS 更新卡片
            ZSONObject zsonObject = new ZSONObject();
            if(dimension == 1){
                zsonObject.put("title","迷你尺寸" + new Random().nextInt());
            }
            if(dimension == 2){
                zsonObject.put("title","小尺寸" + new Random().nextInt());
            }
            if(dimension == 3){
                zsonObject.put("title","中尺寸" + new Random().nextInt());
            }
            if(dimension == 4){
                zsonObject.put("title","大尺寸" + new Random().nextInt());
            }
            FormBindingData formBindingData = new FormBindingData(zsonObject);
            try {
```

```
                updateForm(formId,formBindingData);
            } catch (FormException e) {
                e.printStackTrace();
            }
        }
    },0,2000);
}
```

### 6.3.5 服务卡片跳转事件和消息事件

服务卡片仅支持click通用事件,事件类型分为跳转事件(router)和消息事件(message)。

#### 1. 跳转事件

在index.hml文件中给要触发的控件添加onclick,例如:onclick=routerEvent1,如图6.29所示。

```
<div class="card" onclick="routerEvent1">
    <div class="dim2X2" if="{{dim2X2}}">...</div>
    <div class="dim2X4">...</div>
</div>
```

图6.29 卡片中的单击事件

在index.json文件中的actions用于定义routerEvent1对象,该对象中的action的值为router(路由)或者message(消息),如代码示例6.27所示,这里的事件动作用于调整到bundleName对应的ability页面,可以通过params携带参数。

**代码示例6.27** chapter06\jscard\entry\src\main\js\NewsCard\pages\index\index.json

```json
{
"data": {
"dim2X2": false,
"dim2X4": false
  },
"actions": {
"routerEvent1": {
"action": "router",
"bundleName": "com.harmonyosui.test",
"abilityName": "com.harmonyosui.test.MainAbility",
"params": {
"url": "{{url}}"
      }
    }
  }
}
```

#### 2. 消息事件

消息事件(message)会触发MainAbility中的onTriggerFormEvent()函数,如代码示

例 6.28 所示。

**代码示例 6.28**
```
"actions": {
"sendMessageEvent": {
"action": "message",
"params": {
"p1": "news",
"index": "1"
    }
  }
}
```

在 onTriggerFormEvent()函数中获取消息事件的参数，并更新卡片，如代码示例 6.29 所示。

**代码示例 6.29**
```java
public void onTriggerFormEvent(long formId, String message) {
    HiLog.info(TAG, "onTriggerFormEvent." + message);

    //先获取 message 中的参数
    ZSONObject data = ZSONObject.stringToZSON(message);
    String p1 = data.getString("p1");
    Integer index = data.getIntValue("index");

    ZSONObject zsonObject = new ZSONObject();
    zsonObject.put("index",index);
    FormBindingData formBindingData = new FormBindingData(zsonObject);
    try {
        ((MainAbility)context).updateForm(formId,formBindingData);
    } catch (FormException e) {
        e.printStackTrace();
        HiLog.info(TAG, "更新卡片失败");
    }
}
```

### 3. list 跳转事件

list 组件只能添加一个 onclick，在单击的同时需要获取单击的 list 列表中的某一项数据，如代码示例 6.30 所示。

**代码示例 6.30**
```
<list class = "list" else >
<list-item for = "{{list}}" class = "list-item">
<div class = "div" onclick = "sendRouteEvent">
        ... ...
</div>
</list-item>
</list>
```

单击 list 列表中的某一项,根据不同的项调整到对应的页面,如代码示例 6.31 所示。

代码示例 6.31

```
"actions": {
"routerEvent1": {
"action": "router",
"bundleName": "com.harmonyosui.test",
"abilityName": "com.harmonyosui.test.MainAbility",
"params": {
"id": "$ item.id"
    }
  }
}
```

# 第三篇　JavaScript API篇

ArkUI运行时为上层的ArkUI JS UI和ArkUI ETS UI框架提供了丰富的JavaScript服务接口,这些接口包括访问路由、分布式调度、网络通信、分布式文件系统、多实例管理、硬件的能力。

# 第 7 章 基本功能接口

本章介绍 ArkUI 运行时提供的基础功能接口,包括应用配置、日志、定时器、窗口、动画、路由、弹框、剪贴板等基础功能接口。

## 7.1 页面路由

页面之间跳转是由系统接口通过 Ability 进行跳转的,所以它不是简单的页面 hash 路由,而是通过 FA 的方式进行路由跳转的,如图 7.1 所示。

图 7.1 页面路由

注意：页面路由需要在页面渲染完成之后才能调用，在 onInit 和 onReady 生命周期中页面还处于渲染阶段，因此禁止调用页面路由方法。

### 7.1.1 页面路由用法

下面具体介绍页面路由的使用步骤。

第1步：导入路由模块，代码如下。

```
import router from '@system.router';
```

第2步：选择路由方法。

router.push(OBJECT)跳转到应用内的指定页面，如代码示例7.1所示。

**代码示例 7.1**

```
//在当前页面中
router.push({
  uri: 'pages/routerpage2/routerpage2',
  params: {
      data1: 'message',
      data2: {
          data3: [123,456,789],
          data4: {
              data5: 'message'
          },
      },
  },
});
//在 routerpage2 页面中
console.info('showData1:' + this.data1);
console.info('showData3:' + this.data2.data3);
```

1. router.replace(OBJECT)

用应用内的某个页面替换当前页面，并销毁被替换的页面，如代码示例7.2所示。

**代码示例 7.2**

```
//在当前页面中
router.replace({
  uri: 'pages/detail/detail',
  params: {
      data1: 'message',
  },
});
//在 detail 页面中
console.info('showData1:' + this.data1);
```

2. router.back(OBJECT)

返回上一页面或指定的页面，如代码示例7.3所示。

代码示例 7.3

```
//index 页面
router.push({
  uri: 'pages/detail/detail',
});

//detail 页面
router.push({
  uri: 'pages/mall/mall',
});

//mall 页面通过 back 返回 detail 页面
router.back();
//detail 页面通过 back 返回 index 页面
router.back();
//通过 back 返回 detail 页面
router.back({path:'pages/detail/detail'});
```

### 3. router.clear()

清空页面栈中的所有历史页面,仅保留当前页面作为栈顶页面,代码如下:

```
router.clear();
```

### 4. router.getLength()

获取当前在页面栈内的页面数量,代码如下:

```
var size = router.getLength();
console.log('pages stack size = ' + size);
```

### 5. router.getState()

获取当前页面的状态信息,代码如下:

```
var page = router.getState();
console.log('current index = ' + page.index);
console.log('current name = ' + page.name);
console.log('current path = ' + page.path);
```

## 7.1.2 页面路由动画

可以为页面之间添加转场动画,目前的页面之间的转场动画仅支持进入和退出动画设置,如图 7.2 和图 7.3 所示。

页面转场注意事项如下:

(1) 配置自定义转场时,建议将页面背景色配置为不透明颜色,否则在转场过程中可能会出现衔接不自然的现象。

(2) transition-enter 和 transition-exit 可单独配置，没有配置时使用系统默认的参数。
(3) transition-duration 没有配置的时候使用系统默认值。
(4) transition-enter/transition-exit 说明如下。
- push 场景下：进入页面栈的 Page2.js 应用 transition-enter 描述的动画配置；进入页面栈第二位置的 Page1.js 应用 transition-exit 描述的动画配置。

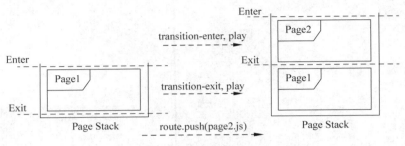

图 7.2　页面路由转场动画(1)

- back 场景下：退出页面栈的 Page2.js 应用 transition-enter 描述的动画配置，并进行倒播；从页面栈第二位置进入栈顶位置的 Page1.js 应用 transition-exit 描述的动画配置，并进行倒播。

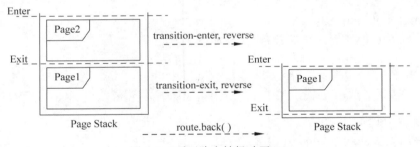

图 7.3　页面路由转场动画(2)

接下来讲解一个转场动画案例。在两个页面中添加转场动画，需要同时在两个页面的 CSS 中添加进入和退出关键帧动画，如代码示例 7.4 所示。

代码示例 7.4　chapter07\basic_service_demo\entry\src\main\js\default\pages\index.css

```css
.container {
    flex-direction: column;
    justify-content: center;
    align-items: center;
    width: 100%;
    height: 100%;
    background-color: red;

    /*自定义页面转场样式*/
```

```css
    transition-enter: present;
    transition-exit: dismiss;
    transition-timing-function: extreme-deceleration;
}

@keyframes present {
    from {
        opacity: 0;
    }

    to {
        opacity: 1;
    }
}

@keyframes dismiss {
    from {
        opacity: 1;
    }

    to {
        opacity: 0;
    }
}
```

这里的进入和退出的过渡动画被添加在组件的根节点上，同时需要给两个页面添加，如图 7.4 所示，页面之间的跳转使用 router.push() 和 router.back() 方法，如代码示例 7.5 所示。

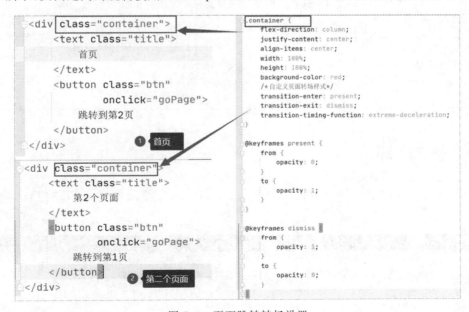

图 7.4　页面跳转转场设置

代码示例 7.5　chapter07\basic_service_demo\entry\src\main\js\default\pages\index.js

```js
goPage() {
    router.push({
        uri:"pages/second/second"
    })
}
```

## 7.2　应用上下文

@system.app 模块封装了应用的上下文信息，可以使用该模块获取当前应用配置文件中声明的信息。

导入模块，代码如下：

```
import app from '@system.app';
```

退出当前 Ability，代码如下：

```
app.terminate()
```

获取当前应用配置文件中声明的信息，如表 7.1 所示，代码如下：

```
app.getInfo()
```

表 7.1　获取当前应用配置文件中声明的信息

| 参数名 | 类型 | 说　　明 |
| --- | --- | --- |
| appID | string | 表示应用的包名，用于标识应用的唯一性 |
| appName | string | 表示应用的名称 |
| versionName | string | 表示应用的版本名称 |
| versionCode | number | 表示应用的版本号 |

## 7.3　日志打印

使用 console.debug│log│info│warn│error(message) 可以打印日志信息，如代码示例 7.6 所示。

代码示例 7.6

```js
var versionCode = 1;
console.info('Hello World. The current version code is ' + versionCode);
console.log(`versionCode: ${versionCode}`);
console.log('versionCode: %d.', versionCode);
```

在 DevEco Studio 的底部，切换到 HiLog 窗口，选择当前的设备及进程，日志级别选择 Info，搜索内容设置为 Hello World。此时窗口仅显示符合条件的日志，效果如图 7.5 所示。

图 7.5　日志打印

说明：console.log()打印的是 debug 级别日志信息。

## 7.4　应用配置

@system.configuration 模块可以获取应用当前的语言和地区，默认与系统的语言和地区同步。

configuration.getLocale()用于获取应用当前的语言和地区，默认与系统的语言和地区同步。

当前语言的国家或地区信息如表 7.2 所示。

表 7.2　当前语言的国家或地区信息

| 参数名 | 类型 | 说明 |
| --- | --- | --- |
| language | string | 语言，例如：zh |
| countryOrRegion | string | 国家或地区，例如：CN |
| dir | string | 文字布局方向，取值范围：<br>ltr：从左到右；<br>rtl：从右到左 |
| unicodeSetting | string | 语言环境定义的 Unicode 语言环境键集，如果此语言环境没有特定键集，则返回空集 |

## 7.5　窗口

@ohos.window 模块用于设置窗口的属性，如图 7.6 所示。

通过 window.getTopWindow()方法可以获取当前窗口，用于获取 window 实例，该方法的返回值 windowClass 是窗口的对象，可以使用该对象设置窗口的属性值。

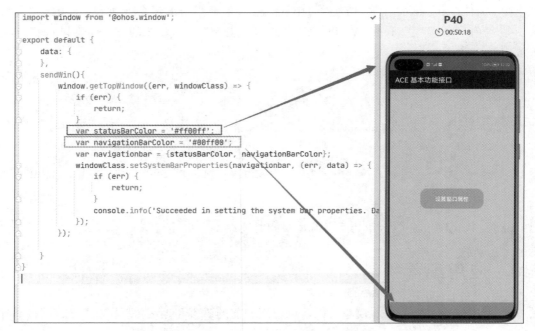

图 7.6 设置窗口的属性

设置窗口内导航条的颜色可以使用 windowClass.setSystemBarProperties()方法，如代码示例 7.7 所示。

代码示例 7.7　chapter07\basic_service_demo\entry\src\main\js\default\pages\win.js

```js
sendWin(){
    window.getTopWindow((err, windowClass) => {
        if (err) {
            return;
        }
        var statusBarColor = '#ff00ff';
        var navigationBarColor = '#00ff00';
        var navigationbar = {statusBarColor, navigationBarColor};
        windowClass.setSystemBarProperties(navigationbar, (err, data) => {
            if (err) {
                return;
            }
            console.info('Succeeded in setting the system bar properties. Data: ' + JSON.stringify(data));
        });
    });
}
```

## 7.6 弹框

系统弹框有两种形式,一种是 Toast,另一种是 Dialog。
下面具体介绍弹框的使用步骤。
步骤 1:导入模块,代码如下。

```js
import prompt from '@system.prompt';
```

步骤 2:方法使用。
prompt.showToast(OBJECT)方法用于显示文本弹框,如代码示例 7.8 所示。

**代码示例 7.8**

```js
prompt.showToast({
  message: 'Message Info',
  duration: 2000,
});
```

prompt.showDialog(OBJECT)方法用于在页面内显示对话框,如代码示例 7.9 所示。

**代码示例 7.9** chapter07\basic_service_demo\entry\src\main\js\default\pages\toast.js

```js
prompt.showDialog({
  title: 'Title Info',
  message: 'Message Info',
  buttons: [
    {
      text: 'button',
      color: '#666666',
    },
  ],
  success: function(data) {
    console.log('dialog success callback,click button : ' + data.index);
  },
  cancel: function() {
    console.log('dialog cancel callback');
  },
});
```

弹框效果如图 7.7 所示。

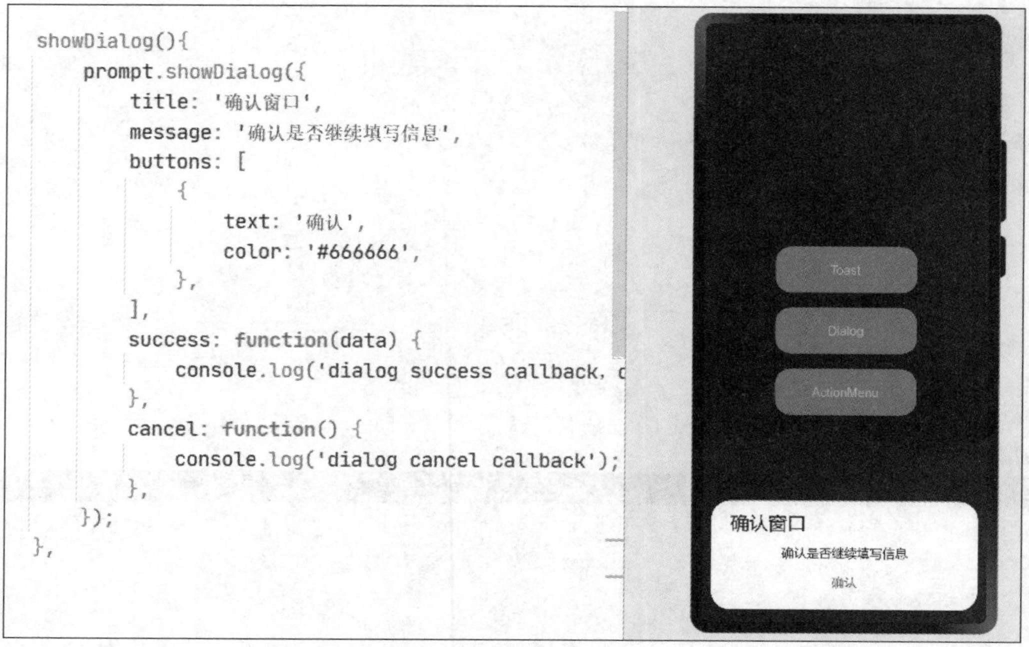
showDialog(){
prompt.showDialog({
title: '确认窗口',
message: '确认是否继续填写信息',
buttons: [
{
text: '确认',
color: '#666666',
},
],
success: function(data) {
console.log('dialog success callback, c
},
cancel: function() {
console.log('dialog cancel callback');
},
});
},
Toast
Dialog
ActionMenu
确认窗口
确认是否继续填写信息
确认

图 7.7 弹框效果

## 7.7 动画

为了提高界面的帧动画执行效率，使用了类似 HTML5 中的请求动画帧，逐帧回调 JS 函数，如表 7.3 所示。

**表 7.3 帧动画函数**

| 函 数 名 | 说 明 | 函 数 参 数 |
|---|---|---|
| requestAnimationFrame | 请求动画帧，逐帧回调 JS 函数 | handler：表示要逐帧回调的函数<br>requestAnimationFrame 函数回调 handler 函数时会在第 1 个参数位置传入 timestamp 时间戳。它表示 requestAnimationFrame 开始去执行回调函数的时刻 |
| | | ...args 附加参数，函数回调时，它们会作为参数传递给 handler |
| cancelAnimationFrame | 取消动画帧，取消逐帧回调请求 | requestId 逐帧回调函数的标识 id |
| createAnimator | 创建动画对象 | duration 动画播放的时长，单位毫秒，默认值为 0 |
| | | easing 动画插值曲线，默认值为'ease' |

续表

| 函 数 名 | 说　　明 | 函数参数 |
| --- | --- | --- |
| createAnimator | 创建动画对象 | delay 动画延时播放时长，单位毫秒，默认值为 0，即不延时 |
| | | fill 动画启停模式，默认值为 none，详情见：animation-fill-mode |
| | | direction 动画播放模式，默认值为 normal，详情见：animation-direction |
| | | iterations 动画播放次数，默认值为 1，设置为 0 时不播放，设置为 −1 时无限次播放 |
| | | begin 动画插值起点，不设置时默认值为 0 |
| | | end 动画插值终点，不设置时默认值为 1 |

动画案例

单击放大组件的宽和高，放大过程中执行动画过渡效果，如代码示例 7.10 所示。

代码示例 7.10　chapter07\basic_service_demo\entry\src\main\js\default\pages\AnimationFrame.hml

```
< div class = "container">
< div class = "Animation"
        style = "height: {{divHeight}}px; width: {{divWidth}}px; background - color: red;"
        onclick = "Show">
</div>
</div>
```

动画如代码示例 7.11 所示。

代码示例 7.11　chapter07\basic_service_demo\entry\src\main\js\default\pages\AnimationFrame.js

```
import Animator from "@ohos.animator";
export default {
    data : {
        divWidth: 200,
        divHeight: 200,
        animator: null
    },
    onInit() {
        var options = {
            duration: 1500,
            easing: 'friction',
            fill: 'forwards',
            iterations: 2,
            begin: 200.0,
            end: 400.0
        };
        this.animator = Animator.createAnimator(options);
    },
```

```
Show() {
    var options1 = {
        duration: 2000,
        easing: 'friction',
        fill: 'forwards',
        iterations: 1,
        begin: 200.0,
        end: 400.0
    };
    this.animator.update(options1);
    var _this = this;
    this.animator.onframe = function(value) {
        _this.divWidth = value;
        _this.divHeight = value;
    };
    this.animator.play();
}
```

## 7.8 剪贴板

剪贴板可实现同一设备的应用程序 A、B 之间可以借助系统剪贴板服务完成简单数据的传递，即应用程序 A 向剪贴板服务写入数据后，应用程序 B 可以从中读取出数据，如图 7.8 所示。

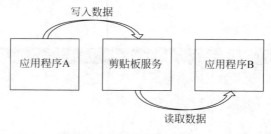

图 7.8 剪贴板读写数据

在使用剪贴板服务时，需要注意以下几点：

（1）只有在前台获取焦点的应用才有读取系统剪贴板的权限（系统默认输入法应用除外）。

（2）写入剪贴板服务中的剪贴板数据不会随应用程序的结束而销毁。

（3）对同一用户而言，写入剪贴板服务的数据会被下一次写入的剪贴板数据所覆盖。

（4）在同一设备内，剪贴板单次传递的内容不应超过 500KB。

### 1. SystemPasteboard

在调用 SystemPasteboard 的接口前，需要先通过 getSystemPasteboard 获取一个 SystemPasteboard 对象。

## 2. setPasteData(pasteData:PasteData): Promise &lt; void &gt;

将数据写入系统剪贴板，如代码示例 7.12 所示。

代码示例 7.12    chapter07\basic_service_demo\entry\src\main\js\default\pages\pasteboard.js

```js
var systemPasteboard = pasteboard.getSystemPasteboard();
var pasteData = pasteboard.createPlainTextData("content");
systemPasteboard.setPasteData(pasteData, (error, data) => {
    if (error) {
        console.error('Failed to set PasteData. Cause: ' + error.message);
        return;
    }
    console.info('PasteData set successfully. ' + data);
});
```

## 3. getPasteData( callback:AsyncCallback &lt; PasteData &gt; ): void

读取系统剪贴板内容，如代码示例 7.13 所示。

代码示例 7.13

```js
var systemPasteboard = pasteboard.getSystemPasteboard();
systemPasteboard.getPasteData((error, pasteData) => {
    if (error) {
        console.error('Failed to obtain PasteData. Cause: ' + error.message);
        return;
    }
    var text = pasteData.getPrimaryText();
});
```

剪贴板如图 7.9 所示。

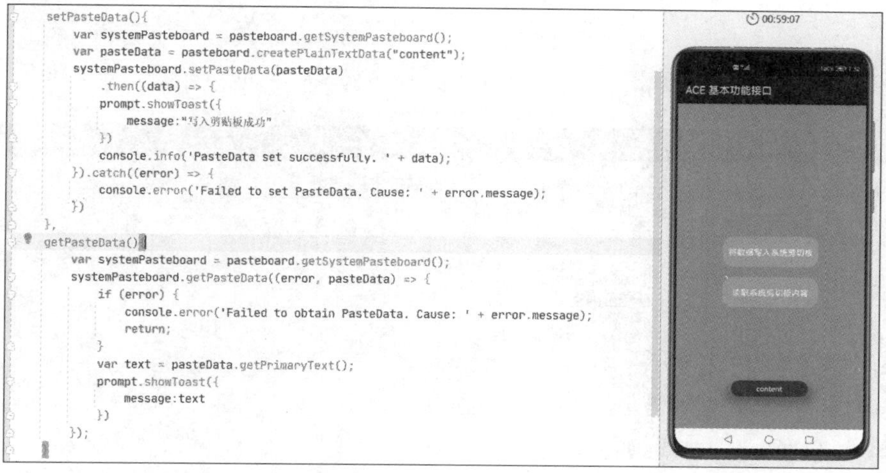

图 7.9    剪贴板

# 第 8 章 网络与媒体接口

本章讲解网络服务和多媒体相关接口,ArkUI 中提供了对 HTTP 协议和 WebSocket 协议调用的支持,同时提供了上传和下载的服务接口,支持应用开发中网络场景的开发。

## 8.1 网络访问

ArkUI 中提供了两种通过 HTTP 协议访问网络数据的方法,分别封装在@system 和 @ohos 的模块中,从 API Version 6 开始,@system.fetch 接口不再维护,推荐使用新接口 @ohos.net.http 发起 http 数据请求。

**注意**:需要开启权限 ohos.permission.INTERNET。默认支持 https,如果要支持 http,则需要在 config.json 文件里增加 network 标签和属性标识 cleartextTraffic:true。

下面我们具体介绍网络访问的使用步骤。

步骤 1:导入模块,代码如下。

```
import http from '@ohos.net.http';
```

步骤 2:配置权限,在 config.json 文件中添加权限,如代码示例 8.1 所示。

**代码示例 8.1  配置权限**

```
"module": {
"abilities": [],
"reqPermissions": [
        {
"name": "ohos.permission.INTERNET"
        }
]
}
```

步骤 3:配置 http 支持,如代码示例 8.2 所示。

**代码示例 8.2**

```
"deviceConfig": {
```

```
        "default": {
        "network": {
        "usesClearText": true
            }
        }
    },
```

步骤 4：调用方法。

通过网络获取数据，如代码示例 8.3 所示。

**代码示例 8.3**

```
import http from '@ohos.net.http';

//每个 httpRequest 对应一个 HTTP 请求任务,不可复用
let httpRequest = http.createHttp();
//用于订阅 http 响应头,此接口会比 request 请求先返回.可以根据业务需要订阅此消息
httpRequest.on('headerReceive', (err, data) => {
    if (!err) {
        console.info('header: ' + data.header);
    } else {
        console.info('error:' + err.data);
    }
});
httpRequest.request(
    //填写 HTTP 请求的 URL 网址,可以带参数也可以不带参数.URL 网址需要开发者自定义.GET 请
    //求的参数可以在 extraData 中指定
"EXAMPLE_URL",
    {
        method: 'POST',                    //可选,默认为'GET'
        //开发者可根据自身业务需要添加 header 字段
        header: {
            'Content-Type': 'application/json'
        },
        //当使用 POST 请求时此字段用于传递内容
        extraData: "data to post",
        readTimeout: 60000,                //可选,默认为 60000ms
        connectTimeout: 60000              //可选,默认为 60000ms
    },(err, data) => {
        if (!err) {
            //data.result 为 http 响应内容,可根据业务需要进行解析
            console.info('Result:' + data.result);
            console.info('code:' + data.responseCode);
            //data.header 为 http 响应头,可根据业务需要进行解析
            console.info('header:' + data.header);
        } else {
            console.info('error:' + err.data);
        }
    }
);
```

## 8.2 WebSocket

使用 WebSocket 建立服务器与客户端的双向连接,需要先通过 createWebSocket()方法创建 WebSocket 对象,然后通过 connect()方法连接到服务器。当连接成功后,客户端会收到 open 事件的回调,之后客户端就可以通过 send()方法与服务器进行通信了。当服务器发信息给客户端时,客户端会收到 message 事件的回调。当客户端不需要此连接时,可以通过调用 close()方法主动断开连接,之后客户端会收到 close 事件的回调。

下面通过具体的步骤介绍如何使用 WebSocket。

步骤 1:导入模块,代码如下:

```
import webSocket from '@ohos.net.webSocket';
```

步骤 2:申请网络权限,需要 ohos.permission.INTERNET 权限,如代码示例 8.4 所示。

**代码示例 8.4**
```
"module": {
"abilities": [],
"reqPermissions": [
    {
"name": "ohos.permission.INTERNET"
    }
   ]
}
```

步骤 3:配置 http 支持,如代码示例 8.5 所示。

**代码示例 8.5**
```
"deviceConfig": {
"default": {
"network": {
"usesCleartext": true
    }
   }
 },
```

步骤 4:创建 WebSocket 服务器端,这里使用 Node.js 中的 ws 模块创建服务器端,如代码示例 8.6 所示。

**代码示例 8.6    \chapter08\ws_server\server.js**
```
var WebSocketServer = require('ws').Server
```

```
    , wss = new WebSocketServer({port: 8080});
wss.on('connection', function(ws) {
    ws.on('message', function(message) {
        console.log('received: %s', message);
    });
    ws.send('websocket 连接成功!这是从 Node WebSocket 服务器端发送的数据!');
});
```

步骤 5:在页面的初始化时调用 WebSocket 连接,如代码示例 8.7 所示。

**代码示例 8.7** chapter08\websocket_app\entry\src\main\js\default\pages\index.js

```
import webSocket from '@ohos.net.webSocket';
export default {
    data: {
    },
    onInit() {
        var defaultIpAddress = "ws://ip:8080";
        let ws = webSocket.createWebSocket();
        ws.on('open', (err, value) => {
            //当收到 on('open')事件时,可以通过 send()方法与服务器进行通信
            ws.send("Hello, server!", (err, value) => {
                if (!err) {
                    console.log("send success");
                } else {
                    console.log("send fail, err:" + JSON.stringify(err));
                }
            });
        });
        ws.on('message', (err, value) => {
            console.log("on message, message:" + value);
            //当收到服务器的'bye'消息时(此消息字段仅为示意,具体字段需要与服务器协商),
            //主动断开连接
            if (value === 'bye') {
                ws.close((err, value) => {
                    if (!err) {
                        console.log("close success");
                    } else {
                        console.log("close fail, err is " + JSON.stringify(err));
                    }
                });
            }
        });
        ws.on('close', (err, value) => {
            console.log("on close, code is " + value.code + ", reason is " + value.reason);
        });
        ws.on('error', (err) => {
            console.log("on error, error:" + JSON.stringify(err));
        });
```

```
            ws.connect(defaultIpAddress, (err, value) => {
                if (!err) {
                    console.log("connect success");
                } else {
                    console.log("connect fail, err:" + JSON.stringify(err));
                }
            });
        }
    }
```

App 运行后,通过 IDE 日志打印出 WebSocket 端返回的数据,如图 8.1 所示。

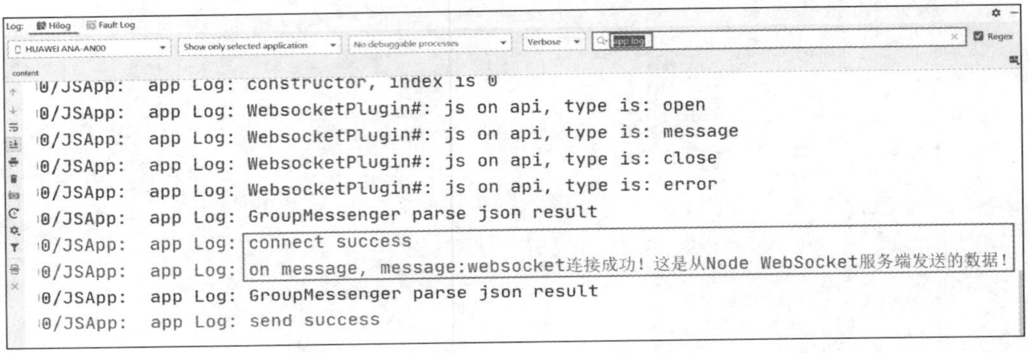

图 8.1　WebSocket 连接成功后打印的信息

## 8.3　上传和下载

从 API Version 6 开始,该接口不再维护,推荐使用新接口 @ohos.request 进行上传和下载,上传代码如代码示例 8.8 所示。

**代码示例 8.8**

```
request.upload({ url: 'https://patch' }).then((data) => {
    this.uploadTask = data;
}).catch((err) => {
    console.error('Failed to request the upload. Cause: ' + JSON.stringify(err));
})
```

下载如代码示例 8.9 所示。

**代码示例 8.9**

```
request.download({ url: 'https://patch' }, (err, data) => {
    if (err) {
        console.error('Failed to request the download. Cause: ' + JSON.stringify(err));
        return;
```

```
        }
        this.downloadTask = data;
    });
```

## 8.4 媒体

媒体模块提供声频播放和控制功能,可以使用该模块播放在线媒体源,下面通过详细步骤介绍如何开发一个在线的 MP3 播放器。

步骤 1:导入模块,代码如下。

```
import media from '@ohos.multimedia.media';
```

步骤 2:申请网络权限,如代码示例 8.10 所示。
录音:ohos.permission.MICROPHONE。
网络:ohos.permission.INTERNET。

代码示例 8.10
```
"module": {
"abilities": [],
"reqPermissions": [
    {
"name": "ohos.permission.INTERNET"
    },
    {
"name": "ohos.permission.MICROPHONE"
    },
    ]
}
```

步骤 3:配置 http 支持,如代码示例 8.11 所示。

代码示例 8.11
```
"deviceConfig": {
"default": {
"network": {
"usesCleartext": true
    }
   }
 },
```

步骤 4:播放音乐。
AudioPlayer 声频播放管理类用于管理和播放声频媒体。在调用 AudioPlayer 的方法前,需要先通过 createAudioPlayer()方法构建一个 AudioPlayer 实例,如代码示例 8.12 所示。

**代码示例 8.12**

```
var audiorecorder = media.createAudioRecorder();
audioplayer.src = 'https://media.xxxx.com/sounds.mp4';
audioplayer.src = 'file://xxx/sounds.mp4';
audioplayer.on('play', () => {
  console.log('Playback starts.');
  audioplayer.pause();
});
audioplayer.play();
```

步骤 5：停止播放音乐，如代码示例 8.13 所示。

**代码示例 8.13**

```
audioplayer.on('pause', () => {
  console.log('Playback paused.');
});
audioplayer.pause();
```

# 第 9 章 分布式能力接口

本章介绍分布式调度能力和分布式文件管理能力接口,该接口包括分布式迁移、拉起、控制和分布式文件操作的能力。

## 9.1 分布式迁移

分布式迁移提供了一个主动迁移接口及一系列页面生命周期回调,以便支持应用迁移到其他设备。注意:如果迁移到的设备上已经运行该 FA,则生命周期 onNewRequest 将被回调。

### 9.1.1 申请分布式迁移权限

在 config.json 文件的 reqPermissions 中添加以下权限申请,如代码示例 9.1 所示。

代码示例 9.1
```
"module": {
"abilities": [],
"reqPermissions": [
    {
"name": "ohos.permission.DISTRIBUTED_DATASYNC"
    }
  ]
}
```

### 9.1.2 通过 FeatureAbility 发起迁移

FeatureAbility.continueAbility()方法为主动进行 FA 迁移的入口,该方法无参数,返回值为 JSON 字符串,指示是否成功,字段如表 9.1 所示。

表 9.1　continueAbility 返回值

| 参数名 | 类型 | 非空 | 说明 |
| --- | --- | --- | --- |
| code | number | 是 | 0:发起迁移成功。<br>非 0:失败,原因见 data |

续表

| 参数名 | 类型 | 非空 | 说明 |
| --- | --- | --- | --- |
| data | object | 是 | 成功：返回 null<br>失败：携带错误信息，类型为 String |

下面通过伪代码示例展示一个备忘录 FA 从发起到完成迁移的过程。发起迁移 FA，如代码示例 9.2 所示。

**代码示例 9.2　发起迁移**

```javascript
import prompt from '@system.prompt'

export default {
  data: {
    continueAbilityData: {
      remoteData1: 'self define continue data for distribute',
      remoteData2: {
        item1: 0,
        item2: true,
        item3: 'inner string'
      },
      remoteData3: [1, 2, 3]
    }
  },
  //shareData 的数据会在 onSaveData 触发时与 saveData 一起传送到迁移目标 FA,并绑定到其
  //shareData 数据段上
  //shareData 的数据可以直接使用 this 访问,如 eg:this.remoteShareData1
  shareData: {
    remoteShareData1: 'share data for distribute',
    remoteShareData2: {
      item1: 0,
      item2: false,
      item3: 'inner string'
    },
    remoteShareData3: [4, 5, 6]
  },
  tryContinueAbility: async function() {
    //应用进行迁移
    let result = await FeatureAbility.continueAbility();
    console.info("result:" + JSON.stringify(result));
  },
  onStartContinuation: function onStartContinuation() {
    //判断当前的状态是否适合迁移
    console.info("trigger onStartContinuation");
    return true;
  },
  onCompleteContinuation: function onCompleteContinuation(code) {
```

```
    //迁移操作完成,code 返回结果
    console.info("trigger onCompleteContinuation: code = " + code);
  },
  onSaveData: function onSaveData(saveData) {
    //将数据保存到 savedData 中进行迁移
    var data = this.continueAbilityData;
    Object.assign(saveData, data)
  }
}
```

迁移到 FA 后会调用 onRestoreData() 方法,恢复缓存的数据,代码如下:

```
onRestoreData: function onRestoreData(restoreData) {
  //收到迁移数据,恢复缓存的数据
  this.continueAbilityData = restoreData;
}
```

下面看一看在迁移过程中这些回调方法的用法和作用。

1) onStartContinuation()

FA 发起迁移时的回调,在此回调中应用可以根据当前状态决定是否迁移。该方法无参数,返回值为 Boolean 类型,true 表示允许进行迁移,false 表示不允许迁移。

2) onSaveData(OBJECT)

保存状态数据的回调,开发者需要往参数对象中填入需迁移到目标设备上的数据。参数如表 9.2 所示。返回值:无。

表 9.2 onSaveData 参数

| 参数名 | 类型 | 非空 | 说明 |
| --- | --- | --- | --- |
| savedData | object | 是 | 可以往其中填入可被序列化的自定义数据 |

**注意**:说明:shareData 中的数据在迁移时会自动迁移到目标设备上。

3) onRestoreData(OBJECT)

恢复发起迁移时对 onSaveData() 方法保存的数据进行回调。参数如表 9.3 所示。返回值:无。

表 9.3 onRestoreData 参数

| 参数名 | 类型 | 非空 | 说明 |
| --- | --- | --- | --- |
| restoreData | object | 是 | 用于恢复应用状态的对象,其中的数据及结构由 onSaveData() 方法决定 |

4) onCompleteContinuation(code)

迁移完成的回调,在调用端被触发,表示应用迁移到目标设备上的结果。参数如表 9.4

所示。返回值：无。

表 9.4　onCompleteContinuation 参数

| 参数名 | 类型 | 非空 | 说明 |
| --- | --- | --- | --- |
| Code | number | 是 | 迁移完成的结果。0：成功。—1：失败 |

**注意**：在 HarmonyOS 中，分布式任务调度平台对搭建 HarmonyOS 的多设备构建的"超级虚拟设备终端"提供了统一的组件管理能力，为应用定义统一的能力基础、接口形式、数据结构、服务描述语言、屏蔽硬件的差异、支持远程启动、远程调用和业务无缝迁移等分布式任务。

## 9.2　分布式拉起

分布式拉起允许拉起一个本地或远程的 FA，拉起时可以传递参数。如果使用 startAbilityForResult，则可以获得 FA 的运行结果。

**注意**：如果设备上已经运行了该 FA，并且 launchType 为 singleton，则生命周期 onNewRequest() 将被回调。

### 9.2.1　申请分布式迁移权限

在 config.json 文件的 reqPermissions 中添加以下权限申请，如代码示例 9.3 所示。

**代码示例 9.3**

```
"module": {
"abilities": [],
"reqPermissions": [
    {
"name": "ohos.permission.DISTRIBUTED_DATASYNC"
    }
  ]
}
```

### 9.2.2　允许以显式的方式拉起远程或本地的 FA

FeatureAbility.startAbility(OBJECT) 方法用于拉起一个 FA，无回调结果，如代码示例 9.4 所示。

**代码示例 9.4**

```
export default {
  start() {
    let actionData = {
      uri: 'www.huawei.com'
    };
    let target = {
```

```
      bundleName: "com.example.harmonydevsample",
      abilityName: "com.example.harmonydevsample.EntryJSApiAbility",
      data: actionData
    };

    let result = await FeatureAbility.startAbility(target);
    let ret = JSON.parse(result);
    if (ret.code == 0) {
      console.log('success');
    } else {
      console.log('cannot start browing service, reason: ' + ret.data);
    }
  }
}

//callee
export default {
  data: {
    contact: "contact information",
    location: "location information"
  }
}
```

**注意**：分布式场景下，data 数据段下的字段名，如果与后台传入的字段同名，则 JS 中的字段名会被覆盖。分布式拉起时传入的数据，会直接挂到应用页面的 data 数据段下，可直接通过 this.XXX 访问。

## 9.2.3 拉起远程带返回值的 FA

可以拉起远程 FA 的能力，并在返回的回调中等待被拉起 FA 的结果，该方法的代码如下：

FeatureAbility.startAbilityForResult(OBJECT)

一个没有定位功能的 FA 可以调用另一个有能力的 FA，调用地图并且获得用户在地图上选择的位置。

和 FeatureAbility.startAbility 的区别在于结果是在被拉起的 FA 消亡后再返回结果。

**注意**：FeatureAbility.startAbilityForResult()需要与 FeatureAbility.finishWithResult()关联使用。

下面的示例展示了一个 ability 如何拉起另一个 ability，并在另一个 ability 退出时获得其中的数据，如代码示例 9.5 所示。

**代码示例 9.5**

```
export default {
  data: {
    startAbilityForResultExplicitResult: 'NA'
```

```
    },
    startAbilityForResultExplicit: async function() {
        var result = await FeatureAbility.startAbilityForResult({
          bundleName: "com.example.harmonydevsample",
          abilityName: "com.example.harmonydevsample.EntryJSApiAbility"
        });
        this.startAbilityForResultExplicitResult = JSON.stringify(result);
    }
}

//callee
export default {
    onShow() {
        let request = {};
        request.result = {
          contact: "contact information",
          location: "location information"
        };
        FeatureAbility.finishWithResult(100, request);
    }
}
```

### 9.2.4 分布式 API 在 FA 中的生命周期

FA 在分布式调用中生命周期的调用顺序如图 9.1 所示。

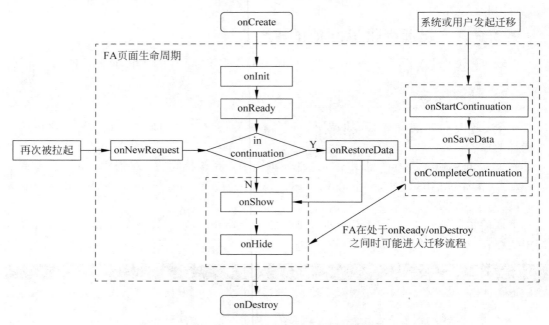

图 9.1 分布式 API 在 FA 的生命周期

## 9.3 文件数据管理

本节介绍文件的存储管理和交互,包括文件的轻量级存储、文件交互、文件管理。

### 9.3.1 轻量级存储

轻量级存储为应用提供 key-value 键值型的文件数据处理能力,支持应用对数据进行轻量级存储及查询。数据存储形式为键值对,键的类型为字符串型,值的存储数据类型包括数字型、字符型、布尔型。

**1. storage 对象**

storage 对象提供获取和修改存储数据的接口。

(1) 通过 store 添加数据,如代码示例 9.6 所示。

代码示例 9.6
```
//同步访问
let data = store.putSync(key, 'defValue');
//异步 promise
const promise = store.put(key, 'test');
promise.then((data) => {
    }).catch((err) => {
});
//异步回调
store.put(key, 'test', function(err, data) {
});
```

(2) 删除给定 key 的存储,如代码示例 9.7 所示。

代码示例 9.7
```
//同步访问
store.deleteSync(key);
//异步回调
store.delete(key, function(err, data) {
});
//异步 promise
const promise = store.delete(key);
promise.then((data) => {
    }).catch((err) => {
});
```

(3) 获取给定 key 的存储,如代码示例 9.8 所示。

代码示例 9.8
```
//同步获取
let ret = store.getSync(key, 'defValue');
```

```javascript
//异步回调获取
store.get(key, 'test', function(err, data) {
    console.info(data);
});
//异步 promise 获取
const promise = store.get(key, 'test');
promise.then((data) => {
   console.info(data);
   }).catch((err) => {
});
```

**2. 文件操作**

导入模块，代码如下：

```javascript
import data_storage from '@ohos.data.storage';
```

读取指定文件，将数据加载到 Storage 实例中，如代码示例9.9所示。

**代码示例9.9**

```javascript
//同步访问
const PATH = '/data/data/com.example.myapplication/{storage_name}';
let store = data_storage.getStorageSync(PATH);
//异步访问
data_storage.getStorage(PATH, function(err, data) {
    store = data;
});
```

从内存中移除指定文件对应的 Storage 单实例，并删除指定文件及其备份文件、损坏文件。在删除指定文件时，应用不允许再使用该实例进行数据操作，否则会出现数据一致性问题，此方法为同步方法，如代码示例9.10所示。

**代码示例9.10**

```javascript
//同步访问
const PATH = '/data/data/com.example.myapplication/{storage_name}';
data_storage.deleteStorageSync(PATH);

//异步访问
data_storage.deleteStorage(PATH, function(err, data) {
});
```

### 9.3.2 文件管理

从 API Version 6 开始，@system.file 接口不再维护，推荐使用新接口@ohos.fileio 进行文件存储管理。

导入文件管理模块，代码如下：

```
import fileio from '@ohos.fileio';
```

在使用文件管理功能对文件/目录进行操作前,需要先获取文件所在的绝对路径,如表 9.5 所示。

表 9.5 文件路径

| 目录类型 | 说明 | 相关接口 |
| --- | --- | --- |
| 内部存储的缓存目录 | 可读写,随时可能清除,不保证持久性。一般用作下载临时目录或缓存目录 | getCacheDir |
| 内部存储目录 | 随应用卸载删除 | getFilesDir |

文件/目录绝对路径=应用目录路径+文件/目录名。

通过上述接口获取应用目录路径 dir,文件名为 xxx.txt,文件所在绝对路径为

```
let path = dir + "xxx.txt"
```

### 1. 目录操作

(1) 打开文件目录,如代码示例 9.11 所示。

代码示例 9.11

```
//以同步方法打开文件目录
let dir = fileio.opendirSync(path);

//以异步方法打开文件目录,使用 promise 形式返回结果
let dir = fileio.opendir(path);

//以异步方法打开文件目录,使用 callback 形式返回结果
fileio.opendir(path, function (err, dir) {
});
```

(2) 创建目录,如代码示例 9.12 所示。

代码示例 9.12

```
//以异步方法创建目录,使用 promise 形式返回结果
fileio.mkdir(fpath)
    .then(function(err) {
    //目录创建成功,do something
    }).catch(function (e){

});

//以异步方法创建目录,使用 callback 形式返回结果
```

```
await fileio.mkdir(fpath, function(err) {
    if (!err) {
        //do something
    }
})

//以同步方法创建目录
fileio.mkdirSync(fpath);
```

(3) 删除目录,如代码示例9.13所示。

**代码示例 9.13**

```
//以同步方法删除目录
fileio.rmdirSync(fpath);
```

### 2. 文件操作

(1) 打开文件,如代码示例9.14所示。

**代码示例 9.14**

```
//以同步方法打开文件
fileio.openSync(fpath);
```

(2) 从文件中读取数据,如代码示例9.15所示。

**代码示例 9.15**

```
//以异步方法从文件读取数据,使用promise形式返回结果
let fd = fileio.openSync(fpath, 0o2);
let buf = new ArrayBuffer(4096);
let res = await fileio.read(fd, buf);
console.log(String.fromCharCode.apply(null, new Uint8Array(res.buffer)));

//以异步方法从文件读取数据,使用callback形式返回结果
let fd = fileio.openSync(fpath, 0o2);
let buf = new ArrayBuffer(4096);
fileio.read(fd, buf, function (err, readOut) {
    if (!err) {
        console.log(String.fromCharCode.apply(null, new Uint8Array(readOut.buffer)))
    }
});

//以同步方法从文件中读取数据
let fd = fileio.openSync(fpath, 0o2);
let buf = new ArrayBuffer(4096);
fileio.readSync(fd, buf);
console.log(String.fromCharCode.apply(null, new Uint8Array(buf)));
```

(3) 删除文件,如代码示例9.16所示。

**代码示例 9.16**

```
//以异步方法删除文件,使用 promise 形式返回结果
await fileio.unlink(fpath);

//以异步方法删除文件,使用 callback 形式返回结果
await fileio.unlink(fpath, function(err) {
    if (!err) {
        //do something
    }
})

//以同步方法删除文件
fileio.unlinkSync(fpath);
```

(4) 将数据写入文件,如代码示例 9.17 所示。

**代码示例 9.17**

```
//以异步方法将数据写入文件,使用 promise 形式返回结果
let fd = fileio.openSync(fpath, 0o102, 0o666);
await fileio.write(fd, "hello, world");

//以异步方法将数据写入文件,使用 callback 形式返回结果
let fd = fileio.openSync(fpath, 0o102, 0o666);
fileio.write(fd, "hello, world", function (err, BytesWritten) {
    if (!err) {
        console.log(BytesWritten)
    }
});

//以同步方法将数据写入文件
let fd = fileio.openSync(fpath, 0o102, 0o666);
fileio.writeSync(fd, "hello, world");
```

# 第 10 章 系统设备接口

通过系统提供的 API 让 JavaScript 可以通过这些接口获取地理位置、传感器、系统应用等相关信息。

## 10.1 消息通知

消息通知提供系统底层的通知能力,首先导入@system.notification,代码如下:

```
import notification from '@system.notification';
```

notification.show(OBJECT)方法用于显示通知,如代码示例 10.1 所示。

代码示例 10.1

```
notification.show({
      contentTitle: 'harmonyos-ui.com',
      contentText: '《鸿蒙操作系统开发入门经典》上线了',
      clickAction: {
           bundleName: 'com.harmonyosui.test',
           abilityName: 'MainAbility',
uri: 'pages/index/index',
      },
});
```

消息通知参数如表 10.1 所示。

表 10.1 消息通知参数

| 参数名 | 类型 | 必填 | 说明 |
| --- | --- | --- | --- |
| contentTitle | string | 否 | 通知标题 |
| contentText | string | 否 | 通知内容 |
| clickAction | ActionInfo | 否 | 通知单击后触发的动作 |

ActionInfo 属性如表 10.2 所示。

表 10.2　ActionInfo 属性

| 参数名 | 类型 | 必填 | 说　　明 |
|---|---|---|---|
| bundleName | string | 是 | 单击通知后要跳转到的应用的 bundleName |
| abilityName | string | 是 | 单击通知后要跳转到的应用的 abilityName |
| uri | ActionInfo | 是 | 要跳转到的 uri，可以是下面的两种格式。<br>页面绝对路径，由配置文件中 pages 列表提供，例如：<br>pages/index/index<br>pages/detail/detail<br>特殊情况，如果 uri 的值是"/"，则跳转到首页 |

## 10.2　地理位置

获取地理位置信息，同样需要用到系统底层能力。

步骤 1：配置权限，在 config.json 文件中添加权限，如代码示例 10.2 所示。

代码示例 10.2

```
"module": {
"abilities": [],
"reqPermissions": [
    {
"name": "ohos.permission.LOCATION"
    }
]
}
```

步骤 2：导入地理位置模块，代码如下。

```
import geolocation from '@system.geolocation';
```

步骤 3：地理模块中的常见方法。

（1）geolocation.getLocation（OBJECT）方法用于获取设备的地理位置，如代码示例 10.3 所示。

代码示例 10.3

```
geolocation.getLocation({
  success: function(data) {
    console.log('success get location data. latitude:' + data.latitude);
  },
  fail: function(data, code) {
    console.log('fail to get location. code:' + code + ', data:' + data);
  },
});
```

（2）geolocation.getLocationType(OBJECT)方法用于获取当前设备支持的定位类型，如代码示例10.4所示。

代码示例 10.4
```
geolocation.getLocationType({
  success: function(data) {
    console.log('success get location type:' + data.types[0]);
  },
  fail: function(data, code) {
    console.log('fail to get location. code:' + code + ', data:' + data);
  },
});
```

（3）geolocation.subscribe(OBJECT)方法用于订阅设备的地理位置信息。多次调用，只有最后一次的调用生效，如代码示例10.5所示。

代码示例 10.5
```
geolocation.subscribe({
  success: function(data) {
    console.log('get location. latitude:' + data.latitude);
  },
  fail: function(data, code) {
    console.log('fail to get location. code:' + code + ', data:' + data);
  },
});
```

（4）geolocation.unsubscribe()方法用于取消订阅设备的地理位置信息，如代码示例10.6所示。

代码示例 10.6
```
geolocation.unsubscribe();
```

（5）geolocation.getSupportedCoordTypes()方法用于获取设备支持的坐标系类型，如代码示例10.7所示。

- 返回值。
- 字符串数组，表示坐标系类型，如[wgs84，gcj02]。

代码示例 10.7
```
var types = geolocation.getSupportedCoordTypes();
```

## 10.3 设备信息

设备信息的获取，对于开发鸿蒙应用来讲，非常重要，这里首先需要导入@system.device模块，注意，获取device信息需要在onShow()方法中调用，如代码示例10.8所示。

代码示例 10.8

```
import device from '@system.device';
```

获取设备方法,如代码示例 10.9 所示。

代码示例 10.9

```
device.getInfo({
  success: function(data) {
    console.log('success get device info brand:' + data.brand);
  },
  fail: function(data, code) {
    console.log('fail get device info code:' + code + ', data: ' + data);
  },
});
```

## 10.4 应用管理

对于已经安装的 HAP 信息,可以通过导入@system.package 模块获取相关应用的信息,这里需要申请权限才可以使用,如代码示例 10.10 所示。

代码示例 10.10

```
import pkg from '@system.package';
```

需要申请权限 ohos.permission.GET_BUNDLE_INFO。

应用管理方法的使用,如代码示例 10.11 所示。

代码示例 10.11

```
pkg.hasInstalled({
  bundleName: 'com.example.bundlename',
  success: function(data) {
console.log('package has installed: ' + data);
  },
  fail: function(data, code) {
console.log('query package fail, code: ' + code + ', data: ' + data);
  },
});
```

## 10.5 媒体查询

鸿蒙 JS 应用框架提供了在 CSS 的媒体查询,同时也可以通过系统接口的方式获取媒体查询信息,这样方便开发者在代码中通过媒体查询进行不同类型媒体的操作。这里首先

需要导入@system.mediaquery 模块,代码如下:

```
import mediaquery from '@system.mediaquery';
```

媒体查询提供匹配媒体的方法及监听回调的方法如下:

(1) mediaquery.matchMedia(condition)方法可根据媒体查询条件创建 MediaQueryList 对象,代码如下:

```
var mMediaQueryList = mediaquery.matchMedia('(max-width: 466)');
```

(2) MediaQueryList.addListener(OBJECT)方法用于向 MediaQueryList 添加回调函数,回调函数应在 onShow 生命周期之前添加,即需要在 onInit 或 onReady 生命周期里添加。

媒体查询需要注意 this 指向的问题,如果需要访问当前实例,则应尽量使用箭头函数。实现方法如代码示例 10.12 所示。

代码示例 10.12

```
import mediaquery from '@system.mediaquery';
export default {
  onReady() {
    var mMediaQueryList = mediaquery.matchMedia('(max-width: 466)');
    function maxWidthMatch(e) {
      if (e.matches) {
        //do something
        //这里无法访问外部 this
      }
    }
    mMediaQueryList.addListener(maxWidthMatch);     //这样写无法访问外部 this
    mMediaQueryList.addListener(e =>{
        //这样访问外部 this
    })
  },
}
```

(3) MediaQueryList.removeListener(OBJECT)方法用于移除 MediaQueryList 中的回调函数,代码如下:

```
query.removeListener(minWidthMatch);
```

## 10.6 振动

通过系统接口提供的振动调用振动方法。

**注意**:地理位置的获取,需要申请权限 ohos.permission.VIBRATE。

首先需要导入@system.vibrator模块，代码如下：

```
import vibrator from '@system.vibrator';
```

vibrator.vibrate(OBJECT)方法中的振动参数如表10.3所示。

表 10.3　振动参数

| 参数名 | 类型 | 必填 | 说明 |
| --- | --- | --- | --- |
| mode | string | 否 | 振动的模式，其中 long 表示长振动，short 表示短振动，默认值为 long |

振动调用的方法如下：

```
vibrator.vibrate({
  mode: 'short',
  success: function(ret) {
    console.log('vibrate is successful');
  },
  fail: function(ret) {
    console.log('vibrate is failed');
  },
  complete: function(ret) {
    console.log('vibrate is completed');
  }
});
```

# 第 11 章 多实例管理

ArkUI JS UI 框架支持多 Ability 实例管理，不同 Ability 实例可绑定不同窗口实例，并能指定不同 JS Component 入口，运行互不影响。不同 Ability 实例的实例名互不相同，在编写应用时，开发人员需调用接口设置实例名。

## 11.1 多实例接口

多实例接口应用通过 ArkUIAbility 类中 setInstanceName()接口设置该实例的实例名称，需要在 super.onStart(Intent)方法前调用此接口。注意：多实例应用的 module.js 字段中有多个实例项，使用时需找到当前实例对应的项。

### 11.1.1 多 Ability 实例管理

实例名称与应用配置文件(config.json)中 module.js 数组下对象的 name 的值对应，若上述字段的值为{实例名称}，则需要在应用 Ability 实例的 onStart()方法中调用此接口并将实例名设置为{实例名称}。若用户未修改实例名，而使用了默认值 default，则无须调用此接口。

通过 DevEco Studio 在 src/main/js 目录创建 JS Component，如图 11.1 所示。

可以创建多个 Page Ability，通过 setInstanceName 绑定创建的 JS Component 的名字。

这样可以在一个应用中有多个 Page Ability，通过这种方式，可以实现轻应用的开发，如代码示例 11.1 所示。

代码示例 11.1

```
public class MainAbility extends AceAbility {
    @Override
    public void onStart(Intent intent) {
      //config.json 配置文件中 ability.js.name 的标签值
        setInstanceName("JSComponentName");
super.onStart(intent);
    }
}
```

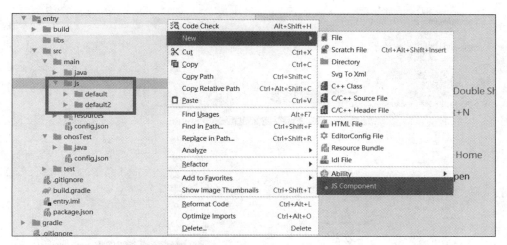

图 11.1 创建 JS Component

通过 setInstanceName 绑定创建好的 JS Component，同时可以创建多个 Page Ability。通过多个 Page Ability 绑定不同的 JS Component 可以实现把一个较大的项目分为多个小的安装项目，这个在后面的游戏场景会用到。

## 11.1.2 多 Ability 之间跳转

一个 JS Ability 管理一个 JS Component 实例，我们可以在这个 JS Ability 内通过 JS 系统模块 @system.router 方法进行组件间的导航，也可以在 JS Ability 之间进行导航跳转。

我们来看个例子。默认情况下，通过模板创建项目，只会创建一个 JS Ability 和一个 default JS Component，如图 11.2 所示。在 java 目录下的 MainAbility.java 文件中管理的 JS Component 是 js 目录下的 default 文件夹，当然可以通过 setInstanceName()方法设置其他的 JS Component 目录。

在 js 目录下新建一个 JS Component，如图 11.3 所示。右击 java 目录下的 package 名称，选择 New → Ability → Empty Page Ability (JS)创建新的 Page Ability。

此时会弹出 Ability 的配置信息，如图 11.4 所示，设置 Page Ability 名称(Page Name)、JS Component Name 及包名(Package Name)，单击 Finish 按钮创建一个新的 Page Ability。

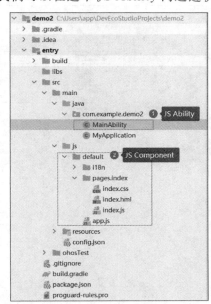

图 11.2 JS Component 目录结构

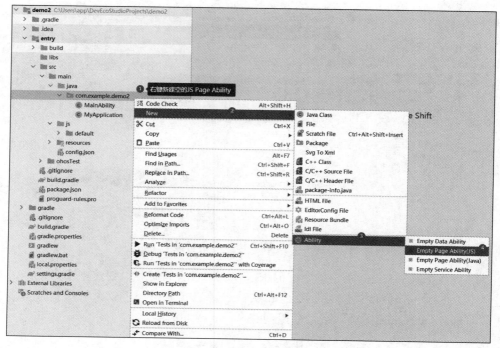

图 11.3 新建一个 JS Component

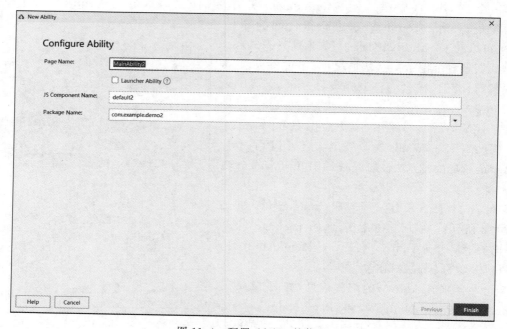

图 11.4 配置 Ability 的信息

创建好一个新的 JS Ability 后，在 js 目录下多了一个 default2 目录，在 java 目录下多了一个 MainAbility2. java 文件，如图 11.5 所示。

创建新的 Ability 后，IDE 会自动在 config. json 配置文件中添加新创建的 Page Ability 的信息，如图 11.6 所示，标注的区域为增加的配置，IDE 会自动给每个新增加的 Ability 添加对应的配置项。

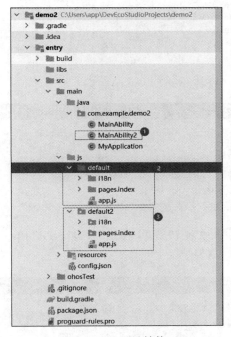

图 11.5　目录结构

```
"abilities": [
  {"name": "com.example.demo2.MainAbility"...},
  {
    "name": "com.example.demo2.MainAbility2",
    "icon": "$media:icon",
    "description": "$string:mainability2_description",
    "label": demo2,
    "type": "page",
    "launchType": "standard"
  }
],
"js": [
  {"name": "default"...},
  {
    "pages": [
      "pages/index/index"
    ],
    "name": "default2",
    "window": {
      "designWidth": 720,
      "autoDesignWidth": false
    }
  }
]
```

图 11.6　Ability 配置文件

创建完成新的 JS Page Ability 后，需要手动设置 JS Ability 对应的 JS Component 名称，打开 java 目录下的 MainAbility2. java 文件，如图 11.7 所示，在 onStart()方法中调用 setInstanceName()方法指定 JS Component 的名称。

```java
import ohos.ace.ability.AceAbility;
import ohos.aafwk.content.Intent;

public class MainAbility2 extends AceAbility {
    @Override
    public void onStart(Intent intent) {
        setInstanceName("default2");
        super.onStart(intent);
    }

    @Override
    public void onStop() { super.onStop(); }

}
```

图 11.7　设置 Ability 管理的 JS Component

接下来，修改 default 和 default2 目录下的 index.hml 页面，为页面添加单击跳转逻辑。
对 default 目录中的 index.hml 文件进行修改，如代码示例 11.2 所示。

**代码示例 11.2　chapter11/default/index.hml**

```html
<div class = "container">
<text class = "title">
    这是第 1 个 JS Component
</text>
<button type = "capsule" onclick = "changePage">跳转到第 2 个 JS Component </button>
</div>
```

对 default2 目录中的 index.hml 文件进行修改，如代码示例 11.3 所示。

**代码示例 11.3　chapter11/default2/index.hml**

```html
<div class = "container">
<text class = "title">
    这是第 2 个 JS Component
</text>
<button type = "capsule" onclick = "changePage">跳转到第 1 个 JS Component </button>
</div>
```

接下来修改 default 目录下的 index.js 文件，添加跳转逻辑，使用 FeatureAbility.startAbility()方法启动指定的 Ability，如代码示例 11.4 所示。

**代码示例 11.4　chapter11/default/index.js**

```js
export default {
    data: {
        title: ""
    },
    onInit() {

    },
    changePage:async function(){
        let actionData = {
            uri: 'www.huawei.com'
        };
        let target = {
            bundleName: "com.example.demo2",
            abilityName: "com.example.demo2.MainAbility2",
            data: actionData
        };

        let result = await FeatureAbility.startAbility(target);
        let ret = JSON.parse(result);
        if (ret.code == 0) {
            console.log('success');
        } else {
```

```
            console.log('cannot start browing service, reason: ' + ret.data);
        }
    }
}
```

修改 default2 目录下的 index.js 文件，添加跳转逻辑，使用 FeatureAbility.startAbility()方法启动指定的 Ability，如代码示例 11.5 所示。

代码示例 11.5　chapter11/default2/index.js

```
export default {
    data: {
        title: ""
    },
    onInit() {
    },
    changePage:async function(){
        let actionData = {
            uri: 'www.huawei.com'
        };
        let target = {
            bundleName: "com.example.demo2",
            abilityName: "com.example.demo2.MainAbility",
            data: actionData
        };

        let result = await FeatureAbility.startAbility(target);
        let ret = JSON.parse(result);
        if (ret.code == 0) {
            console.log('success');
        } else {
            console.log('cannot start browing service, reason: ' + ret.data);
        }
    }
}
```

通过上面的代码就实现了多 JS Ability 的启动跳转。

## 11.2　使用 NPM 安装 JavaScript 模块

在鸿蒙 JS 项目工程目录里可以通过 NPM 安装第三方 Node.js 模块，但是必须是纯 JavaScript 编写的功能库，鸿蒙应用最终通过 C++中定义的组件模块调用执行，因此不能有和浏览器相关的对象，但是一些通用的功能库是可以安装使用的。

这里介绍两个第三方 NPM 库在鸿蒙项目中的用法。

（1）Moment.js JavaScript 日期处理类库

进入 entry 目录,该目录下包含 package.json 文件,在命令行执行的命令如代码示例 11.6 所示。

**代码示例 11.6**

```
npm install moment -- save
#在页面的 JS 代码中通过模块引入
import moment from "moment"          //moment 不需要加路径
```

这样就可以使用 moment 中的方法了,如代码示例 11.7 所示。

**代码示例 11.7**

```
moment().format('MMMM Do YYYY, h:mm:ss a');    //十二月 22 日 2020, 7:39:35 早上
moment().format('dddd');                        //星期二
moment().format("MMM Do YY");                   //12 月 22 日 20
moment().format('YYYY [escaped] YYYY');         //2020 escaped 2020
moment().format();                              //2020-12-22T07:39:35+08:00
```

(2) QRcode.js 文件用于生成二维码,安装代码的命令如代码示例 11.8 所示。

**代码示例 11.8**

```
npm install -- save qrcode
import QRCode from 'qrcode'
```

代码实现如代码示例 11.9 所示。

**代码示例 11.9**

```
var opts = {
    errorCorrectionLevel: 'H',
    type: 'image/jpeg',
    quality: 0.3,
    margin: 1,
color: {
        dark:"#010599FF",
        light:"#FFBF60FF"
    }
}
    QRCode.toCanvas(this.$element("canvas"),
'51itcto.com',opts, function (error) {
        if (error) console.error(error)
      console.log('success!');
})
}
```

**注意**:NPM 安装第三方库,并不是所有的 JS 库都支持,用之前应先测试。

# 第四篇　ArkUI ETS UI篇

2021年，上亿台"富设备"接入鸿蒙操作系统。所谓"富设备"，指的是使用频率高、强交互的设备，例如平板、投影仪、电视机等。

ArkUI ETS UI是专门为鸿蒙"富设备"研发的应用开发框架，相比于ArkUI JS(1.0)框架，ArkUI ETS(2.0)采用了 TypeScript 作为开发语言，ETS(Extended TS)基于标准的 TypeScript 为声明式开发范式做了一些扩展，例如装饰器，尾随闭包等，不过本质上还是 TypeScript，TypeScript 适合大型项目开发，同时在性能方面也有很大提升。

# 第 12 章 ArkUI ETS 开发语言入门

与 ArkUI JS 比较,ArkUI ETS(Extended TS 在 TypeScript 基础上做了扩展)采用 TypeScript 语言开发,TypeScript 语言是专门用来开发复杂 JavaScript 程序的一门预编译脚本语言,TypeScript 是微软开发和维护的一种面向对象的编程语言。它是 JavaScript 的超集,包含了 JavaScript 的所有元素,可以载入 JavaScript 代码运行,并扩展了 JavaScript 的语法。TypeScript 与 JavaScript 的关系如图 12.1 所示。

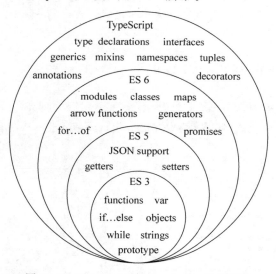

图 12.1 TypeScript 与 JavaScript 的关系图

TypeScript 起源于用 JavaScript 开发的大型项目。由于 JavaScript 语言本身的局限性,难以胜任和维护大型项目开发,因此微软开发了 TypeScript,使其能够胜任开发大型项目。

TypeScript 代码不可以直接运行,最终执行需要通过 TypeScript 编译器或 Babel 转译为 JavaScript 代码,如图 12.2 所示。

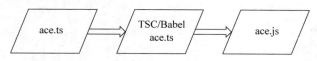

图 12.2 TypeScript 编译器编译 TypeScript 的过程

## 12.1 ArkUI TypeScript 介绍

ArkUI ETS 工程项目的文件使用 ETS 扩展名，如图 12.3 所示。ETS 文件通过 Webpack 中的 babel 插件(ets-loader)编译成原生 JavaScript bundle 文件执行。

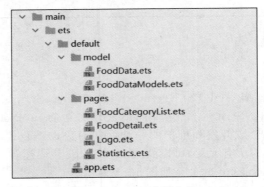

图 12.3 ETS 代码文件

## 12.2 ets-loader 编译 ETS

ets-loader 的作用是为 Webpack 提供用来编译 ETS 脚本文件的转换器(loader)。ets-loader 的位置：\Sdk\ets\(版本号)\build-tools\ets-loader\Webpack.config.js，转换规则如图 12.4 所示。

```
rules: [
  {
    test: /\.(ets|ts)$/,
    use: [
      { loader: path.resolve(__dirname, 'lib/result_process.js') },
      {
        loader: 'ts-loader',
        options: {
          appendTsSuffixTo: [/\.ets$/],
          onlyCompileBundledFiles: true,
          configFile: path.resolve(__dirname, 'tsconfig.json'),
          getCustomTransformers(program) {
            return {
              before: [processUISyntax(program)],
              after: []
            };
          },
          ignoreDiagnostics: IGNORE_ERROR_CODE
        }
      },
      { loader: path.resolve(__dirname, 'lib/pre_process.js') }
    ]
  },
```

图 12.4 ets-loader 编译 ETS 脚本

## 12.3 TypeScript 基础数据类型

TypeScript 支持数据类型声明。TypeScript 编译器在代码编写过程中,帮助开发者检查类型或者语法错误,TypeScript 支持与 JavaScript 几乎相同的数据类型,此外还提供了实用的枚举、元组等类型方便开发者使用。

### 12.3.1 布尔值

最基本的数据类型就是简单的 true/false 值,在 JavaScript 和 TypeScript 里叫作 boolean,布尔类型的定义如代码示例 12.1 所示。

代码示例 12.1

```
let isDone: boolean = false;
```

### 12.3.2 数字

和 JavaScript 一样,TypeScript 里的所有数字都是浮点数。这些浮点数的类型是 number。除了支持十进制和十六进制字面量,TypeScript 还支持 ECMAScript 2015 中引入的二进制和八进制字面量,如代码示例 12.2 所示。

代码示例 12.2

```
let decLiteral: number = 6;
let hexLiteral: number = 0xf00d;
let binaryLiteral: number = 0b1010;
let octalLiteral: number = 0o744;
```

### 12.3.3 字符串

JavaScript 程序的另一项基本操作是处理网页或服务器端的文本数据。像其他语言里的文本数据类型一样,我们使用 string 表示文本数据类型。和 JavaScript 一样,可以使用双引号(")或单引号(')表示字符串,如代码示例 12.3 所示。

代码示例 12.3

```
let name: string = "harmonyos";
name = "openharmony";
```

还可以使用模板字符串,它可以定义多行文本和内嵌表达式。这种字符串被反引号包围(`),并且以 ${ expr } 这种形式嵌入表达式,如代码示例 12.4 所示。

代码示例 12.4

```
let name: string = `Gene`;
```

```
let age: number = 37;
let sentence: string = `Hello, my name is ${ name }.
I'll be ${ age + 1 } years old next month.`;
```

这与下面定义 sentence 的效果相同，如代码示例 12.5 所示。

**代码示例 12.5**

```
let sentence: string = "Hello, my name is " + name + ".\n\n" +
"I'll be " + (age + 1) + " years old next month.";
```

### 12.3.4　数组

TypeScript 像 JavaScript 一样可以操作数组元素。有两种方式可以定义数组。第一种，可以在元素类型后面接上 [ ]，表示由此类型元素组成的一个数组，如代码示例 12.6 所示。

**代码示例 12.6**

```
let list: number[] = [1, 2, 3];
```

第二种方式是使用数组泛型，Array<元素类型>，如代码示例 12.7 所示。

**代码示例 12.7**

```
let list: Array<number> = [1, 2, 3];
```

### 12.3.5　元组

元组类型允许表示一个已知元素数量和类型的数组，各元素的类型不必相同。例如，可以定义一对值分别为 string 和 number 类型的元组，如代码示例 12.8 所示。

**代码示例 12.8**

```
//Declare a tuple type
let x: [string, number];
//Initialize it
x = ['hello', 10];         //OK
//Initialize it incorrectly
x = [10, 'hello'];         //Error
```

当访问一个已知索引的元素时，会得到正确的类型，如代码示例 12.9 所示。

**代码示例 12.9**

```
console.log(x[0].substr(1));     //OK
console.log(x[1].substr(1));     //Error, 'number' does not have 'substr'
```

当访问一个越界的元素时，会使用联合类型替代，如代码示例 12.10 所示。

### 代码示例 12.10

```
x[3] = 'world';              //OK,字符串可以赋值给(string | number)类型
console.log(x[5].toString()); //OK,'string'和'number'都有toString()方法
x[6] = true;                 //Error,布尔不是(string | number)类型
```

## 12.3.6 枚举

enum 类型是对 JavaScript 标准数据类型的一个补充。像 C♯ 等其他语言一样,使用枚举类型可以为一组数值赋予友好的名字,如代码示例 12.11 所示。

### 代码示例 12.11

```
enum Color {Red, Green, Blue}
let c: Color = Color.Green;
```

默认情况下,从 0 开始为元素编号。也可以手动指定成员的数值。例如,将上面的例子改成从 1 开始编号,如代码示例 12.12 所示。

### 代码示例 12.12

```
enum Color {Red = 1, Green, Blue}
let c: Color = Color.Green;
```

或者,全部采用手动赋值,如代码示例 12.13 所示。

### 代码示例 12.13

```
enum Color {Red = 1, Green = 2, Blue = 4}
let c: Color = Color.Green;
```

枚举类型提供的一个便利是可以由枚举的值得到它的名字。例如,知道数值为 2,但是不确定它映射到 Color 里的哪个名字,此时可以查找相应的名字,如代码示例 12.14 所示。

### 代码示例 12.14

```
enum Color {Red = 1, Green, Blue}
let colorName: string = Color[2];
console.log(colorName);        //显示'Green',因为在上面的代码里它的值是 2
```

## 12.3.7 any

有时候,我们会为那些在编程阶段还不清楚类型的变量指定一种类型。这些值可能来自于动态的内容,例如来自用户输入或第三方代码库。这种情况下,我们不希望类型检查器对这些值进行检查而是直接让它们通过编译阶段的检查。此时可以使用 any 类型来标记这些变量,如代码示例 12.15 所示。

代码示例 12.15

```
let notSure: any = 4;
notSure = "maybe a string instead";
notSure = false;              //okay, definitely a boolean
```

在对现有代码进行改写的时候，any 类型是十分有用的，它允许在编译时可选择地包含或移除类型检查。可能会认为 Object 有相似的作用，就像它在其他语言中那样。但是 Object 类型的变量只允许给它赋任意值，而不能够在它上面调用任意的方法，即便它真的有这些方法，如代码示例 12.16 所示。

代码示例 12.16

```
let notSure: any = 4;
notSure.ifItExists();         //okay, ifItExists might exist at runtime
notSure.toFixed();            //okay, toFixed exists (but the compiler doesn't check)

let prettySure: Object = 4;
prettySure.toFixed();         //Error: Property 'toFixed' doesn't exist on type 'Object'
```

当只知道一部分数据的类型时，any 类型也是有用的。例如，有一个数组，它包含了不同的类型的数据，如代码示例 12.17 所示。

代码示例 12.17

```
let list: any[] = [1, true, "free"];
list[1] = 100;
```

### 12.3.8 void

从某种程度上来讲，void 类型像是与 any 类型相反，它表示没有任何类型。当一个函数没有返回值时，通常会见到其返回值的类型是 void，如代码示例 12.18 所示。

代码示例 12.18

```
function warnUser(): void {
    console.log("This is my warning message");
}
```

声明一个 void 类型的变量没有什么大用，因为只能为它赋予 undefined 和 null，如代码示例 12.19 所示。

代码示例 12.19

```
let unusable: void = undefined;
```

### 12.3.9 null 和 undefined

在 TypeScript 里，undefined 和 null 两者各自有自己的类型，分别叫作 undefined 和

null。和 void 相似,它们本身类型的用处不是很大,如代码示例 12.20 所示。

代码示例 12.20
```
//Not much else we can assign to these variables!
let u: undefined = undefined;
let n: null = null;
```

默认情况下 null 和 undefined 是所有类型的子类型。也就是说可以把 null 和 undefined 赋值给 number 类型的变量。

当指定了--strictNullChecks 标记时,null 和 undefined 只能赋值给 void 和它们自身。这能避免很多常见的问题。如果想传入一个 string、null 或 undefined,则可以使用联合类型 string | null | undefined。

注意:应尽可能地使用--strictNullChecks,但在本书中我们假设这个标记是关闭的。

### 12.3.10 never

never 类型表示的是那些永不存在的值的类型。例如,never 类型是那些总是会抛出异常或根本就不会有返回值的函数表达式或箭头函数表达式的返回值的类型,变量也可能是 never 类型,当它们被永不为真的类型保护所约束时。

never 类型是任何类型的子类型,也可以赋值给任何类型,然而,没有类型是 never 的子类型或可以赋值给 never 类型(除了 never 本身之外)。即使 any 也不可以赋值给 never。

下面是一些返回 never 类型的函数,如代码示例 12.21 所示。

代码示例 12.21
```
//返回 never 的函数必须存在无法达到的终点
function error(message: string): never {
    throw new Error(message);
}
//推断的返回值类型为 never
function fail() {
    return error("Something failed");
}
//返回 never 的函数必须存在无法达到的终点
function infiniteLoop(): never {
    while (true) {
    }
}
```

## 12.4 TypeScript 高级数据类型

除了上面的基础数据类型外,TypeScript 中还可以使用一些高级数据类型,如泛型、交叉类型、联合类型。

### 12.4.1 泛型

TypeScript 中引入了 C# 中的泛型,泛型解决类、接口、方法的复用性及对不特定数据类型的支持。

**1. 泛型类**

泛型类可以支持不特定的数据类型,要求传入的参数和返回的参数必须一致,T 表示泛型,具体是什么类型在调用这种方法的时候决定,如代码示例 12.22 所示。

代码示例 12.22

```
//类的泛型
class MyClas <T>{
    public list: T[] = [];
    add(value: T): void {
        this.list.push(value);
    }
    min(): T {
        var minNum = this.list[0];
        for (var i = 0; i < this.list.length; i++) {
            if (minNum > this.list[i]) {
                minNum = this.list[i];
            }
        }
        return minNum;
    }
}
//实例化类并且制定了类的 T,代表的类型是 number
var m1 = new MyClas <number>();
m1.add(1);
m1.add(2);
m1.add(3);
console.log(m1.min());
//实例化类并且制定了类的 T,代表的类型是 string
var m2 = new MyClas <string>();
m2.add('a');
m2.add('b');
m2.add('c');
console.log(m2.min());
```

**2. 泛型接口**

如代码示例 12.23 所示。

代码示例 12.23

```
//泛型接口
interface IConfigFn<T>{
    (value: T): T;
}

function getData<T>(value: T): T{
    return value;
}

var myData: IConfigFn<string> = getData;
console.log(myData('20'));
```

### 3. 定义操作数据库的泛型类

通过泛型类可以定义一个操作数据库的库，支持 MySQL、MsSQL、MongoDB。要求 MySQL、MsSQL、MongoDB 功能一样，都有 add()、update()、delete()、get() 方法，如代码示例 12.24 所示。

代码示例 12.24

```
//定义操作数据库的泛型类
class MysqlAccess<T>{
    add(info: T): boolean {
        console.log(info);
        return true;
    }
}
class MongoAccess<T>{
    add(info: T): boolean {
        console.log(info);
        return true;
    }
}
//想给User表增加数据,定义一个User类和数据库进行映射
class User {
    username: string | undefined;
    pasword: string | undefined;
}
var user = new User();
user.username = "张三";
user.pasword = "123456";
var md1 = new MysqlAccess<User>();
md1.add(user);

//想给Article增加数据,定义一个Article类和数据库进行映射
class Article {
    title: string | undefined;
    desc: string | undefined;
```

```
        status: number | undefined;
        constructor(params: {
            title: string | undefined,
            desc: string | undefined,
            status?: number | undefined
        }) {
            this.title = params.title;
            this.desc = params.desc;
            this.status = params.status;
        }
}

var article = new Article({
    title: "这是文章标题",
    desc: "这是文章描述",
    status: 1
});
var md2 = new MongoAccess<Article>();
md2.add(article);
```

## 12.4.2 交叉类型

交叉类型是将多种类型合并为一种类型。这让我们可以把现有的多种类型叠加到一起成为一种类型，它包含了所需的所有类型的特性。

交叉类型包含 A 的特点，也包含 B 的特点，用伪代码表示就是 A&B。

下面定义了两种类型：Person 和 Student，变量 student 的类型是 Person 和 Student 的交叉类型，student 的类型必须满足两种类型的交叉组合体要求，如代码示例 12.25 所示。

**代码示例 12.25**

```
interface Person {
    name: string
    age: number
}
interface Student {
    school: string
}
const student: Person & Student = {
    name: 'Gavin',
    age: 26,
    school: '清华大学',
}
```

同时 Person & Student 可以使用类型别名，在下面的代码中定义的 StudentInfo 就是 Person&Student 的类型别名，如代码示例 12.26 所示。

代码示例 12.26

```
interface Person {
    name: string
    age: number
}
interface Student {
    school: string
}
type StudentInfo = Person & Student
const student: StudentInfo = {
    name: 'Gavin',
    age: 26,
    school: '清华大学',
}
```

### 12.4.3 联合类型

联合类型，它的类型既可以是 A，也可以是 B，用伪代码表示就是 A|B，如代码示例 12.27 所示。

代码示例 12.27

```
var type:string | number | boolean = '1'
type = 12;
type = true;
```

上面的 type 的类型就是 number 和 boolean 的联合类型，type 的值是这两种类型中的一种，下面的定义是一种字面量类型组合成的一个新的联合类型，如代码示例 12.28 所示。

代码示例 12.28

```
type WorkDays = 1 | 2 | 3 | 4 | 5;

let day: WorkDays = 1;              //ok
day = 5;                            //ok
day = 6;                            //error: Type '6' is not assignable to type WorkDays.
```

字面量联合类型的形式与枚举类型有些类似，所以，如果仅使用数字，则可以考虑是否使用具有表达性的枚举类型。

## 12.5 TypeScript 面向对象特性

TypeScript 增加了类似于 C♯ 语言的面向对象编程，提供了类、接口、抽象类、泛型的支持。

## 12.5.1 类

JavaScript 编程更多还是面向函数编程,在面向对象编程方面支持较弱,虽然在 ES6 后提供了类似 C♯或者 Java 的面向对象编程的特征,但是与 C♯或者 Java 中的面向对象特征差距较大,因此 TypeScript 在语法中加入了完整的面向对象的支持,让熟悉面向对象的开发者可以通过 TypeScript 实现最终的 JavaScript 面向对象的编程体验。

类是对业务领域对象的抽象,类是一张蓝图或一个原型,它定义了特定一类对象共有的变量/属性和方法/函数,对象是面向对象编程中基本的运行实体,类与对象的关系图如图 12.5 所示。

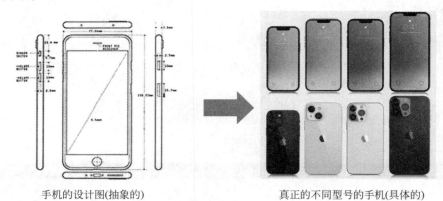

手机的设计图(抽象的)　　　　　　真正的不同型号的手机(具体的)

图 12.5　类与对象的关系

**1. 定义类**

TypeScript 定义类的方式和 ES6 定义类的方式是一样的,在下面的代码中属性和方法前面比 ES6 的类多了一个访问修饰符：private,这里表示该属性和方法是私有的,如代码示例 12.29 所示。

代码示例 12.29

```
class Phone {
    private brandName:string;
    private cpu: string;
    private width : number;
    private height: number;

    constructor(brandName:string,width:number,height:number){
        this.brandName = brandName;
        this.width = width;
        this.height = height;
    }
    private takeCall():void{
        console.log("打电话给……")
    }
}
```

## 2. 访问修饰符

TypeScript 和 Java 类似，可以为类中的属性和方法添加访问修饰符，TypeScript 可以使用 3 种访问修饰符(Access Modifiers)，分别是 public、protected 和 private。

(1) public：公有类型，在当前类里面、子类、类外面都可以访问。

(2) protected：保护类型，在当前类里面、子类里面可以访问，在类外部没法访问。

(3) private：私有类型，在当前类里面可以访问，子类、类外部都没法访问。

注意：如果属性不加修饰符，则默认为公有(public)。

## 3. 存取器

TypeScript 支持通过 getters/setters 来截取对对象成员的访问。它能帮助用户有效地控制对对象成员的访问。在下面的例子中对成员变量 fullName 的访问是通过存储器访问的，可以在 set()方法中添加与权限相关的逻辑来控制对内部成员变量的访问，如代码示例12.30 所示。

代码示例 12.30

```
let passcode = 'password';
class Employee {
    private _fullName: string;
    get fullName(): string {
        return this._fullName;
    }
    set fullName(name: string) {
        if (passcode && passcode === 'password') {
            this._fullName = name;
        } else {
            console.log('授权失败');
        }
    }
}
let employee = new Employee();
employee.fullName = "Gavin Xu";
if (employee.fullName) {
    console.log(employee.fullName)
}
```

## 4. 类的继承

在 TypeScript 中要想实现继承可使用 extends 关键字，只要一旦实现了继承关系，子类中就拥有了父类的属性和方法，而在执行方法过程中，首先从子类开始寻找，如果有，就使用。如果没有，就去父类中寻找。类的继承只能单向继承，如代码示例 12.31 所示。

代码示例 12.31

```
class Person {
    name: string;                    //父类属性，前面省略了 public 关键词
```

```
    constructor(n: string) {         //构造函数,实例化父类的时候触发的方法
        this.name = n;               //使用 this 关键字为当前类的 name 属性赋值
    }
    run(): void {                    //父类方法
        console.log(this.name + "在跑步");
    }
}
class Chinese extends Person {
    age: number;                     //子类属性

    constructor(n: string, a: number) {//构造函数,实例化子类的时候触发的方法
        super(n);                    //使用 super 关键字调用父类中的构造方法
        this.age = a;                //使用 this 关键字为当前类的 age 属性赋值
    }

    speak(): void {                  //子类方法
        super.run();                 //使用 super 关键字调用父类中的方法
        console.log(this.name + "说中文");
    }
}
var c = new Chinese("张三", 28);
c.speak();
```

### 5. 抽象类

TypeScript 中的抽象类：它是提供其他类继承的基类,不能直接被实例化。

用 abstract 关键字定义抽象类和抽象方法,抽象类中的抽象方法不包含具体实现,并且必须在派生类(也就是其子类)中实现,abstract 抽象方法只能放在抽象类里面。

我们常常使用抽象类和抽象方法来定义标准,如代码示例 12.32 所示。

**代码示例 12.32**

```
//动物抽象类,所有动物都会跑(假设),但是吃的东西不一样,所以把吃的方法定义成抽象方法
abstract class Animal {
    name: string;
    constructor(name: string) {
        this.name = name;
    }
    abstract eat(): any;             //抽象方法不包含具体实现,并且必须在派生类中实现
    run() {
        console.log(this.name + "会跑")
    }
}

class Dog extends Animal {
    constructor(name: string) {
        super(name);
    }
    eat(): any {                     //抽象类的子类必须实现抽象类里面的抽象方法
```

```
            console.log(this.name + "啃骨头");
        }
    }

    var d: Dog = new Dog("小狗");d. eat();

    class Cat extends Animal {
        constructor(name: string) {
            super(name);
        }
        eat(): any {              //抽象类的子类必须实现抽象类里面的抽象方法
            console.log(this.name + "吃老鼠");
        }
    }

    var c: Cat = new Cat("小猫");
    c.eat();
```

### 12.5.2 接口

在面向对象的编程中,接口是一种规范的定义,它定义了行为和动作的规范,在程序设计里,接口起到一种限制和规范的作用,编程接口和计算机的各种接口的作用类似,接口定义好后,插头必须完全满足接口标准,这样才可以连接,如图 12.6 所示。

如您的设备接口是标准HDMI接口
本产品适用于您的设备

一般用于MP4、平板电脑、相机等

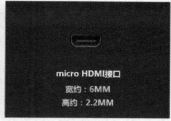
一般用于智能手机、平板电脑等

图 12.6 接口的作用是定义规范

接口定义了某一组类所需要遵守的规范,接口不关心这些类的内部状态数据,也不关心这些类里方法的实现细节,它只规定这批类里必须提供某些方法,提供这些方法的类就可以满足实际需要了。TypeScript 中的接口类似于 C♯ 和 Java 语言中的接口概念,同时还增加了更灵活的接口类型,包括属性、函数、可索引和类等。

在 TypeScript 中定义函数形参{ x,y },如代码示例 12.33 所示。

**代码示例 12.33**

```
function sum ({x, y}: { x: number, y: number}): number {
  return x + y;
}
```

当参数对象的属性比较多时，代码就非常不适合阅读了，这个时候可以使用接口来定义参数的类型，如代码示例12.34所示。

代码示例 12.34

```
interface ISum {
  x: number;
  y: number;
}
function sum ({ x, y }: ISum): number {
  return x + y;
}
```

上面的代码使用接口定义后，可读性得到很好的增强，这就是使用接口带来的好处。除此之外，接口的定义上有非常丰富的用法，下面详细介绍。

1. 可选属性

接口中的属性或者方法可以标记为可选的，和C♯中的可选属性一致，在一个属性后面跟着一个问号（?），用于标记这个属性为可选的，如代码示例12.35所示。

代码示例 12.35

```
interface ISum{
  x: number;
  y?: number;
}
ISum({ x: 0 });
```

2. readonly 属性

接口中的属性可以添加只读标记readonly，添加只读属性后，该属性不可以再赋值了，如代码示例12.36所示。

代码示例 12.36

```
interface IReadonlySum {
  readonly x: number;
  readonly y: number;
}
let p: IReadonlySum = { x: 0, y: 1};
//p.x = 1;
```

如果一旦赋值，则编译器将提示一个错误。

3. 属性检查

接口的作用：限制接口实现对象严格按照接口中定义的规则进行赋值，所以使用接口可以帮助开发者进行属性检查，如代码示例12.37所示。

代码示例 12.37

```
interface ISum{
```

```
    x: number;
    y: number;
}
function create(config: ISum): void {
}
create({ z: 0, x: 0, y: 1 } as ISum)
```

在 JavaScript 中这段代码并不会有错,因为对于对象来讲当传进一个未知的属性时并不是错误,虽然可能会引发潜在的 Bug,但这在 TypeScript 中,这个错误是非常明显的,编译器并不会通过此次编译,除非显式地使用类型断言。

**4. 接口继承**

接口可以继承其他接口,与类的继承使用了相同的关键字,同时支持多重继承,如代码示例 12.38 所示。

代码示例 12.38
```
interface Shape {
    color: string;
}
interface Stroke {
    width: number;
}
interface Square extends Shape, Stroke {
    length: number;
}
var square = <Square>{};
square.color = "blue";
square.length = 10;
square.width = 5.0;
```

在上面的代码中,变量 square 并不是实现了该接口的类,所以不能使用 new 实现,而应使用< Square >{}的写法来创建。

**5. 函数类型**

接口能够描述 JavaScript 中对象拥有的各种各样的外形。除了可以描述带有属性的普通对象外,接口也可以描述函数类型。

为了使用接口表示函数类型,我们需要给接口定义一个调用签名。它就像是一个只有参数列表和返回值类型的函数定义。参数列表里的每个参数都需要名字和类型,如代码示例 12.39 所示。

代码示例 12.39
```
interface IInfo {
  (name: string, age: number): string;
}
let getName1: IInfo = function(name: string, age: number): string {
```

```
    return `${name} ---- ${age}`;
};
console.log(getName1("me", 50)); //me ---- 50
```

#### 6. 索引类型

可索引类型具有一个索引签名,它描述了对象索引的类型,还有相应的索引返回值类型,如代码示例12.40所示。

**代码示例 12.40**

```
interface SomeArray {
    [index: number]: string;
}
let someArray: SomeArray;
someArray = ["string1", "string2"];
let str: string = someArray[0];
console.log(str);
```

#### 7. 类实现(implements)接口

与Java或C♯中的接口规则一致,TypeScript能够用实现(implements)来明确地强制一个类去符合某种契约,如代码示例12.41所示。

**代码示例 12.41**

```
interface Animal {
  name: string;
  eat():void;
}
class Cat implements Animal{
  name: string;
  constructor(name:string){
    this.name = name;
  }
  eat():void{
    console.log(`${this.name}在吃鱼`)
  }
}
class Dog implements Animal{
  name: string;
  constructor(name:string){
    this.name = name;
  }
  eat():void{
    console.log(`${this.name}在啃骨头`)
  }
}
let c = new Cat("小花猫");
c.eat();            //小花猫在吃鱼
```

```
let d = new Dog("小狗");
d.eat();                  //小狗在啃骨头
```

## 12.6 TypeScript 装饰器

装饰器是一种特殊类型的声明,它能够被附加到类、方法、属性或参数上,可以修改类的行为,通俗地讲装饰器就是一种方法,可以注入类、方法、属性或参数上来扩展类、方法、属性或参数的功能。常见的装饰器有类装饰器、方法装饰器、属性装饰器、参数装饰器。

装饰器的写法:普通装饰器(无法传参)、装饰器工厂(可传参),装饰器是 ES7 的标准特性之一。

装饰器执行顺序:属性→方法→方法参数→类。

### 12.6.1 属性装饰器

属性装饰器会被应用到属性描述上,可以用来监视、修改或者替换属性的值。

属性装饰器会在运行时传入下列两个参数:

(1) 对于静态成员来讲是类的构造函数,对于实例成员来讲是类的原型对象。

(2) 成员的名字。

如代码示例 12.42 所示。

**代码示例 12.42**

```
//属性装饰器
function log(params: any) {          //params 就是当前类传递进来的参数
    return function (target: any, attr: any) {
        console.log(target);
        console.log(attr);
        target[attr] = params;
    }
}

class HttpTool {
    @log("http://www.baidu.com")
    public url: any | undefined;

    getData() {
        console.log(this.url);
    }
}

var http = new HttpTool();
http.getData();
```

## 12.6.2　方法装饰器

方法装饰器会被应用到方法描述上,可以用来监视、修改或者替换方法定义。

方法装饰器会在运行时传入下列 3 个参数:

(1) 对于静态成员来讲是类的构造函数,对于实例成员来讲是类的原型对象。

(2) 成员的名字。

(3) 成员的属性描述符。

如代码示例 12.43 所示。

代码示例 12.43

```javascript
function get(params: any) {           //params 就是当前类传递进来的参数
    return function (target: any, methodName: any, desc: any) {
        console.log(target);
        console.log(methodName);
        console.log(desc);
        target.apiUrl = params;
        target.run = function () {
            console.log("run");
        }
    }
}

class HttpTool{
    public url: any | undefined;
    constructor() {
    }
    @get("http://www.harmonyos-ui.com")
    getData() {
        console.log(this.url);
    }
}

var http: any = new HttpTool();
console.log(http.apiUrl);
http.run();
```

## 12.6.3　参数装饰器

参数装饰器表达式会在运行时当作函数被调用,可以使用参数装饰器为类的原型增加一些元素数据,传入下列 3 个参数:

(1) 对于静态成员来讲是类的构造函数,对于实例成员来讲是类的原型对象。

(2) 方法的名字。

(3) 参数在函数参数列表中的索引。

如代码示例12.44所示。

**代码示例 12.44**

```
function logParams(params: any) {
    return function (target: any, methodName: any, paramsIndex: any) {
        console.log(target);
        console.log(methodName);
        console.log(paramsIndex);
        target.apiUrl = params;
    }
}

class HttpTool{
    getData(@logParams("1000") uuid: any) {
        console.log(uuid);
    }
}

var http: any = new HttpTool();
http.getData(123);
console.log(http.apiUrl);
```

### 12.6.4 类装饰器

类装饰器：普通装饰器（无法传参），如代码示例12.45所示。

**代码示例 12.45**

```
function logClass(params: any) {
    console.log(params);          //params 就是当前类
    params.prototype.apiUrl = "apiUrl 是动态扩展的属性";
    params.prototype.run = function () {
        console.log("run 是动态扩展的方法");
    }
}

@logClass
class HttpTool{

}

var http: any = new HttpTool();
console.log(http.apiUrl);
http.run();
```

类装饰器：装饰器工厂（可传参），如代码示例12.46所示。

**代码示例 12.46**

```
function logClass(params: string) {
    return function (target: any) {
```

```
                console.log(target);              //target 就是当前类
                console.log(params);              //params 就是当前类传递进来的参数
                target.prototype.apiUrl = params;
        }
}
@logClass("http://www.harmonyos-ui.com")
class HttpTool{
}
var http: any = new HttpTool();
console.log(http.apiUrl);
```

## 12.7 TypeScript 模块与命名空间

对于大型项目开发来讲,会涉及如何组织和管理代码,TypeScript 使用模块和命名空间来组织代码。

### 12.7.1 模块

模块化是指将一个大的程序文件,拆分成许多小的文件,然后将小文件组合起来。模块化的好处是防止命名冲突、代码可复用及高维护性。

**1. 模块化的语法**

模块功能主要由两个命令构成:export 和 import。

(1) export 命令用于规定模块的对外接口。

(2) import 命令用于输入由其他模块提供的功能。

**2. 模块化的暴露**

方式一:分别暴露,如代码示例 12.47 所示。

**代码示例 12.47**

```
//方式一:分别暴露
export let school = "清华大学";
export function study() {
    console.log("学习鸿蒙 ArkUI!");
}
```

方式二:统一暴露,如代码示例 12.48 所示。

**代码示例 12.48**

```
let school = "清华大学";
function search() {
```

```
        console.log("研究技术");
    }
    export {school, search};
```

方式三：默认暴露，如代码示例12.49所示。

**代码示例 12.49**
```
export default {
    school: "清华大学",
    play: function () {
        console.log("我正在play game!");
    }
}
```

### 3. 模块化的导入

模块导入的方式与ES6中模块导入的方式相同，如代码示例12.50所示。

**代码示例 12.50**
```
//引入 m1.js 模块内容
import * as m1 from "./model/m1";
//引入 m2.js 模块内容
import * as m2 from "./model/m2";
//引入 m3.js 模块内容
import * as m3 from "./model/m3";

m1.study();
m2.search();
m3.default.play();
```

### 4. 解构赋值形式

在导入模块时，通过解构赋值的方式获取对象，如代码示例12.51所示。

**代码示例 12.51**
```
//引入 m1.js 模块内容
import {school, study} from "./model/m1";
//引入 m2.js 模块内容
import {school as s, search} from "./model/m2";
//引入 m3.js 模块内容
import {default as m3} from "./model/m3";

console.log(school);
study();

console.log(s);
search();
```

```
console.log(m3);
m3.play();
```

**注意**：针对默认暴露还可以直接通过代码 import m3 from "./model/m3" 实现。

### 12.7.2 命名空间

命名空间：在代码量较大的情况下，为了避免各种变量命名相冲突，可将相似功能的函数、类、接口等放置到命名空间内，同 Java 的包、.Net 的命名空间一样，TypeScript 的命名空间可以将代码包裹起来，只对外暴露需要在外部访问的对象，命名空间内的对象通过 export 关键字对外暴露。

命名空间和模块的区别如下。

(1) 命名空间：内部模块，主要用于组织代码，避免命名冲突。

(2) 模块：ts 的外部模块的简称，侧重代码的复用，一个模块里可能会有多个命名空间。

如代码示例 12.52 所示。

**代码示例 12.52**

```
namespace A {
    interface Animal {
        name: string;
        eat(): void;
    }
    export class Dog implements Animal {
        name: string;
        constructor(theName: string) {
            this.name = theName;
        }
        eat(): void {
            console.log(`${this.name} 吃狗粮.`);
        }
    }
    export class Cat implements Animal {
        name: string;
        constructor(theName: string) {
            this.name = theName;
        }
        eat(): void {
            console.log(`${this.name} 吃猫粮.`);
        }
    }
}

namespace B {
    interface Animal {
```

```typescript
        name: string;
        eat(): void;
    }
    export class Dog implements Animal {
        name: string;
        constructor(theName: string) {
            this.name = theName;
        }
        eat(): void {
            console.log(`${this.name} 吃狗粮.`);
        }
    }
    export class Cat implements Animal {
        name: string;
        constructor(theName: string) {
            this.name = theName;
        }
        eat(): void {
            console.log(`${this.name} 吃猫粮.`);
        }
    }
}

var cat = new A.Cat("小花");
cat.eat();

var cat2 = new B.Cat("小花");
cat2.eat();
```

# 第 13 章 ArkUI ETS 框架详细讲解

ArkUI ETS UI 框架是专门为富设备和复杂应用开发的一套声明式的 UI 框架，和 ArkUI JS 类 Web 风格的声明式框架进行比较，ArkUI ETS 采用 TypeScript 编写，同时采用声明式的链式调用的写法，在开发效率方面有很大的提升。

声明式 UI 开发框架，采用更接近自然语义的编程方式，让开发者可以直观地描述 UI，不必关心框架如何实现 UI 绘制和渲染，实现极简高效开发。从组件、动效和状态管理 3 个维度来提供 UI 能力，还提供了多能力部署和系统能力接口，真正实现多端代码共享和系统能力的极简调用。

## 13.1 框架特点

ArkUI ETS UI 框架有以下特点：

**1. 开箱即用的组件**

声明式 UI 开发框架提供了丰富的系统预置组件，可以通过链式调用的方式设置系统组件的渲染效果。开发者可以将系统组件组合为自定义组件，通过这种方式将页面组件化为一个个独立的 UI 单元，实现页面不同单元的独立创建、开发和复用，使页面具有更强的工程性。

**2. 丰富的动效接口**

声明式 UI 开发框架提供了 SVG 标准的绘制图形能力，同时开放了丰富的动效接口，开发者可以通过封装的物理模型或者调用动画能力接口实现自定义动画轨迹。

**3. 状态与数据管理**

状态数据管理作为声明式 UI 的特色，通过功能不同的装饰器给开发者提供了清晰的页面更新渲染流程和管道。状态管理包括 UI 组件状态和应用程序状态，两者协作可以使开发者完整地构建整个应用的数据更新和 UI 渲染。

**4. 一次开发多端部署**

声明式 UI 开发框架支持手机、智能手表、智慧屏和车机等不同平台和设备，组件的主题和风格可根据不同设备的特点而进行多态设计，让开发者真正实现不同设备之间的代码共享。

### 5. 系统能力接口

声明式 UI 开发框架还封装了丰富的系统能力接口，开发者可以通过简单的接口调用，实现从 UI 设计到系统能力调用的极简开发。

## 13.2 组件化设计

ArkUI ETS 采用组件化的设计，一切皆组件，开发者可以根据业务需要，把界面拆分成不同颗粒度，可以复用这些组件，下面详细介绍 ArkUI ETS 中组件的定义与详细用法。

### 13.2.1 组件装饰器@Component

组件的定义仿效了 SwiftUI 的语法糖，采用 struct 结构体而不是 class 类的方式定义组件，不过这里的 struct 并不是真正的结构体类型，只是一种 JS 的语法糖，所以本质上和 SwiftUI 是不一样的。

@Component 装饰的 struct 表示该结构体具有组件化能力，能够成为一个独立的组件，这种类型的组件也称为自定义组件。该组件可以组合其他组件，它通过实现 build 方法来描述 UI 结构。

@Entry 添加到组件上，组件被作为页面的入口，页面加载时将被渲染显示。

下面定义一个 MyComponent 组件，该组件在页面加载的时候将被渲染到屏幕上，如图 13.1 和代码示例 13.1 所示。

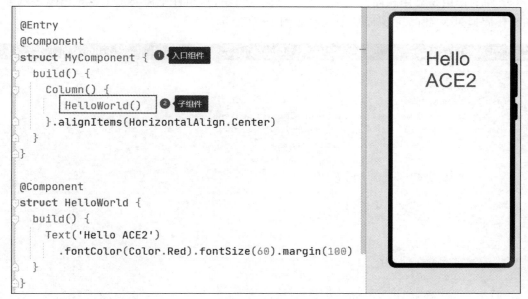

图 13.1　组件的定义

代码示例 13.1　chapter13\02_component\entry\src\main\ets\default\pages\helloworld.ets

```
@Entry
@Component
struct MyComponent {
  build() {
    Column() {
      HelloWorld()
    }.alignItems(HorizontalAlign.Center)
  }
}

@Component
struct HelloWorld {
  build() {
    Text('Hello ArkUI2')
      .fontColor(Color.Red).fontSize(60).margin(100)
  }
}
```

MyComponent 组件中的 build()方法会在初始渲染时执行,此外,当组件中的状态发生变化时,build()方法将被再次执行。

定义一个自定义的组件(HelloWorld.ets),该组件用来显示 Hello ArkUI2 文字,使用自定义组件只需像函数调用一样使用,如在 MyComponent 的 build()方法中声明 HelloWorld()。

组件的作用是复用相同逻辑的 UI,非页面入口组件,组件不需要加上@Entity 装饰器。ETS 代码通过 Webpack 编译后如代码示例 13.2 所示。

代码示例 13.2　.preview\intermediates\res\debug\rich\assets\js\default\pages\helloworld.js

```
(() => {
var __Webpack_exports__ = {};
class MyComponent extends View {
    constructor(compilerAssignedUniqueChildId, parent, params) {
        super(compilerAssignedUniqueChildId, parent);
        this.updateWithValueParams(params);
    }
    updateWithValueParams(params) {
    }
    aboutToBeDeleted() {
        SubscriberManager.Get().delete(this.id());
    }
    render() {
        Column.create();
        Column.debugLine("pages/helloworld.ets(5:5)");
        Column.alignItems(HorizontalAlign.Center);
        let earlierCreatedChild_2 = this.findChildById("2");
        if (earlierCreatedChild_2 == undefined) {
```

```
            View.create(new HelloWorld("2", this, {}));
        }
        else {
            earlierCreatedChild_2.updateWithValueParams({});
            if (!earlierCreatedChild_2.needsUpdate()) {
                earlierCreatedChild_2.markStatic();
            }
            View.create(earlierCreatedChild_2);
        }
        Column.pop();
    }
}
class HelloWorld extends View {
    constructor(compilerAssignedUniqueChildId, parent, params) {
        super(compilerAssignedUniqueChildId, parent);
        this.updateWithValueParams(params);
    }
    updateWithValueParams(params) {
    }
    aboutToBeDeleted() {
        SubscriberManager.Get().delete(this.id());
    }
    render() {
        Text.create('Hello ArkUI2');
        Text.debugLine("pages/helloworld.ets(14:5)");
        Text.fontColor(Color.Red);
        Text.fontSize(60);
        Text.margin(100);
        Text.pop();
    }
}
loadDocument(new MyComponent("1", undefined, {}));

})();
```

## 13.2.2 组件的内部私有状态 @State

@State 装饰器用于修饰组件内部私有状态属性,当组件的状态发生变化时,将会调用所在组件的 build()方法进行 UI 刷新。

@State 状态数据具有以下特征。

(1) 支持多种类型:允许强类型的按值和按引用类型:class、number、boolean、string,以及这些类型的数组,即 Array < class >、Array < string >、Array < boolean >、Array < number >。不允许 object 和 any。

(2) 支持多实例:组件的不同实例的内部状态数据独立。

(3) 内部私有标记为@State 的属性不能直接在组件外部修改。它的生命周期取决于它所在的组件。

（4）需要本地初始化：必须为所有@State变量分配初始值，将变量保持未初始化可能导致框架行为未定义。

（5）创建自定义组件时支持通过状态变量名设置初始值：在创建组件实例时，可以通过变量名显式指定@State状态属性的初始值。

下面通过单击计数的案例（如图13.2所示），介绍@State的用法，在界面上添加一个Text组件和一个Button组件，当单击Button按钮一次时，Text的值累加一次，这里每单击Button按钮一次，界面会显示最新累加的结果，Text组件的值必须是有状态的属性，这样才可以动态地把结果渲染到界面上，如代码示例13.3所示。

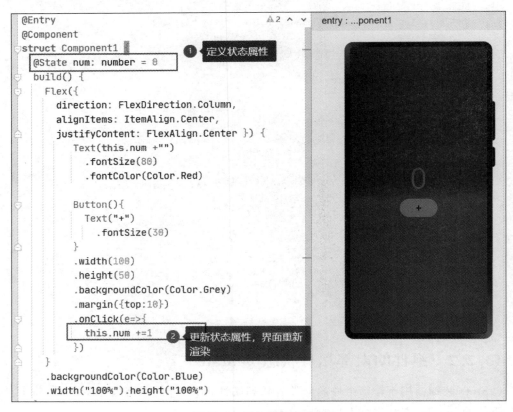

图13.2　组件的状态定义

代码示例13.3　chapter13\02_component\entry\src\main\ets\default\pages\components1.ets

```
@Entry
@Component
struct Component1 {
  @State num: number = 0
  build() {
```

```
Flex({
  direction: FlexDirection.Column,
  alignItems: ItemAlign.Center,
  justifyContent: FlexAlign.Center }) {
    Text(this.num + "")
      .fontSize(80)
      .fontColor(Color.Red)

    Button(){
      Text(" + ")
        .fontSize(30)
    }
    .width(100)
    .height(50)
    .backgroundColor(Color.Grey)
    .margin({top:10})
    .onClick(e =>{
      this.num += 1
    })
  }
  .backgroundColor(Color.Blue)
  .width("100%").height("100%")
}
}
```

代码中的 num 用@State 装饰后，num 变量的变化是响应式的，当 num 值发生变化时，视图被重新渲染，界面 Text 的值不断增加显示。

### 13.2.3 组件的输入和输出属性

组件的内部对外是不可见的，组件交互的边界是组件的输入和输出接口，如图 13.3 所示，ArkUI ETS 中设计了多个双向同步属性，让子组件把内部的属性值传递到组件外部。

图 13.3 组件的输入和输出属性

下面的例子介绍通过组件的属性实现列表项的删除功能，如图 13.4 所示，列表框代表列表项组件的复用，单击每一项的删除按钮后，界面立即显示删除后的列表效果。

这里分为两个组件，外部的主页面组件(Demo6)和内部的列表项组件(Item)，Item 组件包含一个标题和一个删除按钮，标题是组件的输入属性，删除按钮单击后会输出删除事件。

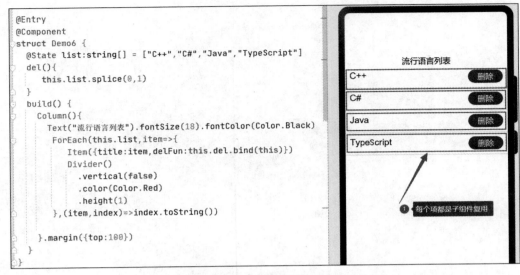

图 13.4 组件的交互案例【删除列表中的一项】

列表项组件属性包含两个输入属性，title 属性和 delFun 删除属性（函数类型），用于接收父组件传入的函数引用，如代码示例 13.4 所示。

代码示例 13.4    chapter13\02_component\entry\src\main\ets\default\pages\component2.ets

```
struct Item {
  //组件输入属性
  public title:string
  //组件输入属性
  public delFun: Function

  build(){
    Flex({
      direction:FlexDirection.Row,
      justifyContent:FlexAlign.SpaceAround
    }){
      Text(this.title).fontSize(20).width("100%")
      Button() {
        Text("删除").fontSize(18).fontColor(Color.White)
      }.width(100).height(30)
      .onClick(e =>{
        this.delFun()
      })

    }.margin(10)
  }
}
```

在父组件 Demo6 中通过 ForEach()方法遍历后给 Item 子组件绑定数据,遍历的数据是 Demo6 组件中的状态属性 list,del()方法实现删除 list 数字中的值,del()方法通过 Item 组件的输入属性传入 Item 组件内部,Item 组件内部调用删除,如代码示例 13.5 所示。

代码示例 13.5　chapter13\02_component\entry\src\main\ets\default\pages\component2.ets

```
@Entry
@Component
struct Demo6 {
  @State list:string[] = ["C++","C#","Java","Typescript"]
  del(){
     this.list.splice(0,1)
  }
  build() {
    Column(){
      Text("流行语言列表").fontSize(18).fontColor(Color.Black)
       ForEach(this.list,item =>{
           Item({title:item,delFun:this.del.bind(this)})
           Divider()
             .vertical(false)
             .color(Color.Red)
             .height(1)
       },(item,index) => index.toString())

    }.margin({top:100})
  }
}
```

## 13.2.4　单向同步父组件状态@Prop

@Prop 具有与@State 相同的语义,但初始化方式不同。@Prop 装饰的变量必须使用其父组件提供的@State 变量进行初始化,允许组件内部修改@Prop 变量,但上述更改不会通知父组件,即@Prop 属于单向数据绑定。

@Prop 状态数据具有以下特征。

(1) 支持简单类型:仅支持简单类型:number、string、boolean。

(2) 私有:仅在组件内访问。

(3) 支持多个实例:一个组件中可以定义多个标有@Prop 的属性。

(4) 创建自定义组件时将值传递给@Prop 变量进行初始化:在创建组件的新实例时,必须初始化所有@Prop 变量,不支持在组件内部进行初始化。

子组件中定义的属性@Prop pData 更新后,不会同步给父组件的@State num 属性,如图 13.5 和代码示例 13.6 所示。

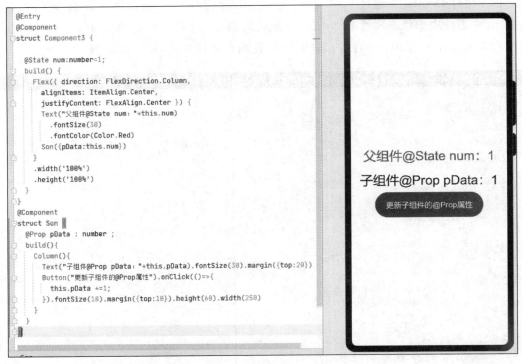

图 13.5　组件的 Prop 属性

代码示例 13.6　chapter13\02_component\entry\src\main\ets\default\pages\component3.ets

```
@Entry
@Component
struct Component3 {

  @State num:number = 1;
  build() {
    Flex({ direction: FlexDirection.Column,
      alignItems: ItemAlign.Center,
      justifyContent: FlexAlign.Center }) {
      Text("父组件@State num:" + this.num)
        .fontSize(30)
        .fontColor(Color.Red)
      Son({pData:this.num})
    }
    .width('100%')
    .height('100%')
  }
}
@Component
struct Son {
```

```
  @Prop pData : number ;
build(){
   Column(){
     Text("子组件@Prop pData:" + this.pData).fontSize(30).margin({top:20})
     Button("更新子组件的@Prop 属性").onClick(()=>{
       this.pData += 1;
     }).fontSize(18).margin({top:10}).height(60).width(250)
   }
 }
}
```

### 13.2.5 双向同步状态@Link

@Link 装饰的变量可以和父组件的@State 变量建立双向数据绑定。

（1）支持多种类型：@Link 变量的值与@State 变量的类型相同，即 class、number、string、boolean 或这些类型的数组。

（2）私有：仅在组件内访问。

（3）单个数据源：初始化@Link 变量的父组件的变量必须是@State 变量。

（4）双向通信：子组件对@Link 变量的更改将同步修改父组件的@State 变量。

（5）创建自定义组件时需要将变量的引用传递给@Link 变量：在创建组件的新实例时，必须使用命名参数初始化所有@Link 变量。@Link 变量可以使用@State 变量或@Link 变量的引用进行初始化。@State 变量可以通过 $ 操作符创建引用。

（6）@Link 变量不能在组件内部进行初始化。

下面的例子，在子组件中定义一个 Button 按钮，当单击子组件的 Button 按钮时，父组件的数字会同步变化，如代码示例 13.7 所示。

**代码示例 13.7** chapter13\02_component\entry\src\main\ets\default\pages\component4.ets

```
@Entry
@Component
struct Component4 {
  @State num :number = 0 ;
  build(){
    Flex({
      direction:FlexDirection.Column,
      justifyContent:FlexAlign.Center,
      alignItems:ItemAlign.Center}){
      Text("父组件@State: " + this.num).fontSize(30)
      SyncSon({synData: $ num})
    }.width("100%")
  }
}

@Component
```

```
struct SyncSon {
  @Link synData:number;
  build(){
    Column(){
      Text("子组件@Link: " + this.synData).fontSize(30)
      Button("同步数据").fontSize(20).onClick(()=>{
        this.synData ++
      })
        .margin({top:20})
        .width(200)
    }
  }
}
```

效果图如图 13.6 所示。

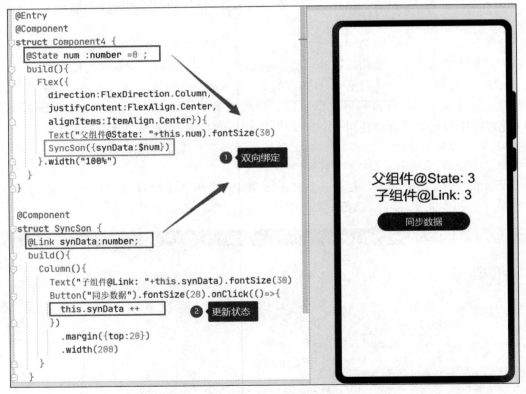

图 13.6　组件的双向状态绑定

## 13.2.6　自定义组件的生命周期函数

自定义组件的生命周期函数用于通知用户该自定义组件的生命周期，这些回调函数是私有的，在运行时由开发框架在特定的时间进行调用，不能从应用程序中手动调用这些回调函数，如表 13.1 所示。

表 13.1 自定义组件的生命周期函数

| 参数名 | 说 明 |
| --- | --- |
| aboutToAppear | 函数在创建自定义组件的新实例后，在执行其 build()函数之前执行。允许在 aboutToAppear 函数中改变状态变量，这些更改将在后续执行 build()函数中生效 |
| aboutToDisappear | 函数在自定义组件析构销毁之前执行。不允许在 aboutToDisappear 函数中改变状态变量，特别是@Link 变量的修改可能会导致应用程序的行为不稳定 |
| onPageShow | 当此页面显示时触发一次，包括路由过程、应用进入前后台等场景，仅@Entry 修饰的自定义组件生效 |
| onPageHide | 当此页面消失时触发一次，包括路由过程、应用进入前后台等场景，仅@Entry 修饰的自定义组件生效 |
| onBackPress | 当用户单击返回按钮时触发，仅@Entry 修饰的自定义组件生效。如果返回 true，则表示页面自己处理返回逻辑，不进行页面路由。如果返回 false，则表示使用默认的返回逻辑。当不返回值会作为 false 处理 |

允许在生命周期函数中使用 Promise 和异步回调函数，例如网络资源获取和定时器设置等，不允许在生命周期函数中使用 async await。

下面的案例（如图 13.7 所示），在组件 aboutToAppear 的函数中定义一个定时器，当页面退出或者切换页面时删除该定时器，如代码示例 13.8 所示。

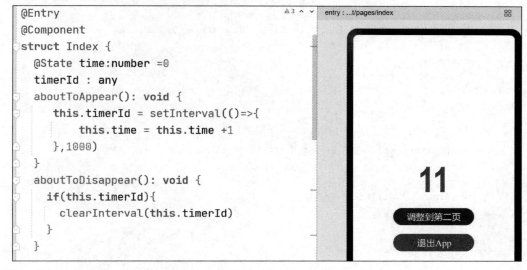

图 13.7 自定义组件的生命周期函数

代码示例 13.8    chapter13\02_component\entry\src\main\ets\default\pages\lifecircle.ets

```
import router from '@system.router';
import app from '@system.app';

@Entry
@Component
struct Index {
  @State time:number = 0
  timerId : any
  aboutToAppear(): void {
    this.timerId = setInterval(() =>{
        this.time = this.time + 1
    },1000)
  }
  aboutToDisappear(): void {
    if(this.timerId){
      clearInterval(this.timerId)
    }
  }

  build() {
    Flex({ direction: FlexDirection.Column,
      alignItems: ItemAlign.Center,
      justifyContent: FlexAlign.Center }) {
      Text('' + this.time)
        .fontSize(80)
        .fontColor(Color.Red)
        .fontWeight(FontWeight.Bold)

      Button() {
        Text("调整到第二页").fontSize(20).fontColor(Color.White)
      }.margin({ top: 20 }).width(180).onClick(e => {
        router.push({
           uri: "pages/second"
        })
      })

      Button() {
        Text("退出App").fontSize(20).fontColor(Color.White)
      }.margin({ top: 20 }).width(180).onClick(e => {
        app.terminate()
      }).backgroundColor(Color.Red)
    }
    .width('100%')
    .height('100%')
  }
}
```

## 13.2.7 跨组件数据传递@Consume 和@Provide

Provide 作为数据的提供方,可以更新其子孙节点的数据,并触发页面渲染,如代码示例 13.9 所示。Consume 在感知到 Provide 数据的更新后,会触发当前 view 的重新渲染,如图 13.8 所示。

**注意**:使用@Provide 和@Consume 要小心,避免循环引用而导致死循环。

代码示例 13.9　chapter13\02_component\entry\src\main\ets\default\pages\component5.ets

```
@Entry
@Component
struct CompA {
  @Provide("n") aNum: number = 0;
  build() {
    Flex({ direction: FlexDirection.Column,
      alignItems: ItemAlign.Center,
      justifyContent: FlexAlign.Center }) {
      CompB()
      Button() {
        Text(`CompA 组件:${this.aNum}`)
          .fontSize(20)
      }.padding(5).width(200)
      .onClick(() => {
        this.aNum += 1;
      })
    }.width("100%").height(500)
    .border({width:5,color:"green",style:BorderStyle.Dashed})
  }
}

@Component
struct CompB {
  build() {
    Column() {
      Text("CompB 组件区域").fontSize(20).margin(5)
      CompC()
    }
    .width(300)
    .height(300)
    .border({width:6,color:"red",style:BorderStyle.Dashed})
  }
}

@Component
struct CompC {
  @Consume("n") num: number;
  build() {
    Column() {
```

```
      Button() {
        Text(`CompC组件: ${this.num}`)
          .fontSize(20)
      }.padding(5).width(200).backgroundColor(Color.Red)
      .onClick(() => {
        this.num += 1;
      })
    }
  }
}
```

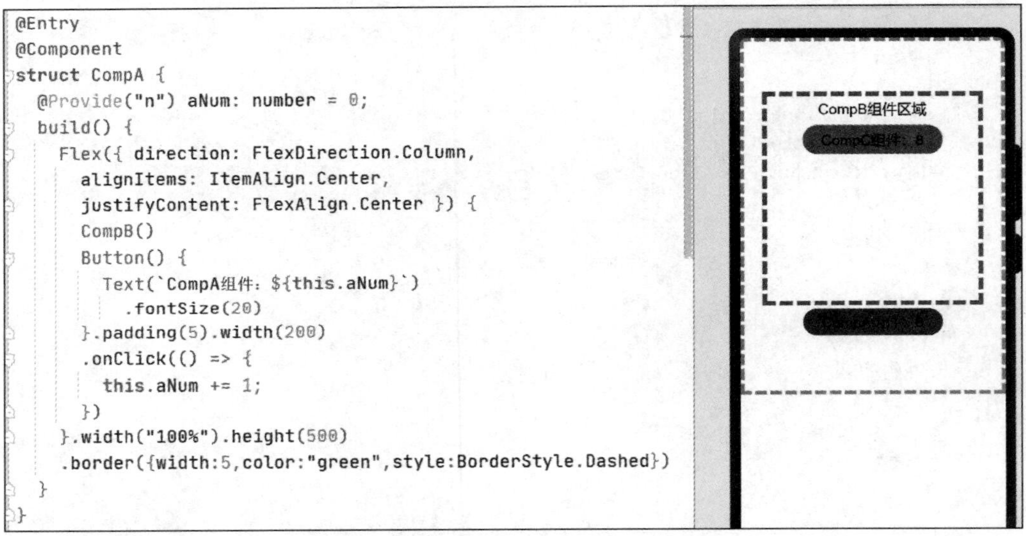

```
@Entry
@Component
struct CompA {
  @Provide("n") aNum: number = 0;
  build() {
    Flex({ direction: FlexDirection.Column,
      alignItems: ItemAlign.Center,
      justifyContent: FlexAlign.Center }) {
      CompB()
      Button() {
        Text(`CompA组件: ${this.aNum}`)
          .fontSize(20)
      }.padding(5).width(200)
      .onClick(() => {
        this.aNum += 1;
      })
    }.width("100%").height(500)
    .border({width:5,color:"green",style:BorderStyle.Dashed})
  }
}
```

图 13.8 跨组件数据传递

### 13.2.8 监听变量状态变更@Watch

应用可以注册回调方法。当一个被@State、@Prop、@Link、@ObjectLink、@Provide、@Consume、@StorageProp 及@StorageLink 中任意一个装饰器修饰的变量改变时,均可触发此回调。@Watch 中的变量一定要使用(" ")进行包装。

下面通过满减案例介绍@Watch 的作用,如图 13.9 所示,当单击增加购买数按钮时,每单击一次按钮,属性 totalPrice 的值就累加 100,这里为 totalPrice 状态属性添加了监听器函数 onTotalUpdated(),当 totalPrice 的值发生变化时,就会触发该监听方法,如代码示例 13.10 所示。

**代码示例 13.10    chapter13\02_component\entry\src\main\ets\default\pages\component6.ets**

```
import prompt from '@system.prompt';
@Entry
@Component
```

```
struct Component6 {
  @State @Watch("onTotalUpdated") totalPrice:number = 0

  //@Watch cb
  onTotalUpdated(propName: string): void {
    if(this.totalPrice > 500){
      prompt.showToast({
        message:"已满 500,送 1000 个积分",
        duration:5000,
        bottom:200
      })
    }
  }

  build() {
    Flex({ direction: FlexDirection.Column,
      alignItems: ItemAlign.Center,
      justifyContent: FlexAlign.Center }) {
      Text(`目前的总金额: ${this.totalPrice}`)
        .fontSize(30)
      Button("增加购买 + 100").margin({top:20}).onClick(() => {
        this.totalPrice += 100
      }).backgroundColor(Color.Red)

    }.width("100%").height("100%")
  }
}
```

图 13.9　监听变量状态变更

## 13.2.9 自定义组件方法 @Builder

@Builder 修饰方法的语法规范与 build()函数一致。为了更加方便地使用语言,因此添加了@Builder 标签,如代码示例 13.11 所示。

代码示例 13.11    chapter13\02_component\entry\src\main\ets\default\pages\component9.ets

```
@Entry
@Component
struct Component9 {
  size: number = 100;

  @Builder SquareText(label: string,color:number) {
    Text(label)
      .fontSize(30)
      .width(1 * this.size)
      .height(1 * this.size)
      .backgroundColor(color)
  }

  @Builder RowOfSquareTexts(label1: string, label2: string) {
    Row() {
      this.SquareText(label1,Color.Red)
      this.SquareText(label2,Color.Blue)
    }
    .width(2 * this.size)
    .height(1 * this.size)
  }

  build() {
    Column() {
      Row() {
        this.SquareText("A",Color.Yellow)
        this.SquareText("B",Color.Gray)
      }
      .width(2 * this.size)
      .height(1 * this.size)

      this.RowOfSquareTexts("C", "D")
    }
    .width(2 * this.size)
    .height(2 * this.size)
  }
}
```

## 13.2.10 统一组件样式 @Extend

@Extend 装饰器可将新的属性函数添加到内置组件上,如 Text、Column、Button 等。

对于一些公共的样式,可以通过@Extend 定义一个扩展函数 comStyle(),如代码示例 13.12 所示。

代码示例 13.12　chapter13\02_component\entry\src\main\ets\default\pages\component10.ets
```
@Extend(Text) function comStyle (color: number) {
  .fontColor(color)
  .fontSize(50)
}

@Entry
@Component
struct Component10 {
  build() {
    Column() {
      Text("第 1 个文本")
        .comStyle(Color.Yellow)
      Text("第 2 个文本")
        .comStyle(Color.Red)
    }
  }
}
```

统一组件样式如图 13.10 所示。

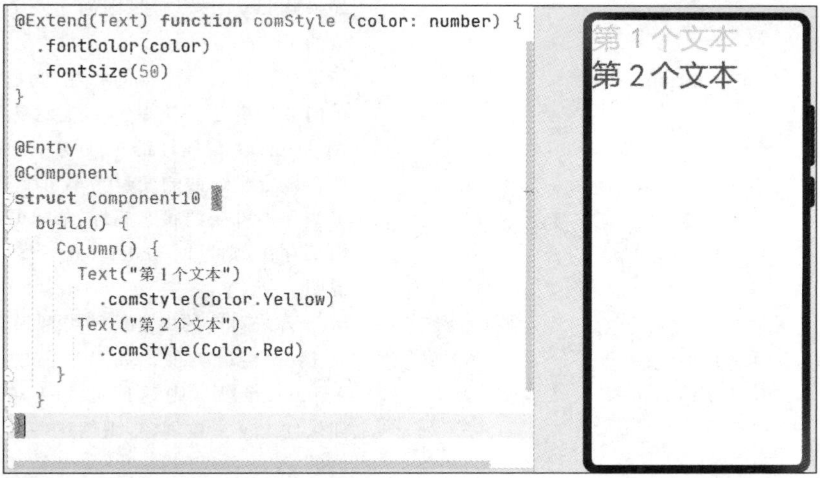

图 13.10　统一组件样式

## 13.3　状态管理仓库

AppStorage 是应用程序中的单例对象,由 UI 框架在应用程序启动时创建。它的目的是为可变应用程序状态属性提供中央存储。AppStorage 包含整个应用程序中需要访问的所有状态属性。只要应用程序保持运行，AppStorage 存储就会保留所有属性及其值，属性值可以通过唯一的键值进行访问。

UI 组件可以通过装饰器将应用程序状态数据与 AppStorage 进行同步。应用业务逻辑的实现也可以通过接口访问 AppStorage。AppStorage 的选择状态属性可以与不同的数据源或数据接收器同步。这些数据源和接收器可以在本地或远程设备上，并具有不同的功能，如数据持久性。这样的数据源和接收器可以独立于 UI 在业务逻辑中实现。默认情况下，AppStorage 中的属性是可变的，AppStorage 还可使用不可变（只读）属性，AppStorage 接口说明如表 13.2 所示。

表 13.2 AppStorage 接口说明

| 参数名 | 类型 | 返回值 | 定义 |
| --- | --- | --- | --- |
| Link | key：string | @Link | 如果存在具有给定键的数据，则返回此属性的双向数据绑定，该双向绑定意味着变量或者组件对数据的更改将同步到 AppStorage，通过 AppStorage 将数据的修改同步到变量或者组件。如果具有此键的属性不存在或属性为只读，则返回 undefined |
| SetAndLink | key：String defaultValue：T | @Link | 与 Link 接口类似。如果当前的 key 保存在 AppStorage 中，则返回此 key 对应的 value。如果此 key 未被创建，则创建一个对应 default 值的 Link 并返回 |
| Prop | key：string | @Prop | 如果存在具有给定键的属性，则返回此属性的单向数据绑定。该单向绑定意味着只能通过 AppStorage 将属性的更改同步到变量或者组件。该方法返回的变量为不可变变量，适用于可变和不可变的状态属性，如果具有此键的属性不存在，则返回 undefined。<br>说明：<br>prop 方法对应的属性值的类型为简单类型 |
| SetAndProp | key：string defaultValue：S | @Prop | 与 Prop 接口类似。如果当前的 key 保存在 AppStorage 中，则返回此 key 对应的 value。如果此 key 未被创建，则创建一个对应 default 值的 Prop 并返回 |
| SetOrCreate | string，newValue：T | boolean | 如果相同名字的属性存在，并且此属性可以被更改，则返回 true，否则返回 false。如果相同名字的属性不存在，则可创建第 1 个赋值为 defaultValue 的属性，不支持 null 和 undefined |
| Delete | key：string | boolean | 删除属性，如果存在，则返回 true。如果不存在，则返回 false |
| Set | string，newValue：T | void | 对已保存的 key 值，替换其 value 值 |

如代码示例13.13所示。

代码示例13.13
```
let link1 = AppStorage.Link('PropA')
let link2 = AppStorage.Link('PropA')
let prop = AppStorage.Prop('PropA')
link1 = 47 //causes link1 == link2 == prop == 47
link2 = link1 + prop //causes link1 == link2 == prop == 94
prop = 1 //error, prop is immutable
```

### 13.3.1 持久化数据管理

PersistentStorage用于管理应用的持久化数据。此对象可以将特定标记的持久化数据连接到AppStorage中,并由AppStorage接口访问对应的持久化数据,或者通过@StorageLink修饰器访问对应key的变量,PersistentStorage接口如表13.3所示。

表13.3 PersistentStorage接口

| 参数名 | 类型 | 返回值 | 定义 |
| --- | --- | --- | --- |
| PersistProp | key: string<br>defaultValue: T | void | 关联命名的属性在AppStorage中变为持久化数据。赋值覆盖流程:<br>(1) 如果此属性存在于AppStorage中,则将Persistent中的数据复写为AppStorage中的属性值。<br>(2) 如果Persistent中有此命名的属性,则使用Persistent中的属性值。<br>(3) 使用defaultValue,不支持null和undefined |
| DeleteProp | key: string | void | 取消双向数据绑定,该属性值将从持久存储中删除 |
| PersistProps | keys: { key: string, defaultValue: any }[] | void | 关联多个命名的属性绑定 |
| Keys | void | Array&lt;string&gt; | 返回所有持久化属性的标记 |

### 13.3.2 环境变量Environment

Environment是框架在应用程序启动时创建的单例对象,它为AppStorage提供了一系列应用程序需要的环境状态属性,这些属性描述了应用程序运行的设备环境。

Environment及其属性是不可变的,所有属性值的类型均为简单类型。

从Environment获取语音环境,代码如下:

```
Environment.EnvProp("accessibilityEnabled", "default");
var enable = AppStorageGet("accessibilityEnabled");
```

accessibilityEnabled 是 Environment 提供的默认系统变量识别符。首先需要将对应系统属性绑定到 AppStorage 中,然后通过 AppStorage 中的方法或者装饰器,访问对应系统属性数据。

### 13.3.3 AppStorage 与组件同步

组件变量与 AppStorage 同步,主要提供@StorageLink 和@StorageProp 装饰器。

1. @StorageLink 装饰器

组件通过@StorageLink(key)装饰的状态变量,与 AppStorage 建立双向数据绑定,key 用于标识 AppStorage 中的属性键值。当创建包含@StorageLink 的状态变量的组件时,该状态变量的值将使用 AppStorage 中的值进行初始化,不允许使用本地初始化。

在 UI 组件中对@StorageLink 的状态变量所做的更改将同步到 AppStorage 中,并从 AppStorage 同步到任何其他绑定实例中,如 PersistentStorage 或其他绑定的 UI 组件。

2. @StorageProp 装饰器

组件通过@StorageProp(key)装饰的状态变量,将于 AppStorage 中建立单向数据绑定,key 用于标识 AppStorage 中的属性键值。当创建包含@StorageProp 的状态变量的组件时,该状态变量的值将使用 AppStorage 中的值进行初始化,不允许使用本地初始化。

AppStorage 中的属性值更改会导致绑定的 UI 组件进行状态更新,如代码示例 13.14 所示。

**代码示例 13.14 逻辑代码**

```
let varA = AppStorage.link('varA')
PersistentStorage.link('varA')
varA = 47
let envLang = AppStorage.prop('languageCode')
```

界面如代码示例 13.15 所示。

**代码示例 13.15　chapter13\02_component\entry\src\main\ets\default\pages\component11.ets**

```
@Entry
@Component
struct Component11 {
  @StorageLink('varA') varA: number = 0
  @StorageProp('languageCode') lang: string = ""
  private label: string = 'count'
  private aboutToAppear() {
    this.label = (this.lang === 'zh') ? '数' : 'Count'
  }
  build(){
    Button(`${this.label}: ${this.varA}`)
    .onClick(() => {this.varA++})
  }
}
```

## 13.4 渲染控制语法

### 13.4.1 条件渲染 if…else…

使用 if/else 进行条件渲染。
(1) if 条件语句可以使用状态变量。
(2) 使用 if 可以使子组件的渲染依赖条件语句。
(3) 必须在容器组件内使用。
(4) 某些容器组件会限制子组件的类型或数量。当将 if 放置在这些组件内时,这些限制将应用于 if 和 else 语句内创建的组件,如当在 Grid 组件内使用 if 时,则仅允许 if 条件语句内使用 GridItem 组件,而在 List 组件内则仅允许 ListItem 组件。

使用 if 条件语句,如代码示例 13.16 所示。

**代码示例 13.16**
```
Column() {
    if (this.count > 0) {
        Text('count is positive')
    }
}
```

使用 if、else if、else 条件语句,如代码示例 13.17 所示。

**代码示例 13.17**
```
Column() {
    if (this.count < 0) {
        Text('count is negative')
    } else if (this.count % 2 === 0) {
        Divier()
        Text('even')
    } else {
    Divider()
    Text('odd')
    }
}
```

### 13.4.2 循环渲染 ForEach

开发框架提供了 ForEach 组件来迭代数组,并为每个数组项创建相应的组件。ForEach 的定义,如代码示例 13.18 所示。

**代码示例 13.18**
```
ForEach(
```

```
    arr: any[], //Array to be iterated
    itemGenerator: (item: any) => void,
    keyGenerator?: (item: any) => string
)
```

## 13.5 动画效果

ArkUI ETS 提供了多种动画的设置,包括属性动画、显式动画、转场动画、路径动画。

### 13.5.1 属性动画

当组件的通用属性发生变化时,可以创建属性动画进行渐变,以此提升用户体验。

下面示例的功能为单击后进行宽和高属性的动画,当按下时进行缩小,当松开后恢复,如代码示例 13.19 所示。

代码示例 13.19　chapter13\02_component\entry\src\main\ets\default\pages\animate1.ets

```
@Entry
@Component
struct Animate1 {
  @State widthSize: number = 100
  @State heightSize: number = 200
  build() {
    Flex({ direction: FlexDirection.Column,
      alignItems: ItemAlign.Center,
      justifyContent: FlexAlign.Center }) {
      Button() {
        Text('按钮')
      }
      .width(this.widthSize)
      .height(this.heightSize)
      .onTouch((event: TouchEvent) => {
        if (event.type === TouchType.Down) {
          this.widthSize = 90
          this.heightSize = 180
        } else if (event.type === TouchType.Up) {
          this.widthSize = 100
          this.heightSize = 200
        }
      })
      //对前面定义的宽和高属性进行动画配置
      .animation({duration: 300, curve: Curve.EaseIn})
    }.width("100%").height("100%")
  }
}
```

## 13.5.2 显式动画

animateTo(value: AnimationOption, event: () => void): void 提供了全局 animateTo 显式动画接口来指定由于闭包代码导致的状态变化而插入过渡效果。event 用于指定显示动效的闭包函数，在闭包函数中导致的状态变化系统会自动插入过渡动画，如代码示例 13.20 所示。

代码示例 13.20　chapter13\02_component\entry\src\main\ets\default\pages\animate2.ets

```
@Entry
@Component
struct Animate2 {

  @State height1:number = 200
  @State width1:number = 200

  build() {
    Flex({ direction: FlexDirection.Column, alignItems: ItemAlign.Center, justifyContent:
FlexAlign.Center }) {
      Text('显式动画')
        .fontSize(50)
        .width(this.width1)
        .height(this.height1)
        .fontWeight(FontWeight.Bold)
        .backgroundColor(Color.Red)
      .onTouch(e =>{
        animateTo({duration: 500, curve: Curve.Ease}, () => {
          this.height1 = 500
          this.width1 = 300
        })
      })
    }
    .width('100%')
    .height('100%')
  }
}
```

## 13.5.3 转场动画

转场动画包含页面间转场、组件内转场、共享元素转场。

### 1. 页面间转场

页面转场通过全局 pageTransition() 方法进行配置转场参数，如表 13.4 所示。

表 13.4　pageTransition() 方法的配置

| 组件名 | 参数 | 参 数 描 述 |
| --- | --- | --- |
| PageTransitionEnter | Object | 页面入场组件，用于自定义当前页面的入场效果 |
| PageTransitionExit | Object | 页面退场组件，用于自定义当前页面的退场效果 |

页面间转场动画的设置可以使用两种方法,如代码示例13.21所示,自定义方式1:将当前页面的入场动画配置为从左侧滑入,将退场配置为缩小加透明度变化。自定义方式2:将当前页面的入场动画配置为淡入,将退场动画配置为缩小。

代码示例13.21

```
//自定义方式1:使用系统提供的多种默认效果(平移、缩放、透明度等)
pageTransition() {
    PageTransitionEnter({duration: 1200})
      .slide(SlideEffect.Left)
    PageTransitionExit({curve: Curve.Linear})
      .translate({x: 100.0, y: 100.0})
      .opacity(0)
}

//自定义方式2:完全自定义转场过程的效果
pageTransition() {
    PageTransitionEnter({duration: 1200, curve: Curve.Linear})
      .onEnter((type: RouteType, progress: number) => {
        this.scale = 1
        this.opacity = progress
      })
      //进场过程中会逐帧触发onEnter()回调,入参为动效的归一化进度(0%~100%)
    PageTransitionExit({duration: 1500, curve: Curve.Ease})
      .onExit((type: RouteType, progress: number) => {
        this.scale = 1 - progress
        this.opacity = 1
      }) //进场过程中会逐帧触发onExit()回调,入参为动效的归一化进度(0%~100%)
}
```

PageTransitionEnter 和 PageTransitionExit 组件支持的属性如表13.5所示。

表13.5 PageTransitionEnter 和 PageTransitionExit 组件支持的属性

| 参数名称 | 参数类型 | 默认值 | 必填 | 参数描述 |
| --- | --- | --- | --- | --- |
| slide | SlideEffect | Right | 否 | 设置转场的滑入效果<br>Left 设置表示入场时从左边滑入,出场时滑到左边<br>Right 设置表示入场时从右边滑入,出场时滑到右边<br>Top 设置表示入场时从上边滑入,出场时滑到上边<br>Bottom 设置表示入场时从下边滑入,出场时滑到下边 |

续表

| 参数名称 | 参数类型 | 默认值 | 必填 | 参数描述 |
|---|---|---|---|---|
| translate | { x?: number, y?: number, z?: number } | - | 否 | 设置页面转场时的平移效果,为入场时起点和退场时终点的值,当和 slide 同时设置时默认生效 slide |
| scale | { x?: number, y?: number, z?: number, centerX?: number, centerY?: number } | - | 否 | 设置页面转场时的缩放效果,为入场时起点和退场时终点的值 |
| opacity | number | 1 | 否 | 设置入场的起点透明度值或者退场的终点透明度值 |

PageTransitionEnter 和 PageTransitionExit 组件支持的事件如表 13.6 所示。

表 13.6 PageTransitionEnter 和 PageTransitionExit 组件支持的事件

| 参数名称 | 参数描述 |
|---|---|
| onEnter(type: RouteType, progress: number) => void | 回调入参为当前入场动画的归一化进度[0-1] |
| onExit(type: RouteType, progress: number) => void | 回调入参为当前退场动画的归一化进度[0-1] |

当页面通过路由进入 second.ets 页面时,会触发 pageTransition()方法,如代码示例 13.22 所示。

代码示例 13.22 chapter13\02_component\entry\src\main\ets\default\pages\second.ets

```
import router from '@system.router';

@Entry
@Component
struct Second {

  pageTransition() {
    PageTransitionEnter({duration: 1200})
      .slide(SlideEffect.Left)
    PageTransitionExit({curve: Curve.Linear})
      .translate({x: 100.0, y: 100.0})
      .opacity(0)
  }
```

```
  build() {
    Flex({ direction: FlexDirection.Column, alignItems: ItemAlign.Center, justifyContent:
FlexAlign.Center }) {
      Text('第二页')
        .fontSize(30)
        .fontWeight(FontWeight.Bold)

      Button(){
        Text("返回").fontSize(20).fontColor(Color.White)
      }.margin({top:20}).width(180).onClick(e =>{
        router.back()
      })
    }
    .width('100%')
    .height('100%')
  }
}
```

### 2. 组件内转场

组件转场主要通过 transition 属性配置转场参数，在组件插入和删除时进行过渡动效，主要用于容器组件子组件在插入和删除时提升用户体验(需要配合 animateTo 才能生效，动效时长、曲线、延时跟随 animateTo 中的配置)。

下面示例功能通过两个 Button 控制另外两个 Button 的出现和消失，并通过 transition 配置了另外两个 Button 出现和消失的过场动画，如代码示例 13.23 所示。

代码示例 13.23  chapter13\02_component\entry\src\main\ets\default\pages\animate3.ets

```
@Entry
@Component
struct Animate3 {
  @State btn1: boolean = false
  @State btn2: boolean = false

  build() {
    Flex({
      direction: FlexDirection.Column,
      alignItems: ItemAlign.Center,
      justifyContent: FlexAlign.Center }) {
      Column() {
        Button() {
          Text('Switch1')
        }.width(100).onClick(() => {
          animateTo({ duration: 1000 }, () => {
            this.btn1 = !this.btn1
          })
        })
```

```
        Button() {
          Text('Switch2')
        }.margin({top:20}).width(100).onClick(() => {
          animateTo({ duration: 1000 }, () => {
            this.btn2 = !this.btn2
          })
        })
      }

      Column() {
        if (this.btn1) {
          //将插入和删除配置为不同的过渡效果
          Button() {
            Image('DEMO1.png')
          }.transition({ type: TransitionType.Insert, opacity: 0 })
          .transition({ type: TransitionType.Delete, scale: { x: 1.0, y: 0.0 } })
        }
        if (this.btn2) {
          //将插入和删除配置为相同的过渡效果
          Button() {
            Image('DEMO1.png')
          }.transition({ rotate: { x: 1, y: 1, angle: 360 }, opacity: 0.0 })
        }
      }
    }
    .width("100%")
    .height("100%")
  }
}
```

### 3. 共享元素转场

共享元素转场支持页面内的转场，如 Row 组件中的元素转场至 List 组件中，也支持页面间的转场，如当前页面的图片转场至下一页面中，如代码示例 13.24 所示。

**代码示例 13.24** chapter13\02_component\entry\src\main\ets\default\pages\animate4.ets

```
@Entry
@Component
struct Animate4 {
  @State scale: number = 1
  @State opacity: number = 1

  build() {
    List() {
      ListItem() {
        Column() {
          Image($r("app.media.Logo"))
```

```
          .width(100)
          .height(100)
          .objectFit(ImageFit.Contain)
          .sharedTransition('id1', { duration: 800, curve: Curve.Linear,
          delay: 100 })
        Navigator({ target: 'pages/third' }) {
          Text('详情')
            .fontSize(30)
            .fontWeight(FontWeight.Bold)
        }
      }
    }
  }
}
```

跳转至另外一个页面,如代码示例 13.25 所示。

**代码示例 13.25    chapter13\02_component\entry\src\main\ets\default\pages\third.ets**

```
@Entry
@Component
struct Third {
  build() {
    Stack() {
      Image($r("app.media.Logo"))
        .width(100)
        .height(100)
        .objectFit(ImageFit.Contain)
        .sharedTransition('id1')
      Text('详情介绍...')
    }
  }
}
```

## 13.5.4  手势处理

ArkUI ETS 中提供了丰富的手势处理能力,如图 13.11 所示,包括以下手势。

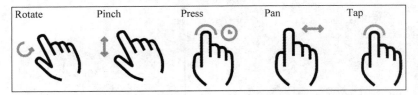

图 13.11  各种不同的手势

(1) Rotate:实现旋转效果,如旋转图片。
(2) Pinch/Zoom:放大、缩小。
(3) Press:长时间按压触发效果。
(4) Pan:左右滑动。
(5) Tap:单击。
(6) GestureGroup:组合手势。

TapGesture,如代码示例 13.26 所示。

**代码示例 13.26**

```
Column() {
Text('Double Click me: ' + this.value)
}
.gesture(
TapGesture({count: 2})
.onAction(() => {
this.value = 'Get Double Tap'
})
)
```

LongPressGesture,如代码示例 13.27 所示。

**代码示例 13.27**

```
Column() {
Text('Get LongPress Repeat Count: ' + this.value)
}.gesture(
LongPressGesture({repeat: true})
    .onAction((event: GestureEvent) => {
if (event.repeat) {
this.value++
}
})
.
(() => {
this.value = 0
})
)
```

PanGesture,如代码示例 13.28 所示。

**代码示例 13.28**

```
Column() {
  Text('Get Pan Gesture: X: ' + this.offsetX + 'Y:' + this.offsetY)
}
.gesture(
  PanGesture()
```

```
    .onActionStart((event: GestureEvent) => {
    console.log('Pan start')
  })
    .onActionUpdate((event: GestureEvent) => {
    this.offsetX = event.offsetX
    this.offsetY = event.offsetY
  })
    .onActionEnd(() => {
    console.log('Pan end')
  })
)
```

PinchGesture,如代码示例 13.29 所示。

**代码示例 13.29**

```
Column() {
  Text('Get Pinch Gesture: Scale Valut is : ' + this.scale)
}
.gesture(
  PinchGesture()
    .onActionStart((event: GestureEvent) => {
    console.log('pinch start')
  })
    .onActionUpdate((event: GestureEvent) => {
    this.scale = event.scale
  })
    .onActionEnd(() => {
    console.log('Pan end')
  })
)
```

RotationGesture,如代码示例 13.30 所示。

**代码示例 13.30**

```
Column() {
  Text('Get Rotation Gesture: angle is ' + this.angle)
}
.gesture(
  RotationGesture()
    .onActionStart((event: GestureEvent) => {
    console.log('Rotation start')
  })
    .onActionUpdate((event: GestureEvent) => {
    this.angle = event.angle
  })
    .onActionEnd(() => {
    console.log('Rotation end')
  })
)
```

组合手势，如代码示例 13.31 所示。

代码示例 13.31
```
Column() {
  Text('Get Rotation Gesture: angle is ' + this.angle)
}
.gesture(
  GestureGroup(GestureMode.Sequence,
    LongPressGesture()
      .onAction((event: GestureEvent) => {
      console.log('LongPress onAction')
    })
      .onActionEnd(() => {
      console.log('LongPress end')
    }),
    PanGesture({ direction: PanDirection.Horizontal })
      .onActionStart(() => {
      console.log('pan start')
    })
      .onActionUpdate(() => {
      console.log('pan update')
    })
  )
    .onCancel(() => {
    console.log('sequence gesture canceled')
  })
)
```

## 13.6 框架结构详细讲解

本章详细讲解 ArkUI ETS UI 框架的项目结构和布局、基础组件的使用等。

### 13.6.1 文件组织

ArkUI ETS FA 应用的 JS 模块（entry/src/main/js/componentName）的典型开发目录结构如图 13.12 所示。

目录结构中文件分类如下：

.ets 结尾的 ETS(Extended TypeScript)文件，这个文件用于描述 UI 布局、样式、事件交互和页面逻辑。

各个文件夹和文件的作用：

(1) app.ets 文件用于管理全局应用逻辑和应用生命周期。

(2) pages 目录用于存放所有组件页面。

(3) common 目录用于存放公共代码文件，例如：自定义组件和公共方法。

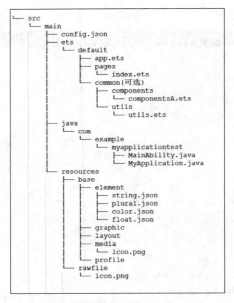

图 13.12　目录结构中的文件分类

（4）resources 目录用于存放资源配置文件，例如：国际化字符串、与资源限定相关的资源和 rawfile 资源等。

资源目录 resources 文件夹位于 src/main 下，此目录下资源文件的详细规范及子目录结构规范可参看资源文件说明。

页面支持导入 TypeScript 和 JavaScript。

### 13.6.2　JS 标签配置

开发框架需要在应用的 config.json 文件中配置相关的 JS 标签，包含实例名称和页面路径。

**1. pages**

定义每个页面入口组件的路由信息，每个页面由页面路径和页面名组成，页面的文件名就是页面名，如代码示例 13.32 所示。

代码示例 13.32

```
{
"pages": [
"pages/index",
"pages/detail"
]
}
```

**注意**：pages 列表中第 1 个页面为应用的首页入口。页面文件名不能使用组件名称，例如：Text.ets、Button.ets 等。每个页面文件中必须包含入口组件。

## 2. window

window 用于配置相关视图显示窗口，支持的属性如表 13.7 所示。

表 13.7 window 的属性

| 参数名 | 默认值 | 说明 |
| --- | --- | --- |
| designWidth | - | 配置视图显示的逻辑宽度，缺省默认值为 720（智能穿戴默认值为 454）。视图显示的逻辑宽度决定了 1px 像素单位大小，如当将 designWidth 配置为 720，并且视图宽度为 1440 物理像素时，1px 为 2 物理像素。详见 13.6.5 节像素单位说明 |

相关配置如代码示例 13.33 所示。

代码示例 13.33

```
"window": {
"designWidth": 720,
"autoDesignWidth": false
}
```

## 3. mode

mode 用于配置 JS Component 的运行类型与风格，支持的属性如表 13.8 所示。

表 13.8 mode 的属性

| 参数名 | 默认值 | 说明 |
| --- | --- | --- |
| type | - | 配置该 JS Component 的运行类型，可选值为<br>pageAbility：以 ability 的方式运行该 JS Component。<br>form：以卡片的方式运行该 JS Component |
| syntax | | 配置该 JS Component 的语法风格，可选值为<br>hml：以 HML/CSS/JS 风格进行编写。<br>ets：以声明式语法风格进行编写 |

相关配置如代码示例 13.34 所示。

代码示例 13.34

```
"mode": {
"syntax": "ets",
"type": "pageAbility"
}
```

不支持同时将 type 类型配置为 form，将 syntax 类型配置为 ets。

### 13.6.3 app.ets

全局应用逻辑和应用生命周期管理如代码示例 13.35 所示。

代码示例 13.35

```
//app.ets
export default {
 onCreate() {
  console.info('Application onCreate')
 },
 onDestroy() {
  console.info('Application onDestroy')
 },
}
```

### 13.6.4 资源访问

在 ets 文件中,可以使用在 resources 目录中定义的资源,如代码示例 13.36 所示。

代码示例 13.36

```
Text($r('app.string.string_hello'))
 .fontColor($r('app.color.color_hello'))
 .fontSize($r('app.float.font_hello'))
}
Text($r('app.string.string_world'))
 .fontColor($r('app.color.color_world'))
 .fontSize($r('app.float.font_world'))
}
//引用 string 资源,$r 的第 2 个参数用于替换 %s
Text($r('app.string.message_arrive', "five of the clock")).fontColor($r('app.color.color_
hello')).fontSize($r('app.float.font_hello'))
}
//plural $r 引用,第 1 个参数用于指定 plural 资源,第 2 个参数用于指定单复数的数量,此处第 3
//个数字为对 %d 的替换
Text($r('app.plural.eat_apple', 5, 5)).fontColor($r('app.color.color_world'))
 .fontSize($r('app.float.font_world'))
}
Image($r('app.media.my_background_image'))        //media 资源的 $r 引用
Image($rawfile('test.png'))                       //rawfile $r 引用 rawfile 目录下的图片
Image($rawfile('newDir/newTest.png'))             //rawfile $r 引用 rawfile 目录下的图片
```

### 13.6.5 像素单位

ArkUI 2 为开发者提供了 4 种像素单位,框架采用 vp 为基准数据单位,并且提供了其他单位数据到 vp 单位转换的方法,如表 13.9 所示。

表 13.9 ArkUI 2 提供的 4 种像素单位

| 参数名 | 说 明 |
| --- | --- |
| px | 屏幕物理像素单位 |
| vp | 与屏幕密度相关的像素,可根据屏幕像素密度转换为屏幕物理像素 |

| 参数名 | 说明 |
|---|---|
| fp | 字体像素,与 vp 类似,适用于屏幕密度变化,可随系统字体大小设置 |
| 1px | 视窗逻辑像素单位,1px 单位为实际屏幕宽度与逻辑宽度的比值。如当将 designWidth 配置为 720 时,在实际宽度为 1440 物理像素的屏幕上,11px 为 2px 大小 |

## 13.7 界面布局

ArkUI ETS UI 提供了一维 Flex 布局和二维 Grid 布局,以及栅格布局和堆叠布局,通过布局容器可以快速构建 App 的界面。

### 13.7.1 Flex 布局

Flex 布局仅适用于手机和平板布局,不支持智慧屏和穿戴设备。ArkUI ETS UI 中的 Flex 布局和 ArkUI JS UI 中的 Flex 布局是一样的,ArkUI ETS 采用了函数式写法,所包含的属性基本是一致的,如图 13.13 所示。

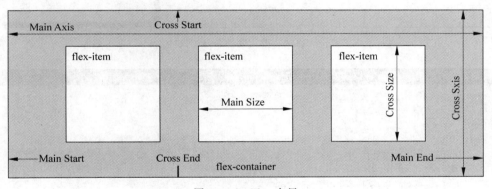

图 13.13　Flex 布局

Flex 布局提供声明式的方法调用并实现弹性布局,如代码示例 13.37 所示。

代码示例 13.37
```
Flex(value:{direction?: FlexDirectionwrap?: FlexWrap,justifyContent?:
FlexAlign,alignItems?: ItemAlign,alignContent?: FlexAlign})
```

下面使用 Flex 布局实现微信底部 Tab 栏的效果,如图 13.14 所示。将 Flex 组件设置为水平方向排列子元素,水平和垂直方向居中对齐,内部子元素直接以空格隔开,如代码示例 13.38 所示。

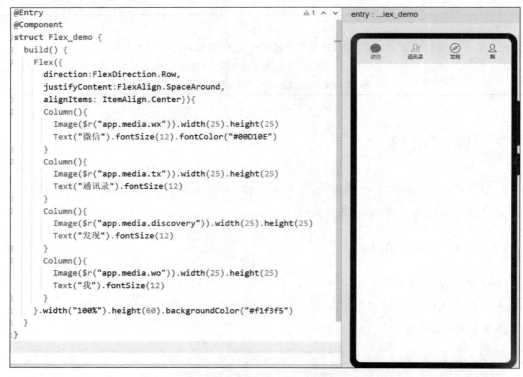

图 13.14　微信底部 Tab 栏效果

代码示例 13.38　chapter13\07_layout\flex_demo.ets

```
@Entry
@Component
struct Flex_demo {
  build() {
    Flex({
      direction:FlexDirection.Row,
      justifyContent:FlexAlign.SpaceAround,
      alignItems: ItemAlign.Center}){
      Column(){
        Image( $ r("app.media.wx")).width(25).height(25)
        Text("微信").fontSize(12).fontColor("#00D10E")
      }
      Column(){
        Image( $ r("app.media.tx")).width(25).height(25)
        Text("通讯录").fontSize(12)
      }
      Column(){
        Image( $ r("app.media.discovery")).width(25).height(25)
        Text("发现").fontSize(12)
```

```
        }
        Column(){
          Image( $ r("app.media.wo")).width(25).height(25)
          Text("我").fontSize(12)
        }
      }.width("100%").height(60).backgroundColor("#f1f3f5")
    }
  }
```

### 13.7.2　Grid 布局

网格布局适用于静态固定尺寸的布局场景。网格容器布局仅支持手机和平板，不支持智慧屏和穿戴设备。

Flex 布局是轴线布局，只能指定"项目"针对轴线的位置，可以看作一维布局，而 Grid 布局则是将容器划分成"行"和"列"，产生单元格，然后指定"项目"所在的单元格，可以看作二维布局，Grid 布局远比 Flex 布局强大，如图 13.15 所示。

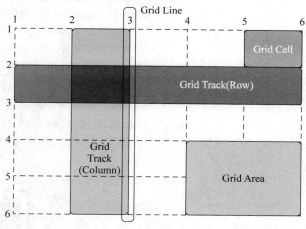

图 13.15　网格容器

网格容器支持的属性如表 13.10 所示。

表 13.10　网格容器支持的属性

| 参数名 | 类型 | 默认值 | 说　　明 |
| --- | --- | --- | --- |
| columnsTemplate | String | '1fr' | 用于设置当前网格布局列的数量，不设置时默认为 1 列。示例，'1fr 1fr 2fr'表示分三列，将父组件允许的宽分为 4 等份，第一列占 1 份，第二列占 1 份，第三列占 2 份 |

续表

| 参数名 | 类型 | 默认值 | 说 明 |
|---|---|---|---|
| rowsTemplate | String | '1fr' | 用于设置当前网格布局行的数量,不设置时默认为1行。示例,'1fr 1fr 2fr'表示分三行,将父组件允许的高分为 4 等份,第一行占 1 份,第二行占 1 份,第三行占 2 份 |
| columnsGap | Length | | 用于设置列与列的间距 |
| rowsGap | Length | | 用于设置行与行的间距 |
| scrollBarWidth | Length | | 用于设置滚动条的宽度 |
| scrollBarColor | Color | | 用于设置滚动条的颜色 |
| scrollBar | BarState | Off | 用于设置滚动条的状态 |

网格子容器(GridItem)支持的属性如表 13.11 所示。

表 13.11 网格子容器(GridItem)支持的属性

| 参数名 | 类型 | 默认值 | 说 明 |
|---|---|---|---|
| rowStart | number | | 用于指定当前元素的起始行号 |
| rowEnd | number | | 用于指定当前元素的终点行号 |
| columnStart | number | | 用于指定当前元素的起始列号 |
| columnEnd | number | | 用于指定当前元素的终点列号 |
| forceRebuild | boolean | false | 用于设置在触发组件 build 时是否重新创建此节点 |

Grid 布局能够完成一些 Flex 布局不太好完成的界面效果,如图 13.16 所示的界面,这是一个不规则的界面效果,使用 Grid 布局可以轻松解决,如代码示例 13.39 所示。

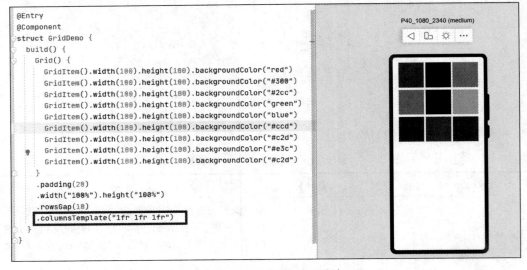

图 13.16 不规则的界面设计

代码示例 13.39　chapter13\07_layout\grid_demo.ets

```
@Entry
@Component
struct GridDemo {
  build() {
    Grid() {
      GridItem().width(100).height(100).backgroundColor("red")
      GridItem().width(100).height(100).backgroundColor("#300")
      GridItem().width(100).height(100).backgroundColor("#2cc")
      GridItem().width(100).height(100).backgroundColor("green")
      GridItem().width(100).height(100).backgroundColor("blue")
      GridItem().width(100).height(100).backgroundColor("#ccd")
      GridItem().width(100).height(100).backgroundColor("#c2d")
      GridItem().width(100).height(100).backgroundColor("#e3c")
      GridItem().width(100).height(100).backgroundColor("#c2d")
    }
    .padding(20)
    .width("100%").height("100%")
    .rowsGap(10)
    .columnsTemplate("1fr 1fr 1fr")
  }
}
```

### 13.7.3　堆叠布局

堆叠布局通过堆叠容器让子组件按照一定的顺序依次入栈，后一个子组件覆盖前一个子组件，如图 13.17 所示。

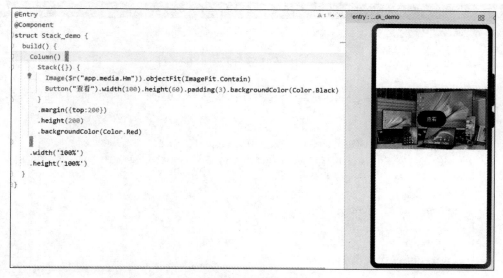

图 13.17　堆叠布局

Stack 构造接口说明如表 13.12 所示。

表 13.12 Stack 构造接口说明

| 参数名 | 类型 | 默认值 | 说明 |
| --- | --- | --- | --- |
| alignContent | Alignment | Center | 设置子组件在容器内的对齐方式 |

Alignment 枚举说明如表 13.13 所示。

表 13.13 Alignment 枚举说明

| 参数名 | 说明 |
| --- | --- |
| TopStart | 顶部起始端 |
| Top | 顶部横向居中 |
| TopEnd | 顶部尾端 |
| Start | 起始端纵向居中 |
| Center | 横向和纵向居中 |
| End | 尾端纵向居中 |
| BottomStart | 底部起始端 |
| Bottom | 底部横向居中 |
| BottomEnd | 底部尾端 |

堆叠的位置,可以通过 Stack 的构造属性 alignContent 设置堆叠的对齐方式,默认为居中堆叠对齐,如代码示例 13.40 所示。

代码示例 13.40　chapter13\07_layout\stack_demo.ets

```
@Entry
@Component
struct Stack_demo {
  build() {
    Flex({ direction: FlexDirection.Column, alignItems: ItemAlign.Center, justifyContent: FlexAlign.Center }) {
      Stack({ alignContent: Alignment.Top }) {
        Image( $ r("app.media.Hm")).objectFit(ImageFit.Contain)
        Button("查看").width(100).height(60).padding(3).backgroundColor(Color.Black)
      }
      .height(200)
      .backgroundColor(Color.Red)
    }
    .width('100 % ')
    .height('100 % ')
  }
}
```

## 13.7.4 栅格布局

栅格布局借鉴了 Bootstrap 中栅格布局的设计,把 HarmonyOS 的设备分为大屏 12 栅格布局、中屏 8 栅格布局和小屏 4 栅格布局,如图 13.18 所示。

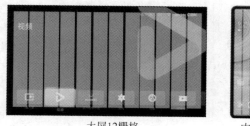

大屏12栅格　　　　中屏8栅格　　小屏4栅格

图 13.18　栅格布局

栅格布局的列宽、列间距由距离最近的 GridContainer 父组件决定。使用栅格属性的组件树上至少需要有 1 个 GridContainer 容器组件,GridContainer 属性如表 13.14 所示。

表 13.14　GridContainer 属性

| 属性名 | 类　　型 | 默认值 | 描　　述 |
| --- | --- | --- | --- |
| useSizeType | {<br>xs?: number \| { span: number, offset: number },<br>sm?: number \| { span: number, offset: number },<br>md?: number \| { span: number, offset: number },<br>lg?: number \| { span: number, offset: number }<br>} | Start | 设置在特定设备宽度类型下的占用列数和偏移列数,<br>span:占用列数;<br>offset:偏移列数。<br>当值为 number 类型时,仅设置列数,当格式如{"span":1, "offset":0}时,指同时设置占用列数与偏移列数。<br>xs:指设备宽度类型为 SizeType.XS 时的占用列数和偏移列数。sm:指设备宽度类型为 SizeType.SM 时的占用列数和偏移列数。md:指设备宽度类型为 SizeType.MD 时的占用列数和偏移列数。lg:指设备宽度类型为 SizeType.LG 时的占用列数和偏移列数 |
| gridSpan | number | 1 | 默认占用列数,指 useSizeType 属性没有设置对应尺寸的列数(span)时,占用的栅格列数。<br>说明:<br>如果设置了栅格 span 属性,则组件的宽度由栅格布局决定 |

续表

| 属性名 | 类型 | 默认值 | 描述 |
|---|---|---|---|
| gridOffset | number | 0 | 默认偏移列数，指 useSizeType 属性没有设置对应尺寸的偏移（offset）时，当前组件沿着父组件 Start 方向偏移的列数，也就是当前组件位于第 n 列。<br>说明：<br>(1) 配置该属性后，当前组件在父组件水平方向的布局不再跟随父组件原有的布局方式，而是沿着父组件的 Start 方向偏移一定位移。<br>(2) 偏移位移＝(列宽＋间距)×列数<br>(3) 设置了偏移(gridOffset)的组件之后的兄弟组件会根据该组件进行相对布局，类似相对布局 |

栅格布局如代码示例 13.41 所示。

代码示例 13.41 chapter13\08_base_components\text_demo.ets

```
@Entry
@Component
struct Grid_container_demo {
  build() {
    GridContainer() {
      Row() {
        Button('G1')
          .gridSpan(2)
          .gridOffset(1)
          .useSizeType({lg: 4})
        Button('G2')
          .gridSpan(2)
          .useSizeType({md: {span: 1, offset: 0}})
      }.width(400)
      .align(Alignment.Center)
    }
    .backgroundColor(Color.Red)
  }
}
```

## 13.8 基础组件

本书中介绍的组件是应用开发中必备的基础组件，其他的组件的用法基本相同，本书不做详细介绍。

### 13.8.1 Text 组件

文本组件,用于呈现一段信息,可以包含 Span 子组件,Text 组件的属性如表 13.15 所示。

表 13.15 Text 组件的属性

| 属性名 | 类型 | 默认值 | 描述 |
| --- | --- | --- | --- |
| textAlign | TextAlign | Start | 设置多行文本的文本对齐方式 |
| textOverflow | {overflow: TextOverflow} | {overflow: TextOverflow.Clip} | 设置文本超长时的显示方式 |
| maxLines | number | Infinity | 设置文本的最大行数 |
| lineHeight | Length | - | 设置文本的文本行高,当设置值不大于 0 时,不限制文本行高,自适应字体大小,当 Length 为 number 类型时单位为 fp |
| decoration | {type: TextDecorationType, color?: Color} | {type: TextDecorationType.None} | 设置文本装饰线样式及其颜色 |
| baselineOffset | Length | - | 设置文本基线的偏移量 |
| textCase | TextCase | TextCase.Normal | 设置文本大小写 |

Text 组件的用法,如图 13.19 所示,Text 组件可以嵌套多个 Span 组件,Text 组件可以设置文字超出范围的截取方式和最大显示的行数。

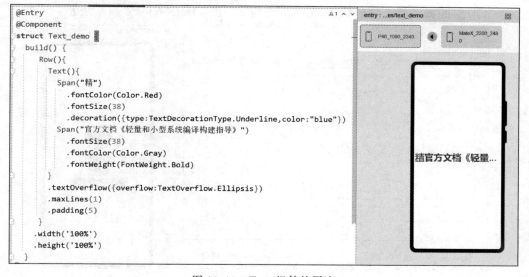

图 13.19 Text 组件的用法

Text 组件的用法如代码示例 13.42 所示。

代码示例 13.42　chapter13\08_base_components\text_demo.ets

```
@Entry
@Component
struct Text_demo {
  build() {
    Row(){
      Text(){
        Span("精")
          .fontColor(Color.Red)
          .fontSize(38)
          .decoration({type:TextDecorationType.Underline,color:"blue"})
        Span("官方文档《轻量和小型系统编译构建指导》")
          .fontSize(38)
          .fontColor(Color.Gray)
          .fontWeight(FontWeight.Bold)
      }
      .textOverflow({overflow:TextOverflow.Ellipsis})
      .maxLines(1)
      .padding(5)
    }
    .width('100%')
    .height('100%')
  }
}
```

### 13.8.2　Button 组件

Button 组件提供了 3 种形状，如图 13.20 所示，通过 ButtonType 枚举设置：

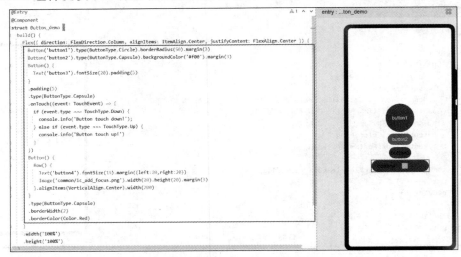

图 13.20　Button 组件的形状和用法

(1) Capsule 胶囊型按钮(圆角默认为高度的一半)。
(2) Circle 圆形按钮。
(3) Normal 普通按钮(默认不带圆角)。

Button 组件的用法如代码示例 13.43 所示。

代码示例 13.43　chapter13\08_base_components\button_demo.ets

```
@Entry
@Component
struct Button_demo {
  build() {
    Flex({ direction: FlexDirection.Column, alignItems: ItemAlign.Center, justifyContent:
FlexAlign.Center }) {
      Button('button1').type(ButtonType.Circle).borderRadius(50).margin(3)
      Button('button2').type(ButtonType.Capsule).backgroundColor('#f00').margin(3)
      Button() {
        Text('button3').fontSize(20).padding(5)
      }
      .padding(5)
      .type(ButtonType.Capsule)
      .onTouch((event: TouchEvent) => {
        if (event.type === TouchType.Down) {
          console.info('Button touch down!');
        } else if (event.type === TouchType.Up) {
          console.info('Button touch up!')
        }
      })
      Button() {
        Row() {
          Text('button4').fontSize(19).margin({left:20,right:20})
          Image('common/ic_add_focus.png').width(20).height(20).margin(3)
        }.alignItems(VerticalAlign.Center).width(200)
      }
      .type(ButtonType.Capsule)
      .borderWidth(2)
      .borderColor(Color.Red)
    }
    .width('100%')
    .height('100%')
  }
}
```

### 13.8.3　Image 组件

图片组件用来渲染展示图片。如果需要使用网络图片,则需要申请 ohos.permission.INTERNET(使用网络图片)权限。Image 组件的属性如表 13.16 所示。

表 13.16 Image 组件的属性

| 属性名 | 类型 | 默认值 | 描述 |
| --- | --- | --- | --- |
| alt | string | - | 加载时显示的占位图。支持本地路径图片资源,也支持通过指定资源类型(type)和资源名称(name)来引用,比如 $ m('image.png') |
| objectRepeat | {overflow: TextOverflow} | {overflow: TextOverflow.Clip} | 设置文本超长时的显示方式 |
| interpolation | - | None | 设置图片的插值效果,仅针对图片放大插值。说明: SVG 类型图源不支持该属性。PixelMap 资源不支持该属性 |
| renderMode | - | Original | 设置图片渲染的模式。说明: SVG 类型图源不支持该属性 |
| sourceSize | { width: number, height: number } | - | 设置图片解码尺寸,将原始图片解码成指定尺寸的图片,number 类型的单位为 px。说明: PixelMap 资源不支持该属性 |

Image 组件的用法如图 13.21 所示。

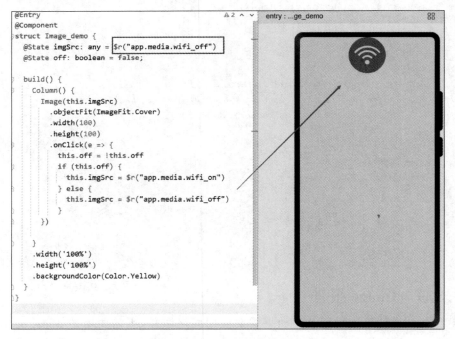

图 13.21 Image 组件的用法

图片案例如代码示例13.44所示。

```
代码示例13.44
@Entry
@Component
struct Image_demo {
  @State imgSrc: any = $r("app.media.wifi_off")
  @State off: boolean = false;

  build() {
    Column() {
      Image(this.imgSrc)
        .objectFit(ImageFit.Cover)
        .width(100)
        .height(100)
        .onClick(e => {
          this.off = !this.off
          if (this.off) {
            this.imgSrc = $r("app.media.wifi_on")
          } else {
            this.imgSrc = $r("app.media.wifi_off")
          }
        })
    }
    .width('100%')
    .height('100%')
    .backgroundColor(Color.Yellow)
  }
}
```

### 13.8.4 List 组件

列表包含一系列相同宽度的列表项。适合连续、多行呈现同类数据，例如图片和文本。ListItem 组件用来展示列表的具体 item，宽度默认充满 List 组件，ListItem 组件必须配合 List 来使用，List 属性如表13.17所示。

表13.17 List 属性

| 属性名 | 类型 | 默认值 | 描述 |
| --- | --- | --- | --- |
| listDirection | Axis.Vertical /Horizontal | Vertical | 设置 List 组件排列方向 |
| divider | { strokeWidth: Length, color?:Color, | - | 用于设置 ListItem 分割线样式，默认无分割线 |

续表

| 属性名 | 类型 | 默认值 | 描 述 |
| --- | --- | --- | --- |
| divider | startMargin?: Length, endMargin?: Length } | - | 用于设置 ListItem 分割线样式,默认无分割线 |
| editMode | boolean | false | 声明当前 List 组件是否处于可编辑模式 |
| edgeEffect | EdgeEffect | Spring | 滑动效果 |
| chainAnimation | boolean | false | 用于设置当前 List 是否启用链式联动动效,开启后列表滑动到顶部和底部拖曳时会有链式联动的效果。<br>链式联动效果：List 内的 list-item 间隔一定距离,在基本的滑动交互行为下,主动对象驱动从动对象进行联动,驱动效果遵循弹簧物理动效。<br>false：不启用链式联动。<br>true：启用链式联动 |

这里通过 List 组件实现一个微信列表页,效果如图 13.22 所示。

图 13.22　List 组件的用法

List 组件的实现如代码示例 13.45 所示。

代码示例 13.45　chapter13\08_base_components\list_demo.ets

```
@Entry
@Component
struct FlexWeChat {

  @State lists:any = [1,2,3,4,5,6,7,8,9,10,11,12]

  build() {
    Flex({
      direction:FlexDirection.Column
    }){
      Flex({justifyContent:FlexAlign.Center,alignItems:ItemAlign.Center}){
        Text("微信(30)").fontSize(20).fontWeight(600)
      }.width("100%").height(60).backgroundColor("#f1f3f5")
      List(){
        ForEach(this.lists,num =>{
          ListItem(){
            Row(){
              Image( $r("app.media.xx")).margin({left:10}).borderRadius(8).width(60).height(60)
              Flex({
                direction:FlexDirection.Row
              }){
                Column(){
                  Text("张三丰-武当派掌门人").
margin({top:20,left:10}).fontSize(18).alignSelf(ItemAlign.Start)
                  Text("最近我的太极拳又修到了80级...")
                    .fontColor("#BDBDBD")
                    .margin({top:10,left:10}).fontSize(14).alignSelf(ItemAlign.Start)
                }.width("100%").height("100%")
                Column(){
                  Text("8月2号").fontColor("#BDBDBD").margin({top:10,left:5}).fontSize
(13).alignSelf(ItemAlign.Start)
                }.width(200).height("100%")
              }
              .margin({top:10,bottom:10,left:10,right:10})
              .width("100%").height("100%")
            }
          }.margin({bottom:1}).width("100%").height(100)
        })
      }
      .height("100%").width("100%").divider({
        strokeWidth: 1, color: "#f1f3f5",
        startMargin: 80, endMargin: 10})

      Flex({
        direction:FlexDirection.Row,
        justifyContent:FlexAlign.SpaceAround,
```

```
          alignItems: ItemAlign.Center}){
        Column(){
          Image($r("app.media.wx")).width(25).height(25)
          Text("微信").fontSize(12).fontColor("#00D10E")
        }
        Column(){
          Image($r("app.media.tx")).width(25).height(25)
          Text("通讯录").fontSize(12)
        }
        Column(){
          Image($r("app.media.discovery")).width(25).height(25)
          Text("发现").fontSize(12)
        }
        Column(){
          Image($r("app.media.wo")).width(25).height(25)
          Text("我").fontSize(12)
        }
      }.width("100%").height(60).backgroundColor("#f1f3f5")
    }
  }
}
```

### 13.8.5　Swiper 组件

Swiper 滑动容器，如图 13.23 所示，提供切换子组件显示的能力，Swiper 构造方法参数如表 13.18 所示。

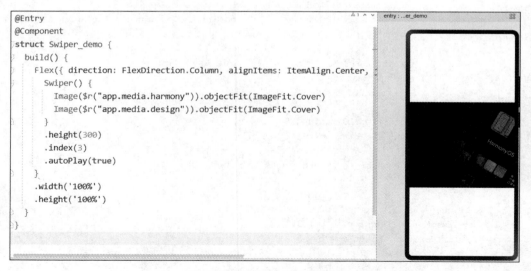

图 13.23　Swiper 组件的用法

表 13.18　Swiper 构造方法参数

| 属性名 | 类型 | 是否必填 | 默认值 | 描述 |
|---|---|---|---|---|
| controller | SwiperController | 否 | Null | 给组件绑定一个控制器,用来控制组件翻页 |

Swiper 属性如表 13.19 所示。

表 13.19　Swiper 属性

| 属性名 | 类型 | 默认值 | 描述 |
|---|---|---|---|
| index | Number | 0 | 设置当前在容器中显示的子组件的索引 |
| autoPlay | boolean | false | 子组件是否自动播放,在自动播放状态下导航点不可操作 |
| interval | Number | 3000 | 使用自动播放时播放的时间间隔,单位为毫秒 |
| indicator | boolean | true | 是否启用导航点指示器 |
| loop | boolean | true | 是否开启循环 |
| duration | number | 400 | 子组件切换的动画时长,单位为毫秒 |
| vertical | boolean | false | 是否为纵向滑动 |
| itemSpace | Length | 0 | 设置子组件与子组件之间的间隙 |
| displayMode | SwiperDisplayMode | Stretch | 设置当 Swiper 容器在主轴上尺寸(水平值描述滑动时为宽度,纵向滑动时为高度)大于子组件时,在 Swiper 里的呈现方式 |

Swiper 组件的用法如代码示例 13.46 所示。

代码示例 13.46　chapter13\08_base_components\swiper_demo.ets

```
@Entry
@Component
struct Swiper_demo {
  build() {
    Flex({ direction: FlexDirection.Column, alignItems: ItemAlign.Center, justifyContent: FlexAlign.Center }) {
      Swiper() {
        Image($r("app.media.harmony")).objectFit(ImageFit.Cover)
        Image($r("app.media.design")).objectFit(ImageFit.Cover)
      }
      .height(300)
      .index(3)
      .autoPlay(true)
    }
    .width('100%')
    .height('100%')
  }
}
```

### 13.8.6　Tabs 组件

Tabs 组件是一种可以通过页签进行内容视图切换的容器组件，每个页签对应一个内容视图（TabContent），即对应一个切换页签的内容视图，如图 13.24 所示，Tabs 属性如表 13.20 所示。

```
@Entry
@Component
struct TabDemo {
  private tabIndex: number = 1;
  private controller: TabsController = new TabsController();
  build() {
    Tabs({
      index: this.tabIndex,
      controller: this.controller
    }) {
      TabContent() {
        Text("1").fontSize(30)
      }
      .backgroundColor(Color.Red)
      .tabBar({ text: "1" })

      TabContent() {
        Text("2").fontSize(30)
      }
      .backgroundColor(Color.Green)
      .tabBar({ text: "2" })
    }
    .height(300)
    .vertical(false)
    .barMode(BarMode.Scrollable)
  }
}
```

图 13.24　Tabs 组件的用法

表 13.20　Tabs 属性

| 属性名 | 类型 | 默认值 | 描述 |
| --- | --- | --- | --- |
| vertical | boolean | 是否为纵向 Tab，默认值为 false | 是否为纵向 Tab，默认值为 false |
| scrollable | boolean | 是否可以通过左右滑动进行页面切换，默认值为 true | 是否可以通过左右滑动进行页面切换，默认值为 true |
| barMode | | TabBar 布局模式 | TabBar 布局模式 |
| barWidth | number | TabBar 的宽度值，不设置时使用系统主题中的默认值 | TabBar 的宽度值，不设置时使用系统主题中的默认值 |
| barHeight | number | TabBar 的高度值，不设置时使用系统主题中的默认值 | TabBar 的高度值，不设置时使用系统主题中的默认值 |
| animationDuration | number | 200 | TabContent 滑动动画时长 |

Tabs 组件的实现如代码示例 13.47 所示。

代码示例 13.47  chapter13\08_base_components\tab_demo.ets
```
@Entry
@Component
struct TabDemo {
  private tabIndex: number = 1;
  private controller: TabsController = new TabsController();
  build() {
    Tabs({
      index: this.tabIndex,
      controller: this.controller
    }) {
      TabContent() {
        Text("1").fontSize(30)
      }
      .backgroundColor(Color.Red)
      .tabBar({ text: "1" })

      TabContent() {
        Text("2").fontSize(30)
      }
      .backgroundColor(Color.Green)
      .tabBar({ text: "2" })
    }
    .height(300)
    .vertical(false)
    .barMode(BarMode.Scrollable)
  }
}
```

## 13.8.7 Scroll 组件

可滚动的容器组件，当子组件的布局尺寸超过父组件的视口时，内容可以滚动。Scroller 控制器类可以通过代码控制 Scroll 的滚动，Scroller 是可滚动容器组件的控制器，可以将 Scroller 实例绑定至容器组件，然后通过它控制容器组件的滚动，目前支持绑定到 List 和 Scroll 组件上，Scroll 属性如表 13.21 所示。

表 13.21  Scroll 属性

| 属性名 | 类型 | 默认值 | 描述 |
| --- | --- | --- | --- |
| scrollable | ScrollDirection | Vertical | 设置滚动方法 |
| scrollBar | BarState | Auto | 设置滚动条状态 |
| scrollBarColor | Color | - | 设置滚动条的颜色 |
| scrollBarWidth | Length | - | 设置滚动条的宽度 |

ScrollDirection 枚举说明如表 13.22 所示。

表 13.22　ScrollDirection 枚举说明

| 属 性 名 | 类　　型 |
| --- | --- |
| Horizontal | 仅支持水平方向滚动 |
| Vertical | 仅支持竖直方向滚动 |
| None | 不可滚动 |

Scroll 组件的用法如图 13.25 所示。

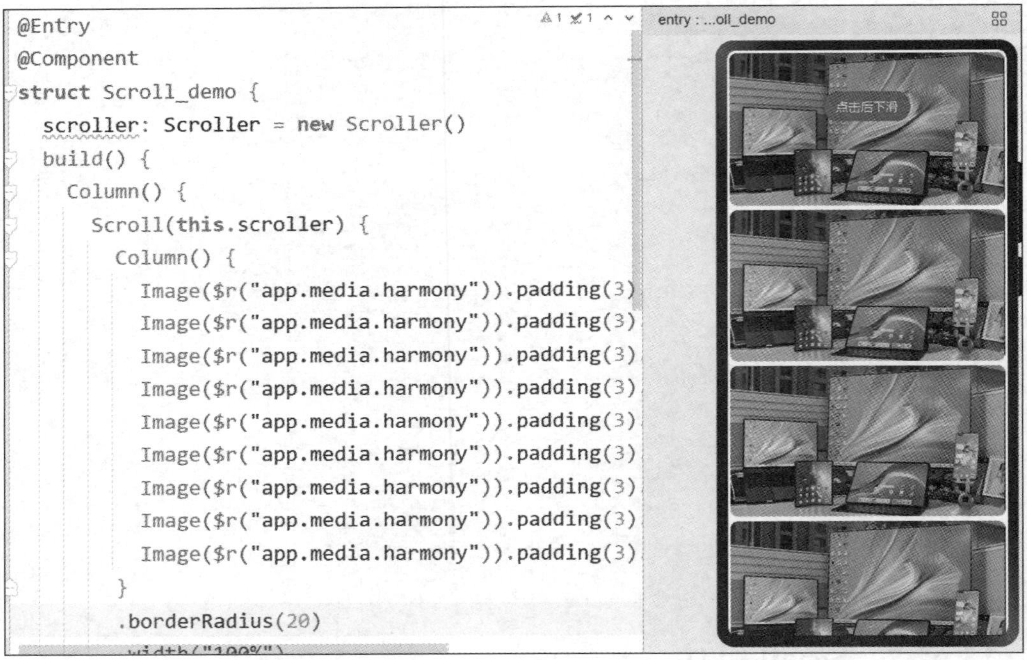

```
@Entry
@Component
struct Scroll_demo {
  scroller: Scroller = new Scroller()
  build() {
    Column() {
      Scroll(this.scroller) {
        Column() {
          Image($r("app.media.harmony")).padding(3)
          Image($r("app.media.harmony")).padding(3)
          Image($r("app.media.harmony")).padding(3)
          Image($r("app.media.harmony")).padding(3)
          Image($r("app.media.harmony")).padding(3)
          Image($r("app.media.harmony")).padding(3)
          Image($r("app.media.harmony")).padding(3)
          Image($r("app.media.harmony")).padding(3)
          Image($r("app.media.harmony")).padding(3)
        }
        .borderRadius(20)
        .width("100%")
```

图 13.25　Scroll 组件的用法

Scroll 组件的实现如代码示例 13.48 所示。

代码示例 13.48　chapter13\08_base_components\scroll_demo.ets

```
@Entry
@Component
struct Scroll_demo {
  scroller: Scroller = new Scroller()
  build() {
    Column() {
      Scroll(this.scroller) {
        Column() {
          Image( $ r("app.media.harmony")).padding(3).borderRadius(10)
```

```
            Image($r("app.media.harmony")).padding(3).borderRadius(10)
            Image($r("app.media.harmony")).padding(3).borderRadius(10)
            Image($r("app.media.harmony")).padding(3).borderRadius(10)
            Image($r("app.media.harmony")).padding(3).borderRadius(10)
            Image($r("app.media.harmony")).padding(3).borderRadius(10)
            Image($r("app.media.harmony")).padding(3).borderRadius(10)
            Image($r("app.media.harmony")).padding(3).borderRadius(10)
            Image($r("app.media.harmony")).padding(3).borderRadius(10)
        }
        .borderRadius(20)
        .width("100%")
    }
    .onScrollEdge((side: Edge) => {
        if (side === Edge.Bottom) {
            console.log('Reach Bottom')
        }
    })

    Row() {
        Button('单击后下滑')
            .onClick(() => {          //单击后下滑100.0距离
                this.scroller
                    .scrollTo({xOffset: 0, yOffset:
                    this.scroller.currentOffset().yOffset + 100})
            }).backgroundColor(Color.Red)
    }
  }
 }
}
```

### 13.8.8 AlertDialog 组件

显示警告弹框组件，可设置文本内容与响应回调，如图13.26所示。

AlertDialog 组件的实现如代码示例13.49所示。

**代码示例13.49   chapter13\08_base_components\show_alert_demo.ets**

```
@Entry
@Component
struct ShowAlertDemo {
  build() {
    Flex({
      justifyContent:FlexAlign.Center,
      alignItems:ItemAlign.Center
    }) {
      Button('单击弹框')
        .onClick(() => {
          AlertDialog.show({
```

```
          title: '删除',
          message: '删除后不可恢复,确认需要删除吗?',
          confirm:
          {
            value: '确认',
            action: () => {
              console.log('Get Alert Dialog handled')
            }
          }})
      })
    }
  }
}
```

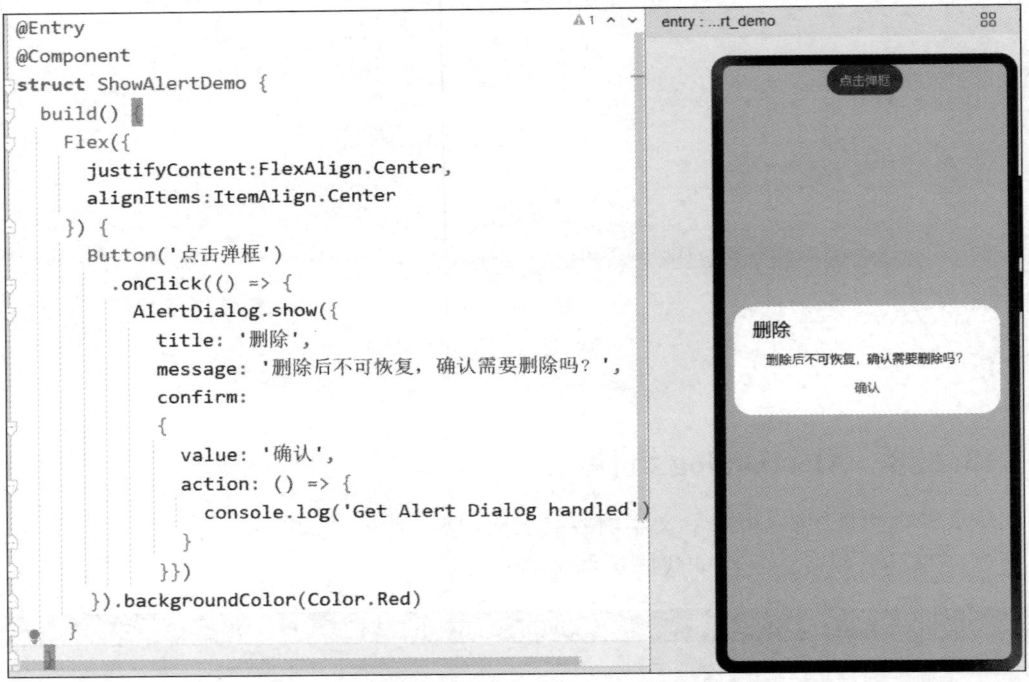
@Entry
@Component
struct ShowAlertDemo {
build() {
Flex({
justifyContent:FlexAlign.Center,
alignItems:ItemAlign.Center
}) {
Button('点击弹框')
.onClick(() => {
AlertDialog.show({
title: '删除',
message: '删除后不可恢复，确认需要删除吗？',
confirm:
{
value: '确认',
action: () => {
console.log('Get Alert Dialog handled')
}
}})
}).backgroundColor(Color.Red)
}
entry : ...rt_demo
点击弹框
删除
删除后不可恢复，确认需要删除吗?
确认

图 13.26　AlertDialog 组件的用法

## 13.8.9　自定义弹框

可以通过 CustomDialogController 构造方法引用创建的自定义的弹框组件，如图 13.27 所示。创建一个自定义组件，如 DialogExample 组件。

自定义弹框组件的实现如代码示例 13.50 所示。

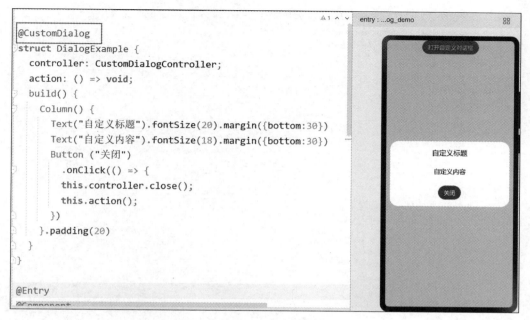

图 13.27　自定义弹框组件的用法

代码示例 13.50　chapter13\08_base_components\text_demo.ets

```
@CustomDialog
struct DialogExample {
  controller: CustomDialogController;
  action: () => void;
  build() {
    Column() {
      Text("自定义标题").fontSize(20).margin({bottom:30})
      Text("自定义内容").fontSize(18).margin({bottom:30})
      Button ("关闭")
        .onClick(() => {
          this.controller.close();
          this.action();
        })
    }.padding(20)
  }
}
```

通过 CustomDialogController 类显示自定义弹框，如代码示例 13.51 所示。

代码示例 13.51　chapter13\08_base_components\custom_dialog_demo.ets

```
@Entry
@Component
struct CustomDialogDemo {
```

```
    dialogController : CustomDialogController = new CustomDialogController({
      builder: DialogExample({action: this.onAccept}),
      cancel: this.existApp,
      autoCancel: true
    });
    onAccept() {
      console.log("onAccept");
    }
    existApp() {
      console.log("Cancel dialog!");
    }
    build() {
      Column() {
        Button("打开自定义对话框")
          .onClick(() => {
            this.dialogController.open()
          })
          .backgroundColor(Color.Red)
      }
      .width("100%")
      .height("100%")
    }
}
```

弹窗容器样式不可自定义,如宽度、圆角、动效等,其中容器宽度固定,高度自适应子节点高度,自定义弹框 builder 的结果作为弹窗容器的子节点。

# 第 14 章 ArkUI ETS UI 开发案例

本章详细讲解 ArkUI ETS UI 开发华为商城 App，本案例中使用的组件都是第 13 章讲解的基础组件的用法，本章分 3 个小节详细介绍如何开发页面中的每个部分，效果如图 14.1 所示。

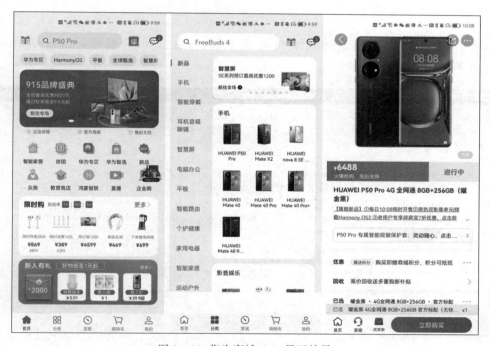

图 14.1　华为商城 App 界面效果

## 14.1　华为商城框架封装

App 开发之前，首先通过分析，把应用中的公共组件进行抽象和封装，公共组件封装的颗粒度以业务复用为目标，这里抽取的公共组件是商城 App 的底部导航。

### 14.1.1 公共组件封装

底部导航组件,效果图如图 14.2 所示,底部导航组件是 App 中非常通用的一种组件,一般可以拆分为两个父子组件嵌套使用,例如把内部的图标和文字封装成一个导航子组件,这样非常方便图标和文字的复用。

导航组件使用 Flex 组件,Flex 组件的 justifyContent 和 alignItems 属性可控制内部子组件的水平排列及对齐方式,这里将 justifyContent 属性设置为 SpaceAround,将垂直方向设置为居中对齐。

图 14.2 底部导航

底部导航的实现如代码示例 14.1 所示,导入 @system.router 模块,router.replace 方法用于导航到不同的页面。

代码示例 14.1 chapter14\huaweiStore\Footer.ets

```
@Component
struct Footer {
  build(){
    Flex({ direction: FlexDirection.Row,
      alignItems: ItemAlign.Center,
      justifyContent: FlexAlign.SpaceAround }){
      Column(){
        Image($r("app.media.main_s")).width(20).height(20).objectFit(ImageFit.Contain)
        Text("首页").fontSize(12)
      }.onClick(e = >{
        router.replace({uri:"pages/index"})
      })
      Column(){
        Image($r("app.media.cate")).width(20).height(20).objectFit(ImageFit.Contain)
        Text("分类").fontSize(12)
      }.onClick(e = >{
        router.replace({uri:"pages/category"})
      })
      Column(){
        Image($r("app.media.faxin")).width(20).height(20).objectFit(ImageFit.Contain)
        Text("发现").fontSize(12)
      }
      Column(){
        Image($r("app.media.cart")).width(20).height(20).objectFit(ImageFit.Contain)
        Text("购物车").fontSize(12)
```

```
        }
        Column(){
          Image( $ r("app.media.personal")).width(20).height(20).objectFit(ImageFit.Contain)
          Text("我的").fontSize(12)
        }
    }.padding(5).width("100%").height(60).backgroundColor("#f1f3f5")
  }
}
```

### 14.1.2 公共数据接口封装

在 App 中调用远程数据接口,通过导入 @ohos.net.http 网络模块,封装网络请求模块,网络模块使用网络权限,需要在 config.json 文件中做权限申请和非加密 HTTP 请求。

在 config.json 文件的 reqPermissions 添加以下权限申请,如代码示例 14.2 所示。

**代码示例 14.2**

```
"module": {
"abilities":[],
"reqPermissions": [
        {
"name": "ohos.permission.INTERNET"
        }
    ]
}
```

默认支持 https,如果要支持 http,需要在 config.json 文件里增加 network 标签,设置属性标识 "cleartextTraffic": true,如代码示例 14.3 所示。

**代码示例 14.3**

```
"deviceConfig": {
"default": {
"network": {
"cleartextTraffic": true
    }
  }
},
```

在项目中创建 api 目录,然后创建 request.ts 文件,在这个 ts 文件中定义网络请求的方法,如代码示例 14.4 所示。

**代码示例 14.4 chapter14\huaweiStore\api\request.ts**

```
import http from '@ohos.net.http';

const baseUrl:string = "http://www.51itcto.com:3000/"

function getData(callback:any) {
```

```
        let httpRequest = http.createHttp();
    httpRequest.request(
        baseUrl,
        {
            header: {
                'Content-Type': 'application/json'
            },
            readTimeout: 60000,
            connectTimeout: 60000
        }, callback)
}

export {getData}
```

上面的代码是标准的 TypeScript 代码,通过 export 暴露外部访问的方法。在 ets 文件中导入 request 模块,在主组件的 aboutToAppear()方法中调用,如代码示例 14.5 所示。

代码示例 14.5 chapter14\huaweiStore\index.ets

```
import {getData} from "../api/request"

@Entry
@Component
struct Index {

  aboutToAppear(){
    getData((err,res) =>{
      if (res.responseCode == 200) {
        console.info(JSON.stringify(res.result))
      }
    })
  }
}
```

## 14.2 商城首页实现

商城首页的效果图如图 14.3 所示,头部和底部固定在上下端位置,中间的部分是可以滚动的,所以最外层使用 Flex 组件布局,宽和高均为 100%,头部和底部高度为 60fp,中间滚动部分使用 Scroll 组件,宽和高均为 100%,如代码示例 14.6 所示。

在开始编写界面代码之前,首先需要对整个界面进行分析,图 14.3 中对每个界面部分进行了标注并使用了不同的布局组件,根据分析的结果,编写页面结构,如代码示例 14.6 所示。

代码示例 14.6 chapter14\huaweiStore\index.ets

```
@Entry
@Component
```

```
struct Index {
  build() {
    Flex({ direction: FlexDirection.Column,
      alignItems: ItemAlign.Center,
      justifyContent: FlexAlign.Center }) {
      Row(){

      }.width("100%").height(60).backgroundColor("#f1f3f5")
      Scroll(){
        Column().height(1200).width("100%")
      }.width("100%").height("100%").backgroundColor(Color.Pink)
      Row().width("100%").height(60).backgroundColor(Color.Yellow)
    }
    .backgroundColor("#f1f3f5")
    .width('100%')
    .height('100%')
  }
}
```

图 14.3 商城首页布局分析

首先,通过 Flex 布局把整个界面分为上、中、下三部分:头部使用 Row 组件,中间使用 Scroll 布局,底部使用 Row 或者 Flex 布局,效果图如图 14.4 所示。

```
Flex({ direction: FlexDirection.Column,
    alignItems: ItemAlign.Center,
    justifyContent: FlexAlign.Center }) {
  Row(){

  }.width("100%").height(60).backgroundColor("#f1f3f5")
  Scroll(){
      Column().height(1200).width("100%")
  }.width("100%").height("100%").backgroundColor(Color.Pink)
  Row().width("100%").height(60).backgroundColor(Color.Yellow)
}
.backgroundColor("#f1f3f5")
.width('100%')
.height('100%')
}
```

图 14.4 商城 App 首页布局结构代码

### 14.2.1 头部组件

首页头部固定在屏幕的顶端，布局效果如图 14.5 所示，Row 布局可以让内部子组件从左往右进行排列。

图 14.5 首页头部搜索栏

为了便于组件复用，头部组件单独创建 Header.ets 文件，头部搜索栏使用 Row 组件模拟，如代码示例 14.7 所示。

代码示例 14.7 chapter14\huaweiStore\Header.ets

```
@Component
struct Header {
  build(){
    Row(){
      Image( $r("app.media.gift"))
        .width(25)
        .height(25)
        .margin({left:20})
```

```
        .objectFit(ImageFit.Contain)
      Row(){
        Image( $ r("app.media.search"))
          .width(20)
          .height(20)
          .margin({left:20})
          .objectFit(ImageFit.Contain)
        Text("p50 pro").fontSize(18).fontColor("#ccc").margin({left:10})
      }
      .width(200).height(40).backgroundColor(Color.White)
      .borderRadius(30).margin({left:20})
      Image( $ r("app.media.cuxiao"))
        .width(25)
        .height(25)
        .margin({left:20})
        .objectFit(ImageFit.Contain)
      Image( $ r("app.media.msg"))
        .width(25)
        .height(25)
        .margin({left:10})
        .objectFit(ImageFit.Contain)
    }.width("100%").height(60).backgroundColor("#f1f3f5")
  }
}
```

在 index.ets 文件中导入 Header.ets 模块,在 Flex 组件中调用 Header 组件,效果图如图 14.6 所示。

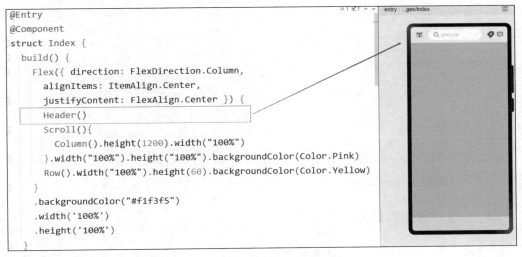

图 14.6  头部组件创建和调用

## 14.2.2 头部滚动

头部滚动效果如图14.7所示,如果滚动区内容超出范围,则可以进行左右滚动,这里使用Scroll组件或者List组件都可以实现滚动效果。选择List组件,可以通过设置List组件的listDirection属性等于Axis.Horizontal实现。

图14.7 头部滚动效果

List组件中的ListItem组件的数据通过组件内的list属性进行绑定,如代码示例14.8所示。

代码示例14.8 chapter14\huaweiStore\index.ets
```
@Component
struct TagList {
  private list: string[] = ["华为专区","P50","P40","HarmonyOS","智慧屏"]
  build(){
    List(){
      ForEach(this.list,item =>{
        ListItem(){
          Text(item).height("100%")
            .fontSize(16).fontColor("#1F2120")
            .margin({left:10,right:10})
        }.margin(5).backgroundColor("#E6E7E9").padding(3).borderRadius(30)
      },(item,index) => index.toString())
    }
    .padding({left:15,right:15})
    .listDirection(Axis.Horizontal)
    .height(50)
    .width("100%")
    .backgroundColor("#f2f3f5")
  }
}
```

这里使用ForEach()方法进行循环,ForEach()方法的第3个参数用来设置组件复用的编号,这里可以通过(item,index) => index.toString()设置,效果图如图14.8所示。

## 14.2.3 轮播广告

轮播的广告区使用Swiper组件实现轮播图片,创建ProductSwiper组件,如代码示例14.9所示。

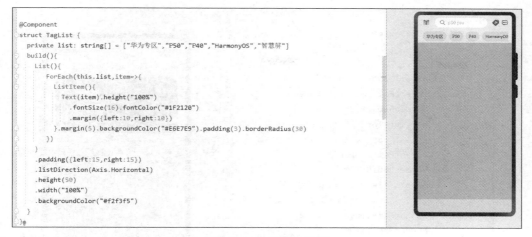

```
@Component
struct TagList {
  private list: string[] = ["华为专区","P50","P40","HarmonyOS","智慧屏"]
  build(){
    List(){
      ForEach(this.list,item=>{
        ListItem(){
          Text(item).height("100%")
            .fontSize(16).fontColor("#1F2120")
            .margin({left:10,right:10})
        }.margin(5).backgroundColor("#E6E7E9").padding(3).borderRadius(30)
      })
    }
    .padding({left:15,right:15})
    .listDirection(Axis.Horizontal)
    .height(50)
    .width("100%")
    .backgroundColor("#f2f3f5")
  }
}
```

图 14.8 头部滚动效果

代码示例 14.9  chapter14\huaweiStore\index.ets

```
@Component
struct ProductSwiper {
  build(){
    Swiper(){
      Image( $ r("app.media.b1"))
        .objectFit(ImageFit.Cover)
        .borderRadius(20)
        .margin({left:15,right:15})
      Image( $ r("app.media.b2"))
        .objectFit(ImageFit.Cover)
        .borderRadius(20)
        .margin({left:15,right:15})
      Image( $ r("app.media.b3"))
        .objectFit(ImageFit.Cover)
        .borderRadius(20)
        .margin({left:15,right:15})
    }
    .padding(5)
    .height(160)
    .width("100%")
    .backgroundColor("#f2f3f5")
  }
}
```

滚动广告效果如图 14.9 所示。

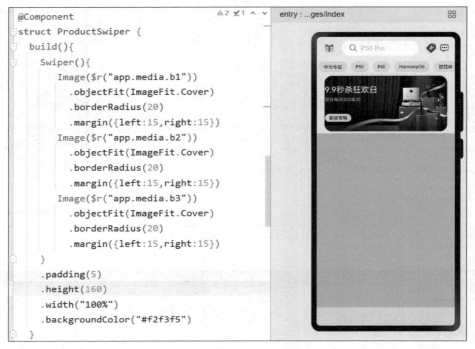

```
@Component
struct ProductSwiper {
  build(){
    Swiper(){
      Image($r("app.media.b1"))
        .objectFit(ImageFit.Cover)
        .borderRadius(20)
        .margin({left:15,right:15})
      Image($r("app.media.b2"))
        .objectFit(ImageFit.Cover)
        .borderRadius(20)
        .margin({left:15,right:15})
      Image($r("app.media.b3"))
        .objectFit(ImageFit.Cover)
        .borderRadius(20)
        .margin({left:15,right:15})
    }
    .padding(5)
    .height(160)
    .width("100%")
    .backgroundColor("#f2f3f5")
  }
}
```

图 14.9 轮播广告

## 14.2.4 导航菜单

首页中间的导航菜单采用两行五列的表格进行设计,如图 14.10 所示。表格布局使用 Grid 组件,将 Grid 组件的 rowsTemplate 属性设置为两行:1fr 1fr。

在 GridMenuList 组件中定义 arr 数组变量,循环绑定给 GridItem 组件,如代码示例 14.10 所示。

**代码示例 14.10　chapter14\huaweiStore\index.ets**

```
@Component
struct GridMenuList {
  private arr:any[] = [
    {"title":"智能家居","img": $r("app.media.hm")},
    {"title":"华为智选","img": $r("app.media.hua")},
    {"title":"教育","img": $r("app.media.jiaju")},
    {"title":"智能家居","img": $r("app.media.jiao")},
    {"title":"企业购","img": $r("app.media.qi")},
    {"title":"新品","img": $r("app.media.zi")},
    {"title":"众筹","img": $r("app.media.zong")},
    {"title":"智能家居","img": $r("app.media.xin")},
    {"title":"直播","img": $r("app.media.bo")},
    {"title":"鸿联","img": $r("app.media.hm")}
```

```
    }
    build(){
        Grid(){
            ForEach(this.arr,item =>{
                GridItem(){
                    Column(){
                        Image(item.img).objectFit(ImageFit.Contain)
                        Text(item.title).fontSize(12).margin({top:3})
                    }.alignItems(HorizontalAlign.Center)
                }
                .width(60)
                .height(50)
                .margin({left:5,right:2,top:2,bottom:2})
                .onClick(e =>{
                    router.push({
                        uri:"pages/Product"
                    })
                })
            },(item,index) => index.toString())
        }
        .padding({left:15,right:15,top:0,bottom:15})
        .height(180)
        .width("100%")
        .backgroundColor("#f1f3f5")
        .rowsTemplate("1fr 1fr")
    }
}
```

图 14.10 表格导航菜单

### 14.2.5 限时购

限时购效果图如图 14.11 所示,采用两行的方式进行布局,每个商品采用 Column 进行布局,也可以采用 Flex 布局。

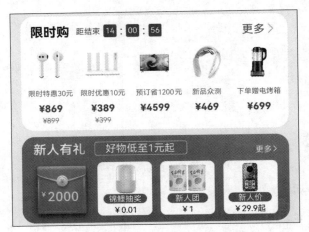

图 14.11 限时购效果

限时购组件的定义如代码示例 14.11 所示。

代码示例 14.11 chapter14\huaweiStore\index.ets

```
@Component
struct BuyList {
  build(){
    Column(){
      Row(){
        Text("限时购").fontWeight(600).fontSize(14).height(50)
        Text("距离结束").fontSize(12).height(50).margin({left:20})
        Text("14").fontSize(10).fontColor(Color.White).padding(3).margin({left:20}).backgroundColor(Color.Red).borderRadius(5)
        Text(":").fontSize(12).padding(5)
        Text("00").fontSize(10).fontColor(Color.White).padding(3).margin({left:2}).backgroundColor(Color.Red).borderRadius(5)
        Text(":").fontSize(12).padding(5)
        Text("28").fontSize(10).fontColor(Color.White).padding(3).margin({left:2}).backgroundColor(Color.Red).borderRadius(5)
        Text("更多>").fontSize(12).height(50).margin({left:50})
      }
      Row(){
        Column(){
          Image($r("app.media.pp")).height(50).objectFit(ImageFit.Cover)
          Text("限时优惠 30 元").fontSize(8)
          Text("￥830 元").fontSize(8).fontWeight(600).fontColor(Color.Red)
          Text("￥860 元").fontSize(8).fontStyle(FontStyle.Italic)
```

```
        }.alignItems(HorizontalAlign.Center).margin(5).width(55).height(100)
      }
      .backgroundColor("#f1f3f5")
    }
}
```

限时购下面的通栏广告部分,如代码示例14.12所示。

代码示例 14.12　chapter14\huaweiStore\index.ets

```
@Component
struct AdImage {
  build(){
    Row(){
      Image($r("app.media.bb")).objectFit(ImageFit.Fill).width("100%").height(100).borderRadius(20)
    }.padding({left:15,right:15}).width("100%").height(100).borderRadius(20)
  }
}
```

## 14.3　商城商品分类页实现

商城商品分类页面,效果如图14.12所示,该页面也是典型的上中下结构,中间的部分分为左边分类滚动区和右边商品滚动区。

图 14.12　商城商品分类效果

首先定义页面的结构布局,这个页面使用 Flex 布局,布局内部的头部使用 Row 布局,底部布局复用公共组件 Footer.ets,中间使用 Row 组件,高度和宽度均设置为 100%,Row 组件内部使用两个 Scroll 组件,分别占据左边和右边滚动区。

商品分类页结构如代码示例 14.13 所示。

代码示例 14.13    chapter14\huaweiStore\category.ets

```
import {Footer} from "./Footer"
import {Header} from "./Header"
@Entry
@Component
struct Category {

  build() {
    Flex({ direction: FlexDirection.Column,
      alignItems: ItemAlign.Start,
      justifyContent: FlexAlign.Center }) {
      Header()
      Row() {
        Scroll() {
          Column() {
          }
        }.width(100).height("100%")

        Scroll() {
        }.width("100%").height("100%").backgroundColor("#f1f3f5")
      }.width("100%").height("100%")
      Footer()
    }
    .backgroundColor("#f1f3f5")
    .width('100%')
    .height('100%')
  }
}
```

完成后的效果如图 14.13 所示。

## 14.3.1  中间左侧分类区

左侧商品分类数据可以通过接口获取,这里直接在组件中定义商品分类数据,代码如代码示例 14.14 所示。

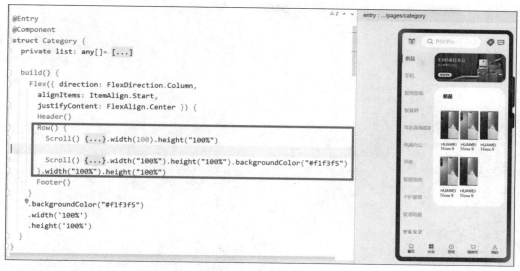

图 14.13　商城商品分类完成效果

代码示例 14.14　chapter14\huaweiStore\category.ets

```
private list: any[] = [
    {
"title": "新品",
"selected": true
    },
    {
"title": "手机"
    },
    {
"title": "智能穿戴"
    },
    {
"title": "智慧屏"
    },
    {
"title": "耳机音箱眼镜"
    },
    {
"title": "计算机办公"
    },
    {
"title": "平板"
    },
    {
"title": "智能路由"
    },
    {
```

```
        "title":"个护健康"
    },
    {
        "title":"家用电器"
    },
    {
        "title":"智能家居"
    },
    {
        "title":"户外运动"
    },
    {
        "title":"出行车品"
    },
    {
        "title":"华为智选"
    }
]
```

将商品分类选中的状态设置为红色,将选中的竖线也设置为红色,如图 14.14 所示。

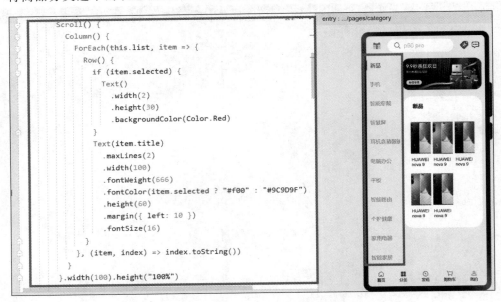

图 14.14 商城商品分类选中效果

商品分类的实现如代码示例 14.15 所示。

**代码示例 14.15　chapter14\huaweiStore\category.ets**

```
@Entry
@Component
```

```
struct Category {
  private list: any[] = [
    {
      "title": "新品",
      "selected": true
    },
    {
      "title": "手机"
    },
    {
      "title": "智能穿戴"
    },
    {
      "title": "智慧屏"
    },
    {
      "title": "耳机音箱眼镜"
    },
    {
      "title": "计算机办公"
    },
    {
      "title": "平板"
    },
    {
      "title": "智能路由"
    },
    {
      "title": "个护健康"
    },
    {
      "title": "家用电器"
    },
    {
      "title": "智能家居"
    },
    {
      "title": "户外运动"
    },
    {
      "title": "出行车品"
    },
    {
      "title": "华为智选"
    }
  ]

  build() {
    Flex({ direction: FlexDirection.Column,
```

```
            alignItems: ItemAlign.Start,
            justifyContent: FlexAlign.Center }) {
        Header()
        Row() {
          Scroll() {
            Column() {
              ForEach(this.list, item => {
                Row() {
                  if (item.selected) {
                    Text()
                      .width(2)
                      .height(30)
                      .backgroundColor(Color.Red)
                  }
                  Text(item.title)
                    .maxLines(2)
                    .width(100)
                    .fontWeight(666)
                    .fontColor(item.selected ? "#f00" : "#9C9D9F")
                    .height(60)
                    .margin({ left: 10 })
                    .fontSize(16)
                }
              }, (item, index) => index.toString())
            }
          }.width(100).height("100%")

          Scroll() {
            Column() {
              ProductSwiper()
              GridProductCard()
            }.width("100%").height("100%").alignItems(HorizontalAlign.Start)
          }.width("100%").height("100%").backgroundColor("#f1f3f5")
        }.width("100%").height("100%")
        Footer()
      }
      .backgroundColor("#f1f3f5")
      .width('100%')
      .height('100%')
    }
  }
```

### 14.3.2 中间右侧商品区

右侧商品区包括顶部的商品轮播展示和商品卡片,商品轮播的实现与首页的轮播实现一样。这里定义两个组件:商品轮播展示组件和商品卡片组件,商品卡片组件要能够复用。

轮播展示组件如代码示例 14.16 所示。

代码示例 14.16　chapter14\huaweiStore\category.ets

```
@Component
struct ProductSwiper {
  build() {
    Swiper() {
      Image( $ r("app.media.b1"))
        .objectFit(ImageFit.Auto)
        .width(240)
        .borderRadius(20)
      Image( $ r("app.media.b2"))
        .objectFit(ImageFit.Contain)
        .width(240)
        .borderRadius(20)
      Image( $ r("app.media.b3"))
        .width(240)
        .objectFit(ImageFit.Contain)
        .borderRadius(20)
    }
    .padding(10)
    .height(120)
    .width("100 % ")
    .backgroundColor(" # f1f3f5")
  }
}
```

商品卡片组件包含一组图片的展示，这里可以采用 Grid 布局容器，代码实现如代码示例 14.17 所示。

代码示例 14.17　chapter14\huaweiStore\category.ets

```
@Component
@Component
struct GridProductCard {
  private products: any[ ] = [
    {
"title": "HUAWEI Nova 9",
"img": $ r("app.media.vv1")
    },
    {
"title": "HUAWEI Mate Pro",
"img": $ r("app.media.vv2")
    },
    {
"title": "HUAWEI Nova 9",
"img": $ r("app.media.vv1")
    },
    {
"title": "HUAWEI Mate Pro",
"img": $ r("app.media.vv2")
```

```
      },{
"title": "HUAWEI Nova 9",
"img": $r("app.media.vv1")
      },
      {
"title": "HUAWEI Mate Pro",
"img": $r("app.media.vv2")
      }
    ]

    build() {
      Column() {
        Row() {
          Text("新品").margin({ left: 30 }).fontSize(16).fontWeight(600)
        }.height(60).width("100%")

        Grid() {
          ForEach(this.products, item = >{
            GridItem(){
              Column(){
                Image(item.img).objectFit(ImageFit.Fill)
                Text(item.title).fontSize(13).margin({top:10})
              }
            }.width(60).height(100).backgroundColor("#f00")
          },(item,index) = > index.toString())
        }
        .padding(20)
        .width("100%")
        .rowsGap(60)
        .columnsTemplate("1fr 1fr 1fr")
        .backgroundColor("#fff")
        .borderRadius(30)
      }
      .margin({ left: 10, right: 10, top: 10, bottom: 10 })
      .borderRadius(30)
      .width(240)
      .height(350)
      .backgroundColor(Color.White)
    }
  }
```

## 14.4 商品详情页实现

商品详情页同样采用上中下结构,顶部的返回栏默认为透明的效果,当页面下拉时慢慢变为不透明的效果,并且固定在屏幕的最上面,中间的商品详细信息是滚动区域,底部的购买栏固定在页面的最底部,如图 14.15 所示。

第14章 ArkUI ETS UI开发案例

图 14.15 商品详情页效果

首先需要定义商品详细页的布局结构,为了便于中间滚动,页面采用 Flex 布局,页面的代码结构如代码示例 14.18 所示。

代码示例 14.18 chapter14\huaweiStore\Product.ets

```
@Entry
@Component
struct Product {
  build() {
    Flex({
      direction: FlexDirection.Column,
      alignItems: ItemAlign.Center,
      justifyContent: FlexAlign.Start }) {
      ItemsSwiper()
      ItemBar()
      ItemTitle()
      ItemSubTitle()
      Scroll() {
        Column(){
          ItemMore()
        }
```

```
        }
        .width("100%")
        .height("100%")
        BuyBar()
      }
      .width('100%')
      .height('100%')
    }
  }
```

### 14.4.1 头部商品图片轮播区

这里轮播图片上面有一个返回栏,返回栏覆盖在轮播图上面,所以这里使用 Stack 布局组件,将 Stack 组件内容的对齐方式设置为 Alignment.Top,如代码示例 14.19 所示。

代码示例 14.19　chapter14\huaweiStore\Product.ets

```
@Component
struct ItemsSwiper {
  build() {
    Column() {
      Stack({ alignContent: Alignment.Top }) {
        Swiper() {
          Image($r("app.media.ww1"))
          Image($r("app.media.ww2"))
          Image($r("app.media.ww3"))
        }

        Row() {
        }.width("100%").height(40).backgroundColor("#f1f3f5").opacity(0.3)
      }
    }
    .width("100%")
    .height(300)
  }
}
```

商品详情页效果如图 14.16 所示。

### 14.4.2 商品价格展示栏

商品价格展示栏采用左右结构,左边为价格,右边为商品的促销状态。结合 Column 和 Row 组件进行布局,如代码示例 14.20 所示。

代码示例 14.20　chapter14\huaweiStore\Product.ets

```
@Component
struct ItemBar {
```

```
build() {
  Row() {
    Column() {
      Text("￥6999").margin({ top: 10, left: 10 }).fontSize(16).fontColor("#fff")
      Text("火爆抢购,先到先得").margin({ top: 2, left: 10 }).fontSize(12).fontColor("#fff")
    }
    .width("70%").height(60).alignItems(HorizontalAlign.Start)

    Row() {
      Text("进行中").fontSize(15).fontColor("#e64566").margin({ left: 20 })
    }.width("30%").height(60).backgroundColor("#fcecef").alignItems(VerticalAlign.Center)
  }.width("100%").height(60).backgroundColor("#FF0D40")
}
```

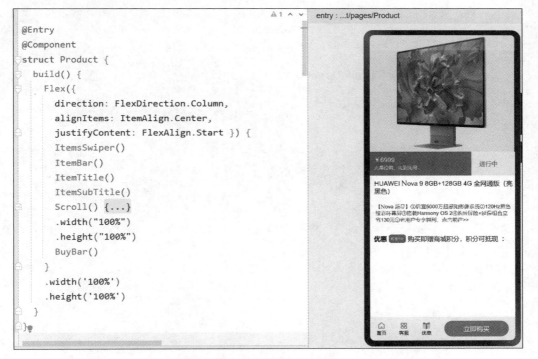

图 14.16　商品详情页轮播效果

## 14.4.3　商品底部购买栏

底部购买栏通过 Flex 布局,居中对齐,将水平方向设置为 FlexAlign.SpaceAround,如代码示例 14.21 所示。

代码示例 14.21　chapter14\huaweiStore\Product.ets

```
@Component
struct BuyBar {
  build() {
    Flex({
      direction: FlexDirection.Row,
      alignItems: ItemAlign.Center,
      justifyContent: FlexAlign.SpaceAround
    }) {
      Column(){
        Image($r("app.media.main")).width(20).height(20).objectFit(ImageFit.Contain)
        Text("首页").fontSize(12)
      }
      Column(){
        Image($r("app.media.cate")).width(20).height(20).objectFit(ImageFit.Contain)
        Text("客服").fontSize(12)
      }
      Column(){
        Image($r("app.media.gift")).width(20).height(20).objectFit(ImageFit.Contain)
        Text("优惠").fontSize(12)
      }
      Button({type:ButtonType.Capsule}){
        Text("立即购买").fontSize(16).fontColor(Color.White)
      }
      .backgroundColor("#FE5945")
      .width(160)
    }
    .width("100%")
    .height(60)
    .backgroundColor("#f1f3f5")
  }
}
```

# 第五篇　OpenHarmony篇

　　OpenHarmony是由开放原子开源基金会（OpenAtom Foundation）孵化及运营的开源项目，目标是面向全场景、全连接、全智能时代，搭建一个智能终端设备操作系统的框架和平台，促进万物互联产业的繁荣发展。

# 第 15 章 OpenHarmony 基础

本章学习 OpenHarmony 的安装编译、烧录及智能家居的应用开发，介绍如何开发 OpenHarmony 应用程序和在 OpenHarmony 系统上运行和卸载 HAP 应用程序。

## 15.1 OpenHarmony 介绍

OpenHarmony 是由开放原子开源基金会（OpenAtom Foundation）孵化及运营的开源项目，目标是面向全场景、全连接、全智能时代，搭建一个智能终端设备操作系统的框架和平台，促进万物互联产业的繁荣发展。

OpenHarmony 支持以下几种系统类型：

**1. 轻量系统类设备（参考内存≥128KB）**

面向 MCU 类处理器，例如 ARM Cortex-M、RISC-V 32 位的设备，硬件资源极其有限，支持的设备最小内存为 128KiB，可以提供多种轻量级网络协议、轻量级的图形框架，以及丰富的 IoT 总线读写部件等。可支撑的产品如智能家居领域的连接类模组、传感器设备、穿戴类设备等。

**2. 小型系统类设备（参考内存≥1MB）**

面向应用处理器，例如 ARM Cortex-A 的设备，支持的设备的最小内存为 1MiB，可以提供更高的安全能力、标准的图形框架、视频编解码的多媒体能力。可支撑的产品如智能家居领域的 IP Camera、电子猫眼、路由器及智慧出行领域的行车记录仪等。

**3. 标准系统类设备（参考内存≥128MB）**

面向应用处理器，例如 ARM Cortex-A 的设备，支持的设备的最小内存为 128MiB，可以提供增强的交互能力、3D GPU 及硬件合成能力、更多控件及动效更丰富的图形能力、完整的应用框架。可支撑的产品如高端的冰箱显示屏。

**4. 大型系统类设备（参考内存≥1GB）**

面向应用处理器，例如 ARM Cortex-A 的设备，参考内存≥1GB，提供完整的兼容应用框架。典型的产品有智慧屏、智能手表等。

## 15.2 OpenHarmony 3.0 LTS 编译与烧录

OpenHarmony 源代码需要在 Linux 系统上进行编译,这里采用 Ubuntu 20.04 LTS 版本进行编译环境的搭建。将 OpenHarmony 的镜像烧录到开发版可以直接在 Windows 系统上操作。

### 15.2.1 编译环境搭建

编译环境的搭建可根据实际情况选择 Docker 方式或工具包方式,如图 15.1 所示。

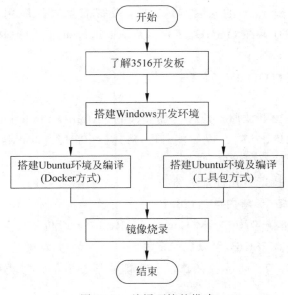

图 15.1 编译环境的搭建

下面介绍在 Ubuntu 系统上直接编译 OpenHarmony 镜像的方法,具体步骤如下。

**1. 安装依赖**

安装的依赖包括 git、git-lfs、python3.8、python3-pip,命令如下:

```
sudo apt-get update && sudo apt-get install binutils git git-lfs gnupg flex bison gperf build-essential zip curl zlib1g-dev gcc-multilib g++-multilib libc6-dev-i386 lib32ncurses5-dev x11proto-core-dev libx11-dev lib32z1-dev ccache libgl1-mesa-dev libxml2-utils xsltproc unzip m4 bc gnutls-bin python3.8 python3-pip ruby
```

这里需要将 Python 软链接设置为 python3.8,命令如下:

```
sudo ln -s /usr/bin/python3.8 /usr/bin/python
```

## 2. 安装工具：repo 和 hc-gen

hc-gen(HDF Configuration Generator)是 HCS 配置转换工具，HCS(HDF Configuration Source)是 HDF 驱动框架的配置描述源码，内容以 Key-Value 为主要形式。它实现了配置代码与驱动代码解耦，便于开发者进行配置管理。

安装码云 repo、hc-gen 工具，可以执行的命令如下：

```
#repo 安装
curl -s https://gitee.com/oschina/repo/raw/fork_flow/repo-py3 > sudo /usr/local/bin/repo
chmod a+x /usr/local/bin/repo
pip3 install -i https://pypi.tuna.tsinghua.edu.cn/simple requests
```

## 3. 安装 Ruby

代码如下：

```
sudo apt-get install ruby-full
```

## 4. 配置 git 相关参数

这里通过 git ssh 下载源代码，设置 git 邮箱和名字，代码如下：

```
git config --global user.name "xxx"
git config --global user.email "xxx@xxx.com"
git config --global credential.helper store
```

## 5. 创建代码目录并拉取代码

代码如下：

```
#3.0 源码下载：
repo init -u https://gitee.com/openharmony/manifest.git -b OpenHarmony-3.0-LTS --no-repo-verify
repo sync -c
repo forall -c 'git lfs pull'
```

## 6. 下载预编译工具

代码如下：

```
./build/prebuilts_download.sh
```

## 7. 运行编译脚本（需自行调整参数）

代码如下：

```
./build.sh --product-name Hi3516DV300 --ccache
```

编译成功后如图 15.2 所示。

```
post_process
=====build Hi3516DV300 successful.
2021-09-29 21:21:23
++++++++++++++++++++++++++++++++++++++++
```

图 15.2　编译成功

### 8. 查看编译结果

命令如下：

```
ls -l out/ohos-arm-release/packages/phone/images/
```

编译结果如图 15.3 所示。

```
++++++++++++++++++++++++++++++++++++++++
xlw@ubuntu:~/OpenHarmony3$ ls -l out/ohos-arm-release/packages/phone/images/
total 327788
-rwxrwxr-x 2 xlw xlw       1240 Sep 29 20:11 Hi3516DV300-emmc.xml
-rw-rw-r-- 1 xlw xlw 1610608640 Sep 29 21:21 system.img
-rwxrwxr-x 2 xlw xlw     240619 Sep 29 20:13 u-boot-hi3516dv300_emmc.bin
-rw-rw-r-- 1 xlw xlw    5729090 Sep 29 20:26 uImage
-rw-rw-r-- 1 xlw xlw   20971520 Sep 29 21:21 updater.img
-rw-rw-r-- 1 xlw xlw 1468006400 Sep 29 21:21 userdata.img
-rw-rw-r-- 1 xlw xlw  268431360 Sep 29 21:21 vendor.img
xlw@ubuntu:~/OpenHarmony3$
```

图 15.3　查看编译的镜像文件列表

### 9. 通过 Samba 共享 Ubuntu 编译镜像

Samba 最大的功能就是可以用于 Linux 与 Windows 系统的文件共享和打印共享，Samba 既可以用于 Windows 与 Linux 之间的文件共享，也可以用于 Linux 与 Linux 之间的资源共享，由于 NFS(网络文件系统)可以很好地完成 Linux 与 Linux 之间的数据共享，因而 Samba 较多地用在了 Linux 与 Windows 之间的数据共享上。

步骤 1：安装命令，命令如下。

```
sudo apt-get install samba
sudo apt-get install samba-common
```

步骤 2：修改 samba 配置文件，命令如下。

```
sudo vim /etc/samba/smb.conf
```

步骤 3：在最后加入的内容如下。

```
[work]
comment = samba home directory
path = /home/User/OpenHarmony3        #需要共享的目录位置
public = yes
```

```
browserable = yes
read only = no
valid users = harmony
create mask = 0777
directory mask = 0777
available = yes
```

步骤4：保存退出后，输入如下命令，设置samba密码。

```
sudo smbpasswd - a harmony        # 添加harmony用户，设置密码
# smbpasswd命令的常用方法
smbpasswd - a                     # 增加用户（要增加的用户必须是系统用户）
smbpasswd - d                     # 冻结用户，也就是这个用户不能再登录了
smbpasswd - e                     # 恢复用户，即解冻用户，让冻结的用户可以再次使用
smbpasswd - n                     # 把用户的密码设置成空
                                  # 要在global中写入null passwords - true
smbpasswd - x                     # 删除用户
```

步骤5：重启samba服务，命令如下。

```
sudo service smbd restart
```

## 15.2.2　标准系统编译和烧录

这里选择 Hi3516DV300 烧录 OpenHarmony 镜像，Hi3516DV300 分别支持"小型系统"和"标准系统"，但是其依赖的编译工具链不同。

Hi3516DV300 作为新一代行业专用 Smart HD IP 摄像机 SoC，集成了新一代 ISP（Image Signal Processor）、H.265 视频压缩编码器，同时集成了高性能 NNIE 引擎，如图 15.4 所示。

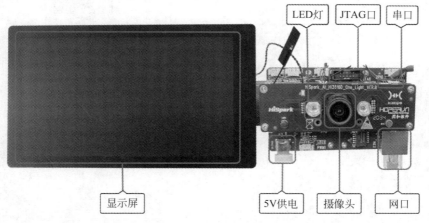

图 15.4　Hi3516 单板正面外观图

烧录镜像可直接在 Windows 系统上操作,这里需要在 Hi3516DV300 官方网站下载驱动程序,需要用到的驱动如图 15.5 所示,驱动的下载网址 http://www.hihope.org/,如表 15.1 所示。

表 15.1　Hi3516DV300 需要用到的驱动列表

| 参数名 | 说　明 |
| --- | --- |
| CH341SER | ch341ser 驱动(USB 转串口驱动工具)是一款很优秀的 USB 到串口的驱动工具 |
| Hi3516-HiTool | Hi3516 烧录工具 |
| HiUSBBurnDirver | USB 烧录的驱动程序 |
| USB-to-Serial Common Port.exe | AI Camera 和 DIY IPC 套件附赠的 USB 串口线中集成了 PL2302 芯片,需要安装此驱动才能识别 |
| usb.reg | Windows 10 操作系统修改注册表 |

安装 HiUSBBurnDriver 驱动程序,Windows 10 操作系统需要修改注册表,否则会报错,如图 15.5 所示。

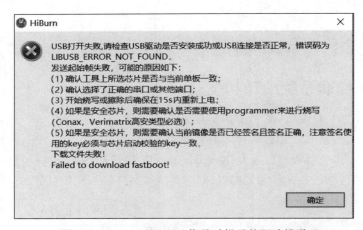

图 15.5　Hi3516DV300 烧录时提示的驱动错误

Windows 10 操作系统修改注册表的方法如下。
(1) 创建一个"文本文档.TXT",文件后缀名修改为.reg,如 usb.reg。
(2) 右击打开已创建的 usb.reg 文件,将如下脚本复制到该文件中,然后保存并关闭。

```
Windows Registry Editor Version 5.00
[HKEY_LOCAL_MACHINE\SYSTEM\CurrentControlSet\Control\usbflags\12D1D0010100]
"SkipBOSDescriptorQuery"=hex:01,00,00,00
"osvc"=hex:00,00
"IgnoreHWSerNum"=hex:01
```

(3) 双击 usb.reg 文件，自动修改注册表文件信息。

使用 HiTool 进行烧录，HiTool 可通过 hihope 网站下载。

### 1. HiTool 将芯片选择为 Hi3516DV300，选择烧录 eMMC

eMMC 分区表文件使用编译完成后对应目录中的 Hi3516DV300-emmc.xml 文件，需要烧录的 bin、img 文件和配置信息会自动设置，如图 15.6 所示。

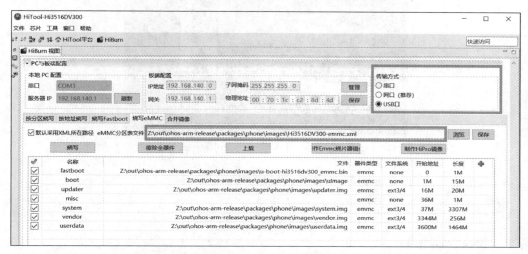

图 15.6　Hi3516DV300 需要用到的驱动列表

### 2. 按住主板上的 update 按键，然后使用 USB 上电

**注意**：因为硬件版本不同，有的 USB 烧录口在板子的左侧，有的在板子的凹槽处，然后等待 HiTool 工具进入烧录模式后，松开 update 按键，update 按键的位置，如图 15.7 所示。

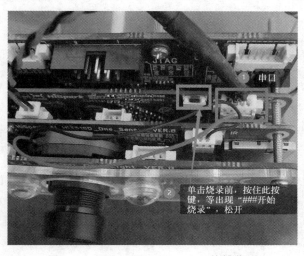

图 15.7　Hi3516DV300 update 按键位置

### 3. 通过 Hi3516DV300 的后置 USB 接口与计算机连接

这里使用 USB 的方式烧录，需要确保上面已经安装了 USB 驱动，并执行了 usb.reg 文件，如图 15.8 所示。

图 15.8　Hi3516DV300 USB 连接位置

### 4. 单击烧录按钮，开始烧录，烧录开始后松开 update 按键

按住 update 按键，单击烧录按钮，确认烧录，如图 15.9 所示。

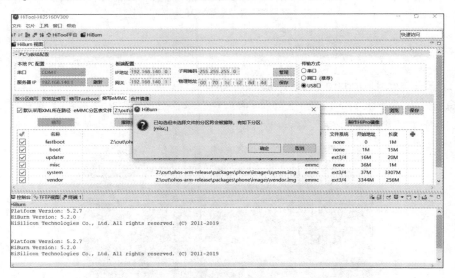

图 15.9　单击烧录

如果之前烧录过系统，可以先单击擦除全器件，等待擦除成功后再重新烧录系统，如图 15.10 所示。

接下来等待 HiTool 工具进入烧录模式，进入烧录模式后松开 update 按键，烧录过程大约需要 1min，烧录完成后，会弹出烧录成功的提示信息，如图 15.11 所示。

烧录完成后，会自动启动系统，并显示桌面，如图 15.12 所示。

第15章 OpenHarmony基础 473

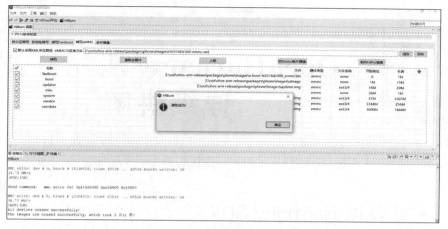

图 15.10　单击擦除全器件

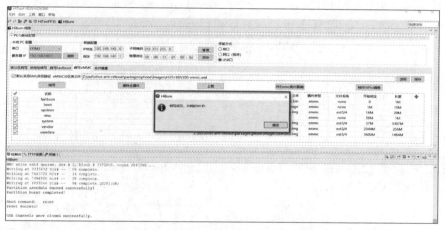

图 15.11　单击擦除全器件

图 15.12　OpenHarmony 的启动界面

# 第 16 章　OpenHarmony 应用开发详细讲解

## 16.1　配置 OpenHarmony SDK

本书中使用的 IDE 是 2021 年 9 月 29 日更新的 IDE 版本,该版本和之前的 IDE 版本的最大差异是支持 ArkUI ETS 项目开发,同时集成了 OpenHarmony SDK,该版本不再需要手动配置 OpenHarmony 的 SDK 了,同时支持 HarmonyOS 和 OpenHarmony 项目开发,版本如图 16.1 所示。

图 16.1　本书使用的 IDE 版本

打开 IDE 设置窗口,在 SDK Manager 里设置 OpenHarmony SDK 和 HarmonyOS SDK 的路径,如图 16.2 所示。

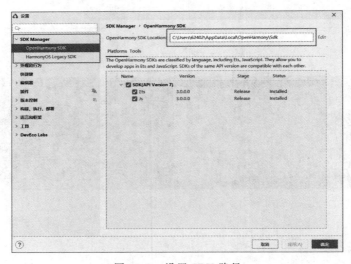

图 16.2　设置 SDK 路径

创建项目后,打开工程目录下的 local.properties 文件,sdk.dir 是 OpenHarmony SDK 路径,hmsdk.dir 是 HarmonyOS SDK 路径,如图 16.3 所示。

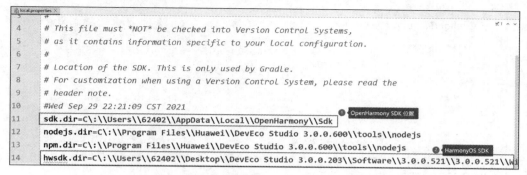

图 16.3　配置项目中的 OpenHarmony SDK 路径

## 16.2　创建 OpenHarmony 工程

创建 OpenHarmony 项目,选择 ArkUI ETS UI 模板创建项目。

### 16.2.1　选择项目模板

选择 Standard 模板生成项目,Standard 模板不包含 Java 代码,如图 16.4 所示。

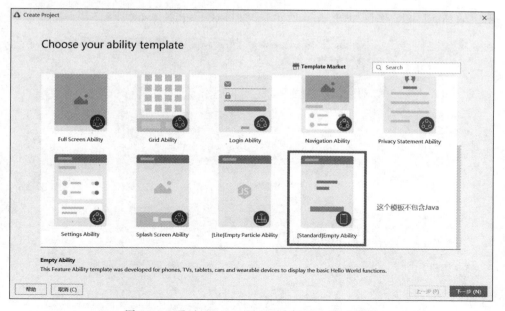

图 16.4　通过 DevEco Studio 选择 Standard 模板

创建 ArkUI JS 项目，如图 16.5 所示。

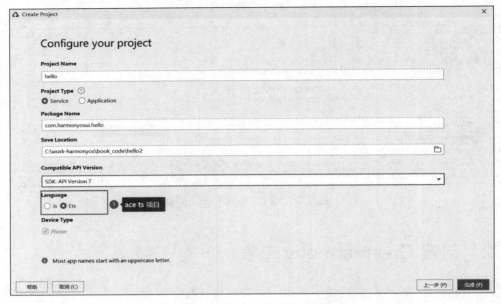

图 16.5　创建 ArkUI JS 项目

## 16.2.2　创建 ArkUI JS 项目

创建的 ArkUI JS 项目如图 16.6 所示，项目目录不包含 Java 代码。

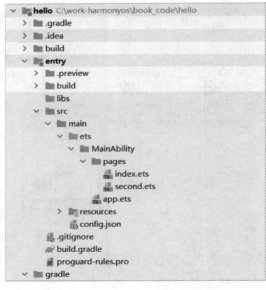

图 16.6　ArkUI JS 项目目录结构

接下来，需要生成签名文件，通过 DevEco Studio 配置签名信息，配置 OpenHarmony 签名的方式和 HarmonyOS 配置签名的方式基本一致，唯一的区别是 profile 文件在本地通过命令行生成即可。

## 16.3 配置 OpenHarmony 应用签名信息

目前的 OpenHarmony 与 HarmonyOS 的证书不通用，所以需要额外进行申请，具体的步骤如图 16.7 所示。

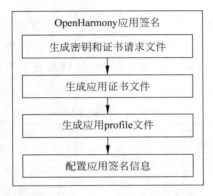

图 16.7　OpenHarmony 应用签名流程

### 16.3.1 生成密钥和证书请求文件

通过 DevEco Studio 来生成密钥文件(.p12 文件)和证书请求文件(.csr 文件)。同时，也可以使用命令行工具的方式来生成密钥文件和证书请求文件。

下面介绍使用 DevEco Studio 生成密钥和证书请求文件，具体步骤如下。

**1. 生成 Key 和 CSR**

在 DevEco Studio 主菜单栏单击构建(Build)→Generate Key and CSR，如图 16.8 所示。

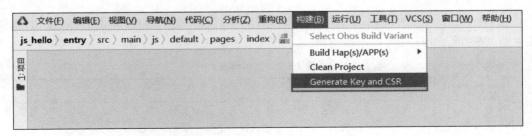

图 16.8　通过 DevEco Studio 生成 Key 和 CSR

## 2. 创建密钥库文件(.p12)

在 Key Store File 中，可以单击 Choose Existing 按钮选择已有的密钥库文件。如果没有密钥库文件，则可单击 New 按钮进行创建，如图 16.9 所示。

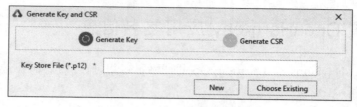

图 16.9　创建密钥库文件

## 3. 填写密钥库信息

在 Create Key Store 窗口中，如图 16.10 所示，填写密钥库信息后，单击 OK 按钮，具体内容如表 16.1 所示。

表 16.1　密钥库信息

| 参数名 | 说明 |
| --- | --- |
| Key Store File | 选择密钥库文件的存储路径 |
| Password | 设置密钥库密码，必须由大写字母、小写字母、数字和特殊符号中的两种以上的字符组合，长度至少为 8 位。需要记住该密码，后续签名配置需要使用 |
| Confirm Password | 再次输入密钥库密码 |

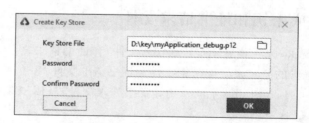

图 16.10　创建密钥库文件

## 4. 填写密钥信息

在 Generate Key 界面中，如图 16.11 所示，继续填写密钥信息后，单击 Next 按钮，具体的参数如表 16.2 所示。

表 16.2　密钥信息

| 参数名 | 说明 |
| --- | --- |
| Alias | 密钥的别名信息，用于标识密钥名称。需要记住该别名，后续签名配置需要使用 |
| Password | 密钥对应的密码，与密钥库密码保持一致，无须手动输入 |
| Validity | 证书有效期，建议设置为 25 年及以上，覆盖应用的完整生命周期 |
| Certificate | 输入证书的基本信息，如组织、城市或地区、国家码等 |

# 第16章 OpenHarmony应用开发详细讲解

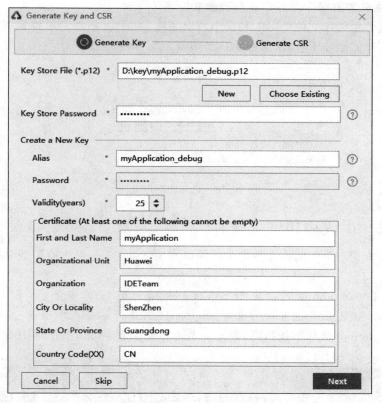

图 16.11 创建密钥库文件

**5. 选择密钥和设置 CSR 文件存储路径(.csr)**

在 Generate CSR 界面，选择密钥和设置 CSR 文件存储路径，如图 16.12 所示。

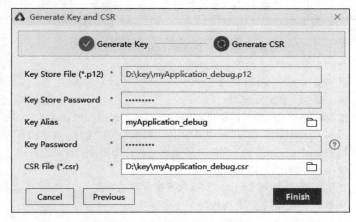

图 16.12 创建密钥库文件

#### 6. 获取生成的密钥库文件和证书请求文件

获取生成的密钥库文件(.p12)和证书请求文件(.csr)。单击 OK 按钮,创建 CSR 文件成功后可以在存储路径下查看,如图 16.13 所示。

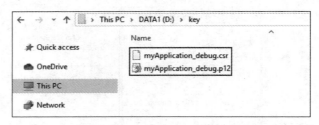

图 16.13　创建密钥库文件

### 16.3.2　生成应用证书文件

使用生成密钥和证书请求文件中生成的证书请求文件,来生成应用签名所需的数字证书文件。

进入 DevEco Studio 安装目录的 Sdk\toolchains\lib 文件夹下(该 SDK 目录只能是 OpenHarmony SDK,配置方法可参考配置 OpenHarmony SDK),打开命令行工具,执行如下命令(如果 keytool 命令不能执行,则应在系统环境变量中添加 JDK 的环境变量)。

```
keytool -gencert -alias "OpenHarmony Application CA" -infile myApplication_ohos.csr
-outfile myApplication_ohos.cer -keystore OpenHarmony.p12 -sigalg SHA384withECDSA
-storepass 123456 -ext KeyUsage:"critical=digitalSignature" -validity 3650 -rfc
```

只需修改输入和输出即可快速生成证书文件。修改-infile:指定证书请求文件 csr 文件路径。修改-outfile:指定输出证书文件名及路径,具体的参数如表 16.3 所示。

表 16.3　生成证书命令参数信息表

| 参数名 | 说　　明 |
| --- | --- |
| alias | 用于签发证书的 CA 私钥别名,OpenHarmony 社区 CA 私钥存于 OpenHarmony.p12 密钥库文件中,该参数不能修改 |
| infile | 证书请求(CSR)文件的路径 |
| outfile | 输出证书文件名及路径 |
| keystore | 签发证书的 CA 密钥库路径,OpenHarmony 密钥库文件名为 OpenHarmony.p12,文件在 OpenHarmony SDK 中 Sdk\toolchains\lib 路径下,该参数不能修改。需要注意,该 OpenHarmony.p12 文件并不是在生成密钥和证书请求文件中生成的.p12 文件 |
| sigalg | 证书签名算法,该参数不能修改 |
| storepass | 密钥库密码,例如密码为 123456,该参数不能修改 |
| ext | 证书扩展项,该参数不能修改 |
| validity | 证书有效期,自定义天数 |
| rfc | 输出文件格式指定,该参数不能修改 |

### 16.3.3 生成应用 Profile 文件

Profile 文件包含 OpenHarmony 应用的包名、数字证书信息、描述应用允许申请的证书权限列表，以及允许应用调试的设备列表（如果应用类型为 Release 类型，则设备列表为空）等内容，每个应用包中均必须包含一个 Profile 文件。

进入 Sdk\toolchains\lib 目录下，打开命令行工具，执行的命令如下：

```
java -jar provisionsigtool.jar sign --in UnsgnedReleasedProfileTemplate.json --out myApplication_ohos_Provision.p7b --keystore OpenHarmony.p12 --storepass 123456 --alias "OpenHarmony Application Profile Release" --sigAlg SHA256withECDSA --cert OpenHarmonyProfileRelease.pem --validity 365 --developer-id ohosdeveloper --bundle-name 包名 --permission 受限权限名(可选) --permission 受限权限名(可选) --distribution-certificate myApplication_ohos.cer
```

参数如表 16.4 所示。

表 16.4 生成应用 Profile 文件命令详情

| 参数名 | 说明 |
| --- | --- |
| provisionsigtool | Profile 文件生成工具，文件在 OpenHarmony SDK 的 Sdk\toolchains\lib 路径下 |
| in | Profile 模板文件所在路径，文件在 OpenHarmony SDK 中 Sdk\toolchains\lib 路径下，该参数不能修改 |
| out | 输出的 Profile 文件名和路径 |
| keystore | 签发证书的密钥库路径，OpenHarmony 密钥库文件名为 OpenHarmony.p12，文件在 OpenHarmony SDK 中 Sdk\toolchains\lib 路径下，该参数不能修改 |
| storepass | 密钥库密码，如密码为 123456，该参数不能修改 |
| alias | 用于签名 Profile 私钥别名，OpenHarmony 社区 CA 私钥存于 OpenHarmony.p12 密钥库文件中，该参数不能修改 |
| sigalg | 证书签名算法，该参数不能修改 |
| validity | 证书有效期，自定义天数 |
| cert | 签名 Profile 的证书文件路径，文件在 OpenHarmony SDK 中 Sdk\toolchains\lib 路径下，该参数不能修改 |
| developer-id | 开发者标识符，自定义一个字符串 |
| bundle-name | 填写应用包名 |
| permission | 可选字段，如果不需要，则可以省去此字段；如果需要添加多个受限权限，则可重复输入。受限权限列表如下：ohos.permission.READ_CONTACTS、ohos.permission.WRITE_CONTACTS |
| distribution-certificate | 在生成应用证书文件中生成的证书文件 |

**注意**：在上面的命令参数中，只需修改下面这 3 个参数，其他参数可保持不变，参数参考如下。

--out，如 c:/xxx/myApplication_ohos_Provision.p7b，

--bundle-name,如 com.harmonyosui.test。
--distribution-certificate,如 C:\xxx\openharmony_test.cer。

### 16.3.4 配置应用签名信息

在真机设备上调试前,需要使用已制作的私钥(.p12)文件、证书(.cer)文件和 Profile(.p7b)文件对调试的模块进行签名。

打开 File→Project Structure,单击 Project→Signing Configs→Debug,在窗口中去除勾选 Automatically generate signing,然后配置指定模块的调试签名信息,如图 16.14 所示,应用签名信息详情如表 16.5 所示。

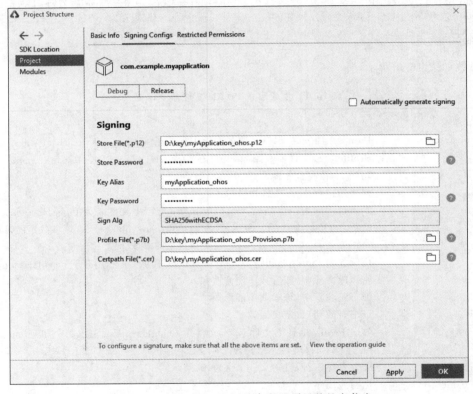

图 16.14 在 DevEco Studio 中配置项目的签名信息

表 16.5 应用签名信息详情

| 参数名 | 说明 |
| --- | --- |
| Store File | 选择密钥库文件,文件后缀为.p12,该文件为在生成密钥和证书请求文件中生成的.p12 文件 |
| Store Password | 输入密钥库密码,该密码应与生成密钥和证书请求文件中填写的密钥库密码保持一致 |

续表

| 参数名 | 说明 |
|---|---|
| Key Alias | 输入密钥的别名信息,应与生成密钥和证书请求文件中填写的别名保持一致 |
| Key Password | 输入密钥的密码,应与 Store Password 保持一致 |
| Sign Alg | 签名算法,固定为 SHA256withECDSA |
| Profile File | 选择在生成应用 Profile 文件中生成的 Profile 文件,文件后缀为.p7b |
| Certpath File | 选择在生成应用证书文件中生成的数字证书文件,文件后缀为.cer |

设置完签名信息后,单击 OK 按钮进行保存,然后可以在工程下的 build.gradle 文件中查看签名的配置信息,如图 16.15 所示。

```
ohos {
    signingConfigs { NamedDomainObjectContainer<SigningConfigOptions> it ->
        debug {
            storeFile file('D:\\key\\myApplication_ohos.p12')
            storePassword '0000001A3D1CCC93694████████████7EEC9570F8C507E39FD80A09CF82885C9ED0539A'
            keyAlias = 'myApplication_ohos'
            keyPassword '0000001A800B548164181████████████8B76DA80AAB57EFB81B4054B15654E3DBB70DD'
            signAlg = 'SHA256withECDSA'
            profile file('D:\\key\\myApplication_ohos_Provision.p7b')
            certpath file('D:\\key\\myApplication_ohos.cer')
        }
    }
}
```

图 16.15　build.gradle 文件中生成的密钥配置信息

编译完成后,OpenHarmony 应用的 HAP 包可以从工程的 build 目录下获取,如图 16.16 所示。

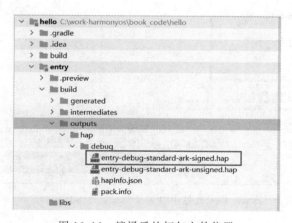

图 16.16　编译后的打包文件位置

## 16.4　推送并将 HAP 安装到开发板/设备

在 OpenHarmony 2.0 开源发布后,官方提供了 HDC 工具,便于在 Windows 上将打包好的应用安装到开发板/设备上,HDC(Harmony Device Connector)是 OpenHarmony 为开

发人员提供的用于设备连接调试的命令行工具，PC 端使用命令行工具 hdc_std（为方便起见，下文统称 hdc），该工具需支持部署在 Windows/Linux/Mac 等系统上与鸿蒙设备（或模拟器）进行连接、调试、通信。PC 端 hdc 工具需要针对以上开发机操作系统平台分别发布相应的版本，设备端 hdc daemon 需跟随设备镜像发布，包括对模拟器进行支持。

hdc 主要由三部分组成。

（1）hdc client 部分：运行于开发机上的客户端，用户可以在开发机命令终端（Windows cmd/linux shell）下请求执行相应的 hdc 命令，运行于开发机器，其他的终端调试 IDE 也包含 hdc client。

（2）hdc server 部分：作为后台进程也运行于开发机器，server 管理 client 和设备端 daemon 之间的通信，包括连接的复用、数据通信包的收发，以及个别本地命令的直接处理。

（3）hdc daemon 部分：daemon 部署于 OpenHarmony 设备端作为守护进程来按需运行，负责处理来自 client 端的请求。

### 16.4.1　OpenHarmony 命令行启动 hdcd

通过串口工具发送 hdcd -t 命令，以此启动 hdc server，hdcd 用于监听来自 client 端的请求，如图 16.17 所示。

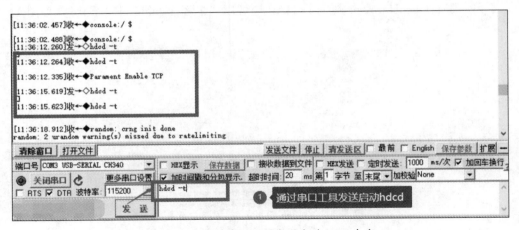

图 16.17　通过串口工具发送启动 hdcd 命令

### 16.4.2　下载 hdc_std 工具

与 HarmonyOS 设备使用的 hdc 工具不同，OpenHarmony 需要 hdc_std 作为调试工具，hdc-std 的下载链接为 https://gitee.com/openharmony/developtools_hdc_standard。

### 16.4.3　配置环境变量（Windows）

将 hdc_std.exe 添加到系统的环境变量中。在 CMD 命令行中输入命令 hdc_std -h 验证环境变量是否添加成功。

### 16.4.4　使用 hdc_std 安装 HAP

使用 USB 连接开发板后，使用命令 hdc_std install <file_path>来安装编译好的 HAP。

首先执行 hdc_std list targets 命令，输出设备的编号，然后执行命令 hdc_std install xx.hap 将 HAP 包安装到设备上，如图 16.18 所示。

图 16.18　在 CMD 命令行执行 hdc_std，将本地的包推送到远程设备

### 16.4.5　Hi3516DV300 的运行

推送成功后，在开发板的桌面上可以看到已创建的项目的图标，效果如图 16.19 所示。

图 16.19　推送成功，桌面显示 App 图标

单击 Icon 图标，显示项目的首页，如图 16.20 所示。

### 16.4.6　hdc_std 连接不到设备

执行 hdc_std list targets 命令后结果为[Empty]，可能的原因有以下几点。

图 16.20　单击桌面 App 图标，进入 App 首页

**1. 设备没有被识别**

在设备管理器中查看是否有 hdc 设备，在通用串行总线设备中会有 HDC Device 信息。如果没有，则表示 hdc 无法连接。此时需要插拔设备，或者烧录最新的镜像。

**2. hdc_std 工作异常**

可以执行 hdc kill 或者 hdc start -r 命令杀掉 hdc 服务或者重启 hdc 服务，然后执行 hdc list targets 命令查看是否已经可以获取设备信息。

**3. hdc_std 与设备不匹配**

如果设备烧录的是最新镜像，则 hdc_std 也需要使用最新版本。由于 hdc_std 会持续更新，所以需要从开源仓 developtools_hdc_standard 中获取最新版本，具体位置在该开源仓的 prebuilt 目录。

# 第 17 章 OpenHarmony "HiSpark 智能赛车"

本章基于 HiSpark 智能赛车套件,在 Hi3861 开发版上搭载 OpenHarmony 1.0 系统,实现多手机遥控比赛的游戏。在游戏控制端使用 JS UI+Java UI 混合开发,实现赛车的操作控制。

## 17.1 鸿蒙 HiSpark 智能赛车游戏介绍

安装 OpenHarmony 鸿蒙操作系统的 WiFi 智能赛车,通过鸿蒙手机 HAP 应用连接,遥控比赛。赛车游戏支持多个鸿蒙智慧赛车加入。

HiSparkWiFi IoT 智能赛车套件基于海思 Hi3861 芯片,支持 HarmonyOS、LiteOS,可实现巡线、避障等功能;通过寻迹模块获取图面轨道数据,运用寻迹算法使智能赛车可以按照固定轨道运行;通过超声波传感器获取周围环境障碍物数据,运用避障实现避障功能;具备 AP 功能,以及支持云平台远程遥控智能赛车;可应用于智能物流、无人车、服务机器人等领域,如图 17.1 所示。

HiSparkWiFi IoT 智能赛车配置的 Hi6861 套件如图 17.2 所示。

图 17.1　HiSpark WiFi IoT 智能赛车套件　　　　图 17.2　HiSpark WiFi IoT 套件

HiSparkWiFi IoT 智能赛车套件的具体参数如表 17.1 所示。

表 17.1　HiSparkWiFi IoT 智能赛车详细套件的具体参数

| 序号 | 类别 | 描述(型号/品名) | 数量 |
|---|---|---|---|
| 1 | WiFi主板 | 型号：HiSpark_WiFi-IoT_Hi3861_CH340G_VER.A | 1 |
| 2 | 通用底板 | 型号：HiSpark_WiFi-IoT_EXB_VER.A | 1 |
| 3 | 显示板 | 型号：HiSpark_WiFi-IoT_OLED_VER.A | 1 |
| 4 | NFC板 | 型号：HiSpark_WiFi-IoT_NFC_VER.A | 1 |
| 5 | 机器人板 | 型号：HiSpark_WiFi-IoT_Robot_VER.A | 1 |
| 6 | 智能小车底盘 | 型号：2WD；尺寸：21.4cm×15cm | 1 |
| 7 | TT电机 | 黄色；3~6V；单轴；1:48 | 2 |
| 8 | TT电机固定支架 | 亚克力电机支架 | 4 |
| | | 螺丝：M3×30 | 4 |
| | | 螺帽：M3 | 4 |
| 9 | 橡胶轮胎 | 黄色；65mm×27mm | 2 |
| 10 | 万向轮 | 直径1英寸；白色PP材质 | 1 |
| 11 | 寻迹模块 | 红外寻迹传感器 | 2 |
| 12 | 寻迹连接线 | 2.54mm间距；3PIN；母对母双头并排带外壳连接线 | 2 |
| 13 | 舵机+舵机配件 | 微型SG90舵机 | 1 |
| | | 舵机支架：单端×1；双端×1；四端×1 | 3 |
| | | 支架螺丝：M2×4 | 1 |
| | | 自攻螺丝：M2×8 | 2 |
| | | 自攻螺丝：M2×6 | 2 |
| 14 | 舵机支架 | SG90舵机支架 | 1 |
| | | 螺丝：M3×10 | 2 |
| | | 螺帽：M3 | 2 |
| | | 螺丝：M2×10 | 2 |
| | | 螺帽：M2 | 2 |
| 15 | 超声波模块 | 型号：HC-SR04；电压：3.3~5V | 1 |
| 16 | 超声波支架 | 超声波固定支架 | 1 |
| | | 螺丝：M2×6 | 2 |
| | | 螺帽：M2 | 2 |
| 17 | 超声波连接线 | 4PIN对4PIN；PH2.0mm转杜邦2.54mm | 1 |
| 18 | NFC板连接线 | 6PIN对6PIN；1.27mm；反向异面：20cm | 1 |
| 19 | 电池盒 | 锂电池(18650)盒；两节；串联；带2.54mm插头 | 1 |
| | | 干电池(5号)盒；四节；串联；带2.54mm插头 | 1 |
| 20 | 配件袋 | 螺帽：M3 | 6 |
| | | 螺丝：M3×6 | 24 |
| | | 螺柱：M3×8+6 | 4 |
| | | 螺柱：M3×30 | 10 |

## 17.2 HiSpark 智能赛车端实现

在 HiSpark 智能赛车搭载的 OpenHarmony 系统中添加 racecar 应用程序，racecar 程序运行在赛车端的系统上，用来监听遥控端的输入，并调用赛车上的电机驱动程序，驱动赛车朝不同方向行驶，本节涉及 C 语言开发，读者可以忽略这部分内容。

### 17.2.1 HiSpark 赛车配置 WiFi 网络

下面通过 SSCOM 5.13.1 串口/网络调试器工具设置 OpenHarmony WiFi 网络，如图 17.3 所示。

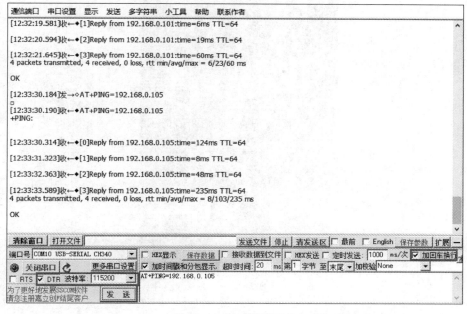

图 17.3 串口调试工具

在串口终端中，依次执行如下 AT 命令，启动 STA 模式，连接指定 AP 热点，并开启 DHCP 功能，命令如下：

```
AT + STARTSTA                      # 启动 STA 模式
AT + SCAN                          # 扫描周边 AP
AT + SCANRESULT                    # 显示扫描结果
AT + CONN = "SSID",,2,"PASSWORD"   # 连接指定 AP,其中 SSID/PASSWORD 为待连接的热点名称和密码
AT + STASTAT                       # 查看连接结果
AT + DHCP = wlan0,1                # 通过 DHCP 向 AP 请求 wlan0 的 IP 地址
```

查看 WLAN 模组与网关联通是否正常，命令如下：

```
AT + IFCFG                    # 查看模组接口 IP
AT + PING = X.X.X.X          # 检查模组与网关的联通性,其中 X.X.X.X 需替换为实际的网关地址
```

### 17.2.2 HiSpark 赛车电机驱动

让赛车跑起来,需要控制驱动赛车的电机,电机的驱动板如图 17.4 所示,目前电机驱动芯片使用的是 L9110S 芯片。

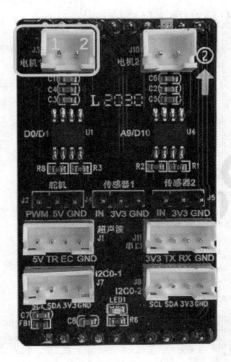

图 17.4 驱动板及 L9110S 芯片

L9110S 芯片的电路如图 17.5 所示。

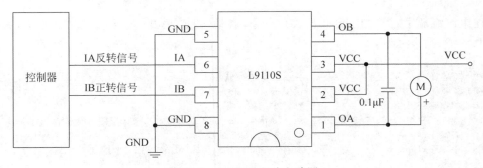

图 17.5 L9110S 芯片电路图

参考图 17.6，要控制电机，芯片至少需要两路 PWM 信号，一路用于控制正转，另一路用于控制反转。由于赛车有两个轮子，需要两个电机，所以一共需要 4 路 PWM 信号。

目前 L9110S 芯片用来控制电机的 4 路 PWM，如表 17.2 所示。

表 17.2　L9110S 芯片 PWM 信号

| 电机 | GPIO 口 | PWM 通道 |
| --- | --- | --- |
| 电机 1 | GPIO 0 | PWM 3 |
| | GPIO 1 | PWM 4 |
| 电机 2 | GPIO 9 | PWM 0 |
| | GPIO 10 | PWM 1 |

知道了 PWM 通道和对应的 GPIO 口，我们就可以开始编程了。

首先初始化 PWM，如代码示例 17.1 所示。

代码示例 17.1　初始化 PWM

```c
void pwm_init(void)
{
    GpioInit();
    //引脚复用
    IoSetFunc(WIFI_IOT_IO_NAME_GPIO_0, WIFI_IOT_IO_FUNC_GPIO_0_PWM3_OUT);
    IoSetFunc(WIFI_IOT_IO_NAME_GPIO_1, WIFI_IOT_IO_FUNC_GPIO_1_PWM4_OUT);
    IoSetFunc(WIFI_IOT_IO_NAME_GPIO_9, WIFI_IOT_IO_FUNC_GPIO_9_PWM0_OUT);
    IoSetFunc(WIFI_IOT_IO_NAME_GPIO_10, WIFI_IOT_IO_FUNC_GPIO_10_PWM1_OUT);
    //初始化 PWM
    PwmInit(WIFI_IOT_PWM_PORT_PWM3);
    PwmInit(WIFI_IOT_PWM_PORT_PWM4);
    PwmInit(WIFI_IOT_PWM_PORT_PWM0);

    PwmInit(WIFI_IOT_PWM_PORT_PWM1);
}
```

控制赛车前进、后退、左转、右转、停止的函数，如代码示例 17.2 所示。

代码示例 17.2　racecar_c/car_model.c

```c
void pwm_stop(void)
{
    //先停止 PWM
    PwmStop(WIFI_IOT_PWM_PORT_PWM3);
    PwmStop(WIFI_IOT_PWM_PORT_PWM4);
    PwmStop(WIFI_IOT_PWM_PORT_PWM0);
    PwmStop(WIFI_IOT_PWM_PORT_PWM1);
}
```

```c
//前进
void pwm_forward(void)
{
//先停止 PWM
  PwmStop(WIFI_IOT_PWM_PORT_PWM3);
  PwmStop(WIFI_IOT_PWM_PORT_PWM4);
  PwmStop(WIFI_IOT_PWM_PORT_PWM0);
  PwmStop(WIFI_IOT_PWM_PORT_PWM1);

//启动 A 路 PWM
  PwmStart(WIFI_IOT_PWM_PORT_PWM3, 750, 1500);
  PwmStart(WIFI_IOT_PWM_PORT_PWM0, 750, 1500);
}

//后退
void pwm_backward(void)
{
   //先停止 PWM
  PwmStop(WIFI_IOT_PWM_PORT_PWM3);
  PwmStop(WIFI_IOT_PWM_PORT_PWM4);
  PwmStop(WIFI_IOT_PWM_PORT_PWM0);

  PwmStop(WIFI_IOT_PWM_PORT_PWM1);

   //启动 A 路 PWM
  PwmStart(WIFI_IOT_PWM_PORT_PWM4, 750, 1500);
  PwmStart(WIFI_IOT_PWM_PORT_PWM1, 750, 1500);
}

//左转
void pwm_left(void)
{
  //先停止 PWM
  PwmStop(WIFI_IOT_PWM_PORT_PWM3);
   PwmStop(WIFI_IOT_PWM_PORT_PWM4);
  PwmStop(WIFI_IOT_PWM_PORT_PWM0);
  PwmStop(WIFI_IOT_PWM_PORT_PWM1);
   //启动 A 路 PWM
  PwmStart(WIFI_IOT_PWM_PORT_PWM3, 750, 1500);
}

//右转
void pwm_right(void)
{
  //先停止 PWM
  PwmStop(WIFI_IOT_PWM_PORT_PWM3);
  PwmStop(WIFI_IOT_PWM_PORT_PWM4);
  PwmStop(WIFI_IOT_PWM_PORT_PWM0);
  PwmStop(WIFI_IOT_PWM_PORT_PWM1);
```

```c
//启动 A 路 PWM
  PwmStart(WIFI_IOT_PWM_PORT_PWM0, 750, 1500);
}
```

接下来，需要修改 vendor\hisi\hi3861\hi3861\build\config\usr_config.mk 文件，把 PWM 功能打开，增加一行代码 CONFIG_PWM_SUPPORT=y，如图 17.6 所示。

增加第 43 行代码，开启 PWM_SUPPORT 支持。

```
33 # CONFIG_I2C_SUPPORT is not set
34 # CONFIG_I2S_SUPPORT is not set
35 # CONFIG_SPI_SUPPORT is not set
36 # CONFIG_DMA_SUPPORT is not set
37 # CONFIG_SDIO_SUPPORT is not set
38 # CONFIG_SPI_DMA_SUPPORT is not set
39 # CONFIG_UART_DMA_SUPPORT is not set
40 # CONFIG_PWM_SUPPORT is not set
41 # CONFIG_PWM_HOLD_AFTER_REBOOT is not set
42 CONFIG_I2C_SUPPORT=y
43 CONFIG_PWM_SUPPORT=y
44 CONFIG_AT_SUPPORT=y
45 CONFIG_FILE_SYSTEM_SUPPORT=y
46 CONFIG_UART0_SUPPORT=y
47 CONFIG_UART1_SUPPORT=y
48 # CONFIG_UART2_SUPPORT is not set
49 # end of BSP Settings
```

图 17.6 开启 PWM_SUPPORT 支持

如果不开启 PWM，则在编译时会出现以下错误，如图 17.7 所示。

```
-lwifi -lwifi_flash -lwifiiot_app -lwifiservice -lwpa --end-group
riscv32-unknown-elf-ld: ohos/libs/libhal_iothardware.a(hal_wifiiot_pwm.o): in f
unction `.L0 ':
hal_wifiiot_pwm.c:(.text.HalPwmInit+0x16): undefined reference to `hi_pwm_set_c
lock'
riscv32-unknown-elf-ld: hal_wifiiot_pwm.c:(.text.HalPwmInit+0x24): undefined re
ference to `hi_pwm_init'
riscv32-unknown-elf-ld: hal_wifiiot_pwm.c:(.text.HalPwmStart+0x12): undefined r
eference to `hi_pwm_start'
riscv32-unknown-elf-ld: hal_wifiiot_pwm.c:(.text.HalPwmStop+0x12): undefined re
ference to `hi_pwm_stop'
scons: *** [output/bin/Hi3861_wifiiot_app.out] Error 1
```

图 17.7 未开启 PWM_SUPPORT 支持时编译会报错

### 17.2.3　HiSpark 赛车操作控制

我们在赛车上面简单地编写一个 UDP 程序，监听 50001 端口。这里使用的通信格式是 JSON，赛车收到 UDP 数据后，解析 JSON 格式数据，并根据命令执行相应的操作，例如前进、后退、左转、右转等，C 语言代码如代码示例 17.3 所示。

代码示例 17.3　racecar_c/udp_model.c

```c
#include "hi_wifi_api.h"
#include "lwip/ip_addr.h"
#include "lwip/netifapi.h"
```

```c
#include "lwip/sockets.h"
#include <stdio.h>

#include <unistd.h>
#include "ohos_init.h"
#include "cmsis_os2.h"
#include "cJSON.h"
#include "car_test.h"

char recvline[1024];
void udp_thread(void *pdata)
{
    int ret;
    struct sockaddr_in servaddr;
    cJSON *recvjson;
    pdata = pdata;
    int sockfd = socket(PF_INET, SOCK_DGRAM, 0);

    //服务器 ip port
    bzero(&servaddr, sizeof(servaddr));
    servaddr.sin_family = AF_INET;
    servaddr.sin_addr.s_addr = htonl(INADDR_ANY);
    servaddr.sin_port = htons(50001);

    printf("udp_thread \r\n");
    bind(sockfd, (struct sockaddr *)&servaddr, sizeof(servaddr));

    while(1)
    {
        struct sockaddr_in addrClient;
        int sizeClientAddr = sizeof(struct sockaddr_in);

        memset(recvline, sizeof(recvline), 0);
        ret = recvfrom(sockfd, recvline, 1024, 0, (struct sockaddr *)&addrClient,(socklen_t *)&sizeClientAddr);

        if(ret > 0)
        {
            char *pClientIP = inet_ntoa(addrClient.sin_addr);

            printf("%s - %d(%d) says:%s\n", pClientIP, ntohs(addrClient.sin_port), addrClient.sin_port, recvline);

            //进行 JSON 解析
            recvjson = cJSON_Parse(recvline);

            if(recvjson != NULL)
            {
```

```c
                if(cJSON_GetObjectItem(recvjson, "cmd")->valuestring != NULL)
                {
                    printf("cmd : %s\r\n", cJSON_GetObjectItem(recvjson, "cmd")->valuestring);

                    if(strcmp("forward", cJSON_GetObjectItem(recvjson, "cmd")->valuestring) == 0)
                    {
                        set_car_status(CAR_STATUS_FORWARD);
                        printf("forward\r\n");
                    }

                    if(strcmp("backward", cJSON_GetObjectItem(recvjson, "cmd")->valuestring) == 0)
                    {
                        set_car_status(CAR_STATUS_BACKWARD);
                        printf("backward\r\n");
                    }

                    if(strcmp("left", cJSON_GetObjectItem(recvjson, "cmd")->valuestring) == 0)
                    {
                        set_car_status(CAR_STATUS_LEFT);
                        printf("left\r\n");
                    }

                    if(strcmp("right", cJSON_GetObjectItem(recvjson, "cmd")->valuestring) == 0)
                    {
                        set_car_status(CAR_STATUS_RIGHT);
                        printf("right\r\n");
                    }

                    if(strcmp("stop", cJSON_GetObjectItem(recvjson, "cmd")->valuestring) == 0)
                    {
                        set_car_status(CAR_STATUS_STOP);
                        printf("stop\r\n");
                    }

                    if(strcmp("led_off", cJSON_GetObjectItem(recvjson, "cmd")->valuestring) == 0)
                    {
                        set_car_status(CAR_STATUS_STOP);
                        printf("led_off\r\n");
                    }

                    if(strcmp("led_on", cJSON_GetObjectItem(recvjson, "cmd")->valuestring) == 0)
                    {
                        set_car_status(CAR_STATUS_STOP);
                        printf("led_on\r\n");
```

```c
                }
            }
            if(cJSON_GetObjectItem(recvjson, "mode")->valuestring != NULL)
            {
                if(strcmp(" step", cJSON_GetObjectItem(recvjson, "mode")->valuestring) == 0)
                {
                    set_car_mode(CAR_MODE_STEP);
                    printf("mode step\r\n");
                }

                if(strcmp(" alway", cJSON_GetObjectItem(recvjson, "mode")->valuestring) == 0)
                {
                    set_car_mode(CAR_MODE_ALWAY);
                    printf("mode alway\r\n");
                }
            }
            cJSON_Delete(recvjson);
        }
    }
}

void start_udp_thread(void)
{
    osThreadAttr_t attr;
    attr.name = "wifi_config_thread";
    attr.attr_bits = 0U;
    attr.cb_mem = NULL;
    attr.cb_size = 0U;
    attr.stack_mem = NULL;
    attr.stack_size = 2048;
    attr.priority = 36;

    if (osThreadNew((osThreadFunc_t)udp_thread, NULL, &attr) == NULL) {
        printf("[LedExample] Falied to create LedTask!\n");
    }
}
```

## 17.3 将赛车控制模块添加到鸿蒙源码并编译

这里采用 OpenHarmony 1.0 系统进行烧录，下载 OpenHarmony1.0 源码，把 racecar 代码复制到 applications\sample\wifi-iot\app 目录中，使用 python build.py wifiiot 命令编译，再通过 hiburn 把系统烧录到 Hi3861 上。

## 17.3.1 添加赛车控制模块代码

将赛车控制模块 racecar 代码添加到 applications\sample\wifi-iot\app 目录中,如图 17.8 所示。

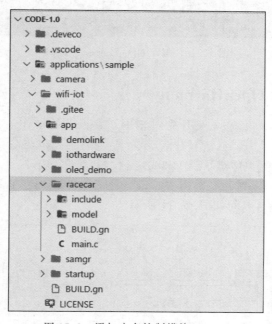

图 17.8 添加小车控制模块 racecar

接下来,修改 BUILD.gn 文件,设置启动 racecar 模块,如图 17.9 所示。

```
import("//build/lite/config/component/lite_component.gni")
lite_component("app") {
    features = [
        "racecar",
    ]
}
```

图 17.9 设置启动 racecar 模块

## 17.3.2 编译 OpenHarmony 源码

在串口中监听系统运行的信息,为了不被系统运行的日志影响可查看调试信息,这里编译命令不要带-b debug 参数,命令如下:

```
python build.py wifiiot
```

编译成功后的提示信息如图 17.10 所示。

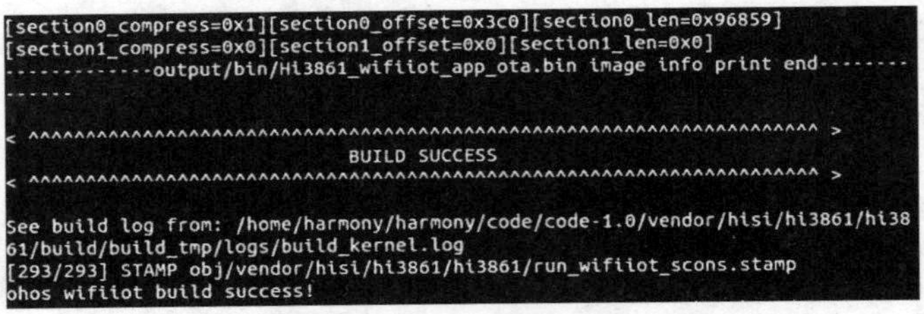

图 17.10　编译 OpenHarmony 源码

### 17.3.3　烧录 OpenHarmony

烧录是指将编译后的程序文件下载到芯片开发板上的动作,为后续的程序调试提供基础。

使用 HiBurn 将.bin 文件烧录到 Hi3861 开发板,步骤如下:

(1) 在 Windows 10 系统执行前需要右击"属性"→解除锁定,否则系统默认会报安全警告,不允许执行。双击后,界面如图 17.11 所示。

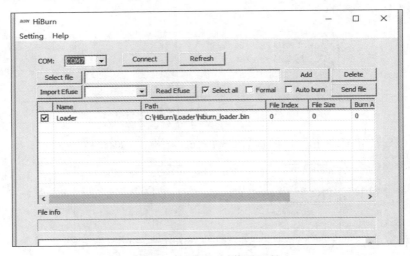

图 17.11　HiBurn 烧录工具

(2) 单击界面左上角的 Setting→Com settings 进入串口参数设置界面,如图 17.12 所示。

(3) 在串口参数设置界面,Baud 为波特率,默认为 115200,可以选择 921600、2000000 或者 3000000(实测最快支持的值),其他参数保持默认,单击"确定"按钮保存,如图 17.13 所示。

(4) 根据设备管理器,选择正确的 COM 口,例如笔者的开发板的端口是 COM8,如果在打开程序之后才插串口线,则可以单击 Refresh 按钮刷新串口下拉列表框的可选项,如图 17.14 所示。

# 第17章 OpenHarmony "HiSpark智能赛车"

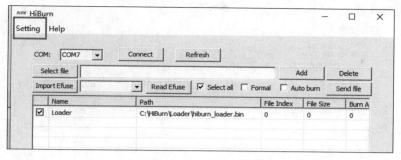

图 17.12 Hi3861 产品参数(1)

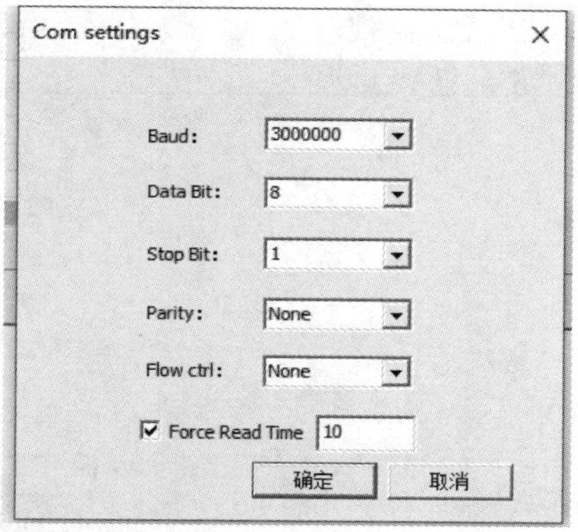

图 17.13 Hi3861 产品参数(2)

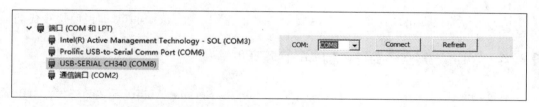

图 17.14 Hi3861 产品参数(3)

(5) 单击 Select file 按钮后会弹出文件选择对话框,选择编译生成的 allinone.bin 文件,这个 bin 文件其实是由多个 bin 文件合并而成的文件,从命名上也能看出来,例如,笔者选择的 z:\harmonyos\openharmony\out\wifiiot\Hi3861_wifiiot_app_allinone.bin。

(6) 勾选 Auto burn,自动下载多个 bin 文件,到这里,配置完毕,如图 17.15 所示。

(7) 单击 Connect 按钮,连接串口设备,这时 HiBurn 会打开串口设备,并尝试开始烧录,需要确保没有其他程序占用串口设备(烧录之前可能正在用超级终端或串口助手查看串口日志,需要确保其他软件已经关闭了当前使用的串口)。

(8) 复位设备,按开发板的 RESET 按键。

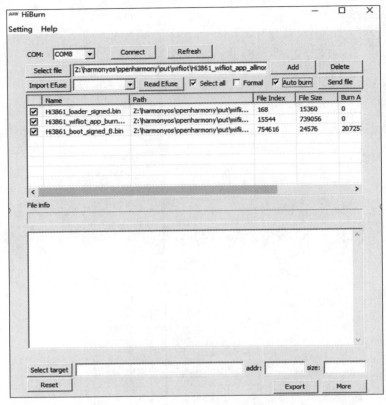

图 17.15　Hi3861 产品参数(4)

(9) 等待输出框出现 3 个========及上方均出现 successful，即说明烧录成功。

## 17.4　鸿蒙 HAP 端控制赛车实现

通过鸿蒙 JS UI 开发赛车的控制端，结合 Java 端 Service Ability 给赛车发送 UDP 命令，实现对赛车的操作，如图 17.16 所示。

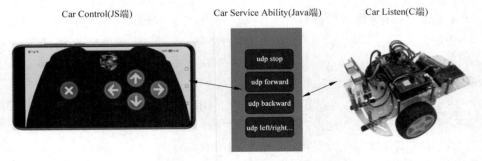

图 17.16　JS UI 端与赛车端调用关系

## 17.4.1 赛车控制手柄界面实现

通过 JS UI 实现赛车控制端,效果如图 17.17 所示,控制端通过上、下、左、右键控制赛车的前进方向。

图 17.17　JS UI 端赛车遥控器

这里通过 fixed 定位的方式实现图 17.17 的赛车控制界面,如代码示例 17.4 所示。

代码示例 17.4　racecar_js/js/default/index.hml

```html
<div class="container">
  <div class="logo">
    <image class="logo_btn" id="logo"></image>
  </div>
  <div class="start">
    <image class="{{startIcon}}" @click="start"></image>
  </div>
  <div class="up">
    <image class="up_btn"></image>
  </div>
  <div class="down">
    <image class="down_btn"></image>
  </div>
  <div class="left">
    <image class="left_btn"></image>
  </div>
  <div class="right">
    <image class="right_btn"></image>
  </div>
</div>
```

界面样式实现如下所示,这里给按钮添加激活样式,给单击操作添加交互的效果,当单击开始按钮后,最上面的车标开始旋转,表示车处于激活状态。当单击停止按钮后,最上面的车标停止旋转,表示车处于停止状态,如代码示例 17.5 所示。

代码示例 17.5　racecar_js/js/default/index.css

```css
.container {
    flex-direction: column;
    justify-content: center;
    align-items: center;
    background-image: url("/common/bj.jpg");
    background-size: 100%;
    background-repeat: no-repeat;
    object-fit: contain;
}

.logo {
    position: fixed;
    left: 45%;
    top: 25px;
    width: 80px;
    height: 80px;
}

.logo_btn {
    background-image: url("/common/logo.png");
    background-repeat: no-repeat;
    background-size: 100%;
    width: 80px;
    height: 80px;
    border-radius: 40px;
    object-fit: cover;
}

.start {
    position: fixed;
    left: 150px;
    top: 150px;
    width: 80px;
    height: 80px;
}

.stop_btn {
    background-image: url("/common/stop.png");
    background-repeat: no-repeat;
    background-size: 100%;
    object-fit: cover;
    border-radius: 40px;
}

.stop_btn:active {
    border-width: 10px;
    border-color: red;
}
```

```css
.start_btn {
    background-image: url("/common/start.png");
    background-repeat: no-repeat;
    background-size: 100%;
    object-fit: cover;
    border-radius: 40px;
}

.start_btn:active {
    border-width: 10px;
    border-color: red;
}


.up {
    position: fixed;
    right: 200px;
    top: 100px;
    width: 80px;
    height: 80px;

}

.up_btn:active {
    border-width: 10px;
    border-color: red;
}

.up_btn {
    background-image: url("/common/up.png");
    background-repeat: no-repeat;
    background-size: 100%;
    object-fit: cover;
    border-radius: 40px;
}

.down {
    position: fixed;
    right: 200px;
    top: 200px;
    width: 80px;
    height: 80px;
}

.down_btn {
    background-image: url("/common/down.png");
    background-repeat: no-repeat;
    background-size: 100%;
    object-fit: cover;
```

```css
        border-radius: 40px;
    }

    .down_btn:active {
        border-width: 10px;
        border-color: red;
    }

    .left {
        position: fixed;
        right: 100px;
        top: 150px;
        width: 80px;
        height: 80px;
    }

    .left_btn {
        background-image: url("/common/right.png");
        background-repeat: no-repeat;
        background-size: 100%;
        object-fit: cover;
        border-radius: 40px;
    }

    .left_btn:active {
        border-width: 10px;
        border-color: red;
    }

    .right {
        position: fixed;
        right: 300px;
        top: 150px;
        width: 80px;
        height: 80px;
    }

    .right_btn {
        background-image: url("/common/left.png");
        background-repeat: no-repeat;
        background-size: 100%;
        object-fit: cover;
        border-radius: 40px;
    }

    .right_btn:active {
        border-width: 10px;
        border-color: red;
    }
```

## 17.4.2 将赛车控制手柄设置为横屏模式

因为控制手柄比较宽,所以需要将界面设置为横屏模式,这里只需将 config.json 文件中的 abilities 中的 orientation 属性值修改为 landscape。为了更好地体验,这里需要把头部标题去掉,设置为窗口透明,如代码示例 17.6 所示。

代码示例 17.6    racecar_js/js/default/config.json

```json
{
"skills": [
    {
"entities": [
"entity.system.home"
    ],
"actions": [
"action.system.home"
    ]
  }
  ],
"name": "com.charjedu.ptgamebook.racecar_js.MainAbility",
"icon": "$media:icon",
"description": "$string:mainability_description",
"label": "$string:app_name",
"type": "page",
"orientation": "landscape",
"launchType": "standard",
"metaData": {
"customizeData": [
    {
"name": "hwc-theme",
"value": "androidhwext:style/Theme.Emui.Light.NoTitleBar",
"extra": ""
    }
  ]
 }
},
```

这里的 metaData 属性是用来设置 Abiltiy 主题的,使用 NoTitleBar 主题,将不显示头部标题栏,代码如下:

```json
"metaData": {
"customizeData": [
    {
"name": "hwc-theme",
"value": "androidhwext:style/Theme.Emui.Light.NoTitleBar",
"extra": ""
    }
  ]
}
```

### 17.4.3　Java 端通过 Service Ability 发送指令

Java 端通过远程 Service Ability 发送 UDP 指令，远程代理对象 CarRemoteObject 的 onRemoteRequest() 方法用于处理 JS 端发送过来的操作码，如 100 表示启动、101 表示停止、102～105 分别表示上、下、左、右的赛车操作码，如代码示例 17.7 所示。

代码示例 17.7　racecar_js/java/CarServiceAbility.java

```java
public class CarServiceAbility extends Ability {
    private static final HiLogLabel LABEL_LOG = new HiLogLabel(3, 0xD001100, "Demo");

    private CarRemoteObject carRemoteObject = new CarRemoteObject();

    @Override
    public void onStart(Intent intent) {
        HiLog.error(LABEL_LOG, "CarServiceAbility::onStart");
        super.onStart(intent);
    }

    @Override
    public IRemoteObject onConnect(Intent intent) {
        return carRemoteObject.asObject();
    }

    class CarRemoteObject extends RemoteObject implements IRemoteBroker {

        public CarRemoteObject() {
            super("hisparck CarRemoteObject");
        }

        @Override
        public IRemoteObject asObject() {
            return this;
        }

        @Override
        public boolean onRemoteRequest(int code, MessageParcel data, MessageParcel reply,
MessageOption option) throws RemoteException {
            String carip = "192.168.0.105";
            int port = 50001;
            Map<String, String> cmdMsg;
            switch (code){
                //启动
                case 100: {
                    cmdMsg = new HashMap<>();
                    cmdMsg.put("cmd", "forward");
                    //mode 为 step 时会自动停下
                    cmdMsg.put("mode", "step");
                    CommonTools.sendMsg(carip, port, cmdMsg);
```

```
            break;
        }
        //停止
        case 101: {
            break;
        }
        //前进
        case 102: {
            cmdMsg = new HashMap<>();
            cmdMsg.put("cmd", "forward");
            //mode 为 step 时会自动停下
            cmdMsg.put("mode", "step");
            CommonTools.sendMsg(carip, port, cmdMsg);
            break;
        }
        //后退
        case 103: {
            cmdMsg = new HashMap<>();
            cmdMsg.put("cmd", "backward");
            cmdMsg.put("mode", "step");
            CommonTools.sendMsg(carip, port, cmdMsg);
            break;
        }
        //向左
        case 104: {
            cmdMsg = new HashMap<>();
            cmdMsg.put("cmd", "left");
            cmdMsg.put("mode", "step");
            CommonTools.sendMsg(carip, port, cmdMsg);
            break;
        }
        //向右
        case 105: {
            cmdMsg = new HashMap<>();
            cmdMsg.put("cmd", "right");
            cmdMsg.put("mode", "step");
            CommonTools.sendMsg(carip, port, cmdMsg);
            break;
        }
    }
    return true;
    }
  }
}
```

在上面的代码中使用 CommonTools.sendMsg()方法向赛车发送操作命令,这种方法有 3 个参数:carip 是指赛车的 IP 地址,port 是赛车用来介绍命令的 UDP 端口号,cmdMsg

是向赛车发送的命令信息,cmdMsg 包括 cmd 命令和运动模式(mode)两个配置,cmd 命令值是赛车设置好的(forward：向前,backward：向后,left：向左,right：向右),如代码示例 17.8 所示。

代码示例 17.8　racecar_js/java/utils/CommonTools.java

```java
public static void sendMsg(final String ipAddress,final int linkPort, final Map<String,String> msgMap) {
    new Thread(() -> {
        ZSONObject address = new ZSONObject();
        try {
            for(Map.Entry<String,String> entry:msgMap.entrySet()){
                address.put(entry.getKey(), entry.getValue());
            }
        } catch (Exception e) {
            e.printStackTrace();
        }
        try {
            InetAddress targetAddress = InetAddress.getByName(ipAddress);
            DatagramSocket socket = new DatagramSocket();
            DatagramPacket packet = new DatagramPacket(address.toString().getBytes(), address.toString().length(), targetAddress, linkPort);
            socket.send(packet);
        } catch (IOException e) {
            e.printStackTrace();
        }
    }).start();
}
```

### 17.4.4　赛车控制手柄界面逻辑实现

JS 端逻辑实现是通过 FeatureAbility.callAbility()方法实现 PA 端调用和发送驱动指令的,赛车控制手柄的界面逻辑实现的步骤如下。

#### 1. 开启分布式调用权限

调用 Service Ability,需要开启远程调用权限,在 config.json 文件中申请远程调用权限,配置如代码示例 17.9 所示。

代码示例 17.9　racecar_js/js/config.json

```json
"reqPermissions": [
  {
"name": "ohos.permission.READ_USER_STORAGE"
  },
  {
"name": "ohos.permission.DISTRIBUTED_DATASYNC"
  },
  {
```

```
  "name": "ohos.permission.servicebus.ACCESS_SERVICE"
  },
  {
  "name": "ohos.permission.servicebus.BIND_SERVICE"
  },
  {
  "name": "ohos.permission.DISTRIBUTED_DEVICE_STATE_CHANGE"
  },
  {
  "name": "ohos.permission.GET_DISTRIBUTED_DEVICE_INFO"
  },
  {
  "name": "ohos.permission.GET_BUNDLE_INFO"
  }
]
```

在 MainAbility.java 文件中获取权限，如代码示例 17.10 所示。

**代码示例 17.10    racecar_js/java/MainAbility.java**

```java
public class MainAbility extends AceAbility {
    @Override
    public void onStart(Intent intent) {
        super.onStart(intent);
        requestPermission();
    }

    //获取权限
    private void requestPermission() {
        String[] permission = {
"ohos.permission.CAMERA",
"ohos.permission.READ_USER_STORAGE",
"ohos.permission.WRITE_USER_STORAGE",
"ohos.permission.DISTRIBUTED_DATASYNC",
"ohos.permission.MICROPHONE",
"ohos.permission.GET_DISTRIBUTED_DEVICE_INFO",
"ohos.permission.KEEP_BACKGROUND_RUNNING",
"ohos.permission.NFC_TAG"};
        List<String> applyPermissions = new ArrayList<>();
        for (String element : permission) {
            if (verifySelfPermission(element) != 0) {
                if (canRequestPermission(element)) {
                    applyPermissions.add(element);
                }
            }
        }
        requestPermissionsFromUser(applyPermissions.toArray(new String[0]), 0);
    }
```

```java
        @Override
    public void onStop() {
        super.onStop();
    }
}
```

### 2. JS 端调用 Service Ability 实现指令发送

JS 端调用 Service Ability 是通过 FeatureAbility.callAbility()方法实现 PA 端调用和发送驱动指令的,这里封装了 FeatureAbility.callAbility()方法,如代码示例 17.11 所示。

**代码示例 17.11　racecar_js/js/default/common/utils.js**

```javascript
callAbility:async function(code){
    var action = {};
    action.bundleName = 'com.charjedu.ptgamebook';
    action.abilityName = 'com.charjedu.ptgamebook.racecar_js.CarServiceAbility';
    action.messageCode = code;
    action.data = {};
    action.abilityType = 0;
    action.syncOption = 0;
    var result = await FeatureAbility.callAbility(action);
    var ret = JSON.parse(result);
    if (ret.code == 0) {
        console.info('plus result is:' + JSON.stringify(ret.abilityResult));
    } else {
        console.error('plus error code:' + JSON.stringify(ret.code));
    }
},
```

修改 index.hml 文件,给控制按钮添加@click=runCar(code)方法,runCar()方法调用 utils.callAbility,以便将不同的编码发送给 PA,PA 接受 code 后将 UDP 包发送给赛车,如代码示例 17.12 所示。

**代码示例 17.12　racecar_js/js/default/index.js**

```javascript
import utils from "../../common/utils.js"
import prompt from '@system.prompt';

export default {
    data: {
        startIcon: "start_btn",
        status: false,
        aniPoster: ""
    },
    onShow() {
        var options = {
            duration: 3000,
            easing: 'linear',
```

```
            fill: 'forwards',
            iterations: 'Infinity',
        };
        var frames = [
            {
                transform: {
                    rotate: '0deg'
                }
            },
            {
                transform: {
                    rotate: '360deg'
                }
            }
        ];
        this.aniPoster = this. $ element('logo').animate(frames, options);
    },
    start() {
        this.status = !this.status;
        if (this.status) {
            this.aniPoster.play();
            this.startIcon = "stop_btn"
        } else {
            this.aniPoster.cancel();
            this.startIcon = "start_btn"
        }
    },
    runCar: async function(code){
        await utils.callAbility(code)
    }
}
```

## 17.5 本章小结

本章通过开发一款 JS 端应用控制 HiSpark 智能赛车实现赛车的游戏,让开发者通过游戏掌握鸿蒙混合开发的技巧。

本章涉及智能赛车端的代码是使用 C 语言编写的,JavaScript 开发读者可以选择跳过此小节,如果读者希望亲自尝试开发,则可以在本书提供的源码下载网址下载 racecar_c 的源码和已经编译好的鸿蒙操作系统文件,读者在下载鸿蒙操作系统相关文件后可通过 HiBurn 烧录到开发板上。

# 第六篇 提高篇

本篇介绍 OpenHarmony 中开源代码的实现,以及其依赖的第三方开源的 JS 引擎和渲染引擎,通过本篇学习,有利于对 OpenHarmony 有更深的理解。

# 第 18 章 轻鸿蒙端 JavaScript 框架

OpenHarmony 在 IoT 设备上使用 JerryScript JS 引擎,用来运行 JavaScript 编写的应用程序,代码仓库的网址为 https://gitee.com/openharmony/third_party_jerryscript。

## 18.1 JerryScript 轻量级引擎

JerryScript 是由三星开发的一款 JavaScript 引擎,是为了让 JavaScript 开发者能够构建物联网应用。物联网设备在 CPU 性能和内存空间上有着严重的制约,因此,三星设计了 JerryScript 引擎,它能够运行在小于 64KB 的内存上,并且全部代码能够存储在不足 200KB 的只读存储(ROM)上。

JerryScript 的主要特征有下几点:
(1) 完全符合 ECMAScript 5.1 标准。
(2) 为 ARM Thumb-2 编译时,二进制文件的大小为 160KB。
(3) 大量优化以降低内存消耗。
(4) 使用 C99 编写,以实现最大的可移植性。
(5) 快照支持将 JavaScript 源代码预编译为字节码。
(6) 成熟的 C API,易于嵌入应用程序中。

### 18.1.1 编译 JerryScript

在 Ubuntu 系统编译 JerryScript,这里需要 Ubuntu 18.04+以上版本。

**1. 安装编译环境**

命令如下:

```
sudo apt-get install gcc gcc-arm-none-eabi cmake cppcheck vera++ python
```

**2. 通过 git 下载 JerryScript 源码**

命令如下:

```
git clone https://github.com/jerryscript-project/jerryscript.git
cd jerryscript
```

### 3. 编译 JerryScript 源码

命令如下：

```
python tools/build.py
```

编译成功后的效果如图 18.1 所示。

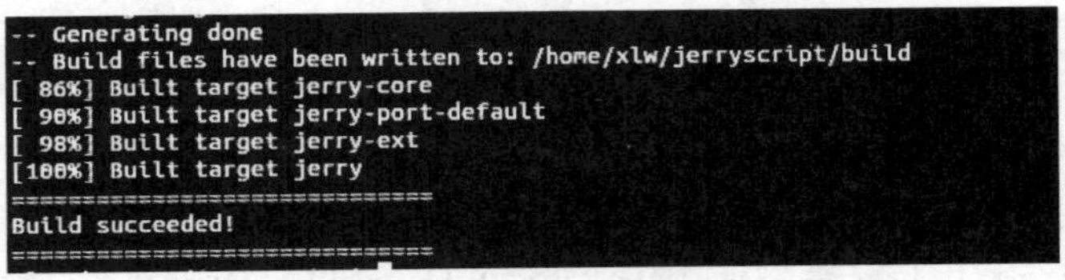

图 18.1 编译 JerryScript 成功后的提示

生成一个新的文件夹：build，如图 18.2 所示，jerry 在 build 里面的 bin 目录下。

图 18.2 编译生成的目录列表

## 18.1.2 运行 JerryScript

在 bin 目录下，执行 ./jerry 命令，进入 JerryScript 交互式命令行，效果如图 18.3 所示。

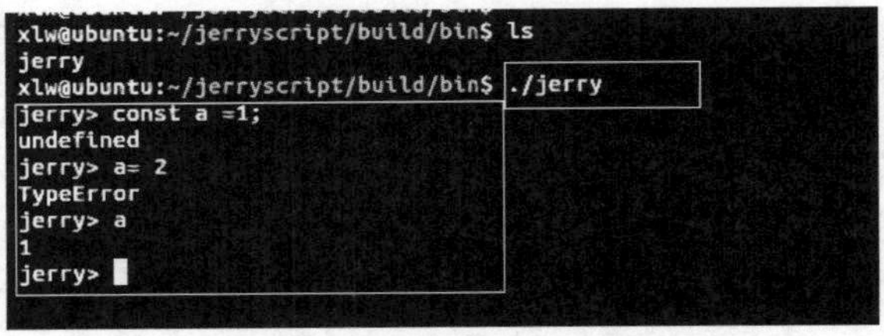

图 18.3 在 JerryScript 的交互式命令行执行命令

## 18.2 轻量级 JS 核心开发框架

Ace_engine_lite 框架子系统包括 JS 数据绑定框架（JS Data binding）、JS 运行时（JS runtime）和 JS 框架（JS Framework），如图 18.4 所示，Ace_engine_lite 框架包括以下部分：

1. **JS Data binding**

JS 数据绑定框架使用 JavaScript 语言提供的一套基础的数据绑定能力。

2. **JS runtime**

JS 运行时用以支持 JS 代码的解析和执行。

3. **JS Framework**

JS 框架部分使用 C++ 语言提供 JS API 和组件的框架机制。

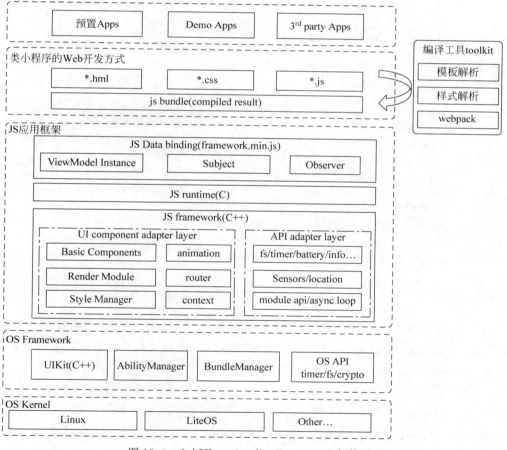

图 18.4 ArkUI_engine_lite Framework 架构图

### 18.2.1 JS Framework

Ace_engine_lite UI 框架使用 JavaScript 语言实现了一套简单的数据劫持框架,称为 runtime-core,源代码的目录结构如图 18.5 所示,通过数据劫持实现了界面上的组件与数据的分离,实现数据驱动式界面开发。

```
/foundation/ace/ace_engine_lite/frameworks/packages
└─ runtime-core
   ├─ .babelrc              # babel配置文件
   ├─ contribution.md
   ├─ .editorconfig         # IDE配置文件
   ├─ .eslintignore         # ESLint配置文件,可以设置不进行ESLint扫描的目录或文件
   ├─ .eslintrc.js          # ESLint配置文件,可以配置扫描规则
   ├─ .gitignore
   ├─ package.json          # NPM包管理文件
   ├─ package-lock.json     # NPM依赖版本锁定文件
   ├─ .prettierrc           # 代码格式化规则配置文件
   ├─ scripts               # 编译脚本存放目录
   │  ├─ build.js           # 编译脚本
   │  └─ configs.js         # Rollup配置文件
   ├─ .size-snapshot.json
   └─ src                   # 源代码
      ├─ core               # ViewModel核心实现目录
      │  └─ index.js
      ├─ index.js
      ├─ observer           # 数据劫持部分代码实现目录
      │  ├─ index.js
      │  ├─ observer.js
      │  ├─ subject.js
      │  └─ utils.js
      ├─ profiler           # profiler目录
      │  └─ index.js
      └─ __test__           # 测试用例目录
         └─ index.test.js
```

图 18.5 runtime core 源码结构图

可以下载源代码,通过 npm 命令编译,编译命令如下:

```
npm run build
```

JS 应用框架所集成的 JS 引擎仅支持 ES5.1 语法,runtime-core 源代码是使用 ES6 语法编写的,因此选择使用 rollup 作为打包工具,配合 babel 实现对语法进行降级处理。在命令行中执行 npm run build 后,会在 build 目录下输出打包结果。

JavaScript runtime-core 采用类似 Vue.js 2.0 框架设计模式,如图 18.6 所示,通过 Object.defineProperty 的 getter 和 setter,并结合观察者模式实现数据绑定。

查看 OpenHarmony 开源代码目录:ace_engine_lite/frameworks/packages/runtime-core/src/observer/observer.js。

与 Vue 2.0 中的 Observer 模式基本类似,代码如下:

# 第18章 轻鸿蒙端JavaScript框架

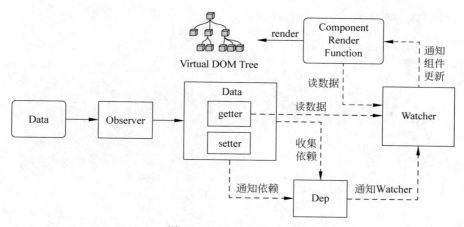

图 18.6 MVVM 模式图

```
import { ObserverStack, canObserve } from './utils';

/**
 * Observer constructor
 * @param {ViewModel} context execute context of callback
 * @param {Function} getter getter function
 * @param {Function} callback callback function
 * @param {Object} meta meta data that Observer object don't care about
 */
export function Observer(context, getter, callback, meta) {
  this._ctx = context;
  this._getter = getter;
  this._fn = callback;
  this._meta = meta;
  this._lastValue = this._get();
}

Observer.prototype._get = function() {
  try {
    ObserverStack.push(this);
    return this._getter.call(this._ctx);
  } finally {
    ObserverStack.pop();
  }
};

Observer.prototype.update = function() {
  const lastValue = this._lastValue;
  const nextValue = this._get();
  const context = this._ctx;
  const meta = this._meta;
```

```
    if (nextValue !== lastValue || canObserve(nextValue)) {
      this._fn.call(context, nextValue, lastValue, meta);
      this._lastValue = nextValue;
    }
  };

Observer.prototype.subscribe = function(subject, key) {
    const detach = subject.attach(key, this);
    if (typeof detach !== 'function') {
      return void 0;
    }
    if (!this._detaches) {
      this._detaches = [];
    }
    this._detaches.push(detach);
  };

Observer.prototype.unsubscribe = function() {
    const detaches = this._detaches;
    if (!detaches) {
      return void 0;
    }
    while (detaches.length) {
      detaches.pop()();
    }
  }
```

### 18.2.2 组件绑定实现

JS 端组件,如< text >、< div > 的 XML 标签组件,都对应一个绑定到 JerryScript 上的 C++ Component 类,如 TextComponent 和 DivComponent 等,网址如下:

```
https://gitee.com/openharmony/ace_engine_lite/blob/OpenHarmony-3.0-LTS/frameworks/src/core/components/text_component.h
```

JS 中以@system 为前缀的 built-in 模块,还提供了 JS 中可用的 Router/Audio/File 等平台能力(参见 ohos_module_config.h 文件)。

### 18.2.3 路由实现

与前端框架 Vue 中的 VueRouter 模块的实现原理不同,ace_engine_lite 中的路由是在运行时深度定制的(参见 router_module.cpp、js_router.cpp 和 js_page_state_machine.cpp 文件)。

JS 框架中的路由执行过程如下:

(1) 在 JS 中调用切换页面的原生方法,如使用 router.push()方法便可进入 C++模块中执行。

(2) C++中根据页面URI路径(如pages/index)加载页面JS,新建页面状态机实例,将其切换至Init状态。

(3) 在新状态机的Init过程中,调用JS引擎去Eval新页面的JS代码,获得新页面的ViewModel。

(4) 再将路由参数附加到ViewModel上,销毁旧状态机及其上的JS对象。

### 18.2.4　图形绘制层

图形库(graphic_lite)提供了UIView这个C++控件基类,其中有一系列(如OnClick/OnLongPress/OnDrag)的虚函数。基本每种JS中可用的原生Component类,对应于一种UIView的子类。

除了各种定制化View之外,它还开放了一系列形如DrawLine / DrawCurve / DrawText等命令式的绘制方法。

这个图形库具备名为GFX的GPU加速模块,但它目前似乎只有象征性的FillArea矩形单色填充能力。

### 18.2.5　渲染流程

ace_engine_lite框架的运行渲染流程大致如下:

(1) 如果对JS中的data:{ }数据进行修改,就会触发JS依赖追踪。

(2) JS依赖追踪回调会触发原生函数,更新C++中的Component组件状态。

(3) Component更新其绑定的UIView子类状态,再触发图形库更新。

(4) 图形库更新内存中的像素状态,完成绘制。

# 第 19 章 富鸿蒙端 JavaScript 框架

富鸿蒙 JS 框架采用更强大的 JS 引擎,如在智慧屏端和手表端使用 QuickJS 引擎,在手机、平板上使用 Google V8 引擎。

## 19.1 QuickJS 引擎

QuickJS 是一个轻量且可嵌入的 JavaScript 引擎,它支持 ES2019 规范,包括 ES module、异步生成器及 proxies。除此之外,还支持可选的数学扩展,例如大整数(BigInt)、大浮点数(BigFloat)和运算符重载。

QuickJS 引擎的主要特性如下。

(1) 轻量且方便嵌入:QuickJS 只包含了一些 C 语言文件,没有额外的依赖,运行一个简单的 hello world 程序只需 190KiB 的 x86 代码。

(2) 拥有启动时间极短的快速解释器:在单核的台式 PC 上,运行 ECMAScript 测试套件的 56000 个测试大约在 100s 内完成。一个 runtime 实例的完整生命周期在不到 300ms 内完成。

(3) 几乎完整的 ES2019 支持,包括 ES module、异步生成器和完整的 Annex B 支持(传统的 Web 兼容性)。

(4) 完全通过了 ECMAScript 测试套件的测试。

(5) 可将 JavaScript 源码编译为没有外部依赖的可执行文件。

(6) 基于引用计数的 GC(以减少内存使用并具有确定性行为)。

(7) 数学扩展:BigInt、BigFloat、运算符重载、bigint mode 和 math mode。

(8) 使用 JavaScript 实现的具有上下文着色功能(Contextual Colorization)的命令行解释器。

(9) 包含使用 C 语言库封装的轻量级内置标准库。

### 19.1.1 安装基础编译环境

在 Ubuntu 系统的命令行中执行的命令如下:

```
sudo apt-get install -y build-essential gcc-multilib
```

## 19.1.2 通过 Git 下载 QuickJS 源码

这里直接在 GitHub 上下载最新版本的 QuickJS 源码,命令如下:

```
git clone https://github.com/quickjs-zh/QuickJS
```

## 19.1.3 编译 QuickJS

使用 root 身份执行 make install 可以将编译的二进制文件和支持文件安装到/usr/local,执行的命令如下:

```
cd quickjs && make && make install
```

## 19.1.4 编译验证 JS

通过 qjs 可进入 quickjs 环境,-h 表示获取帮助,-q 表示退出环境。

qjs 是一个纯解释执行的 JS 引擎。

新建一个 js 脚本 hello.js,内容为 console.log('hello world!'),在 js 目录下执行的代码如下:

```
qjs hello.js
```

输出: hello world!

## 19.2 Google V8 引擎

ArkUI JS 框架在手机、平板上使用 Google V8 引擎来执行 JavaScript 代码。V8 引擎代码包含在 ace_ace_engine→frameworks→bridge→declarative_frontend→engine 目录下,如图 19.1 所示。

V8 是 Google 基于 C++ 编写的开源高性能 JavaScript 与 WebAssembly 引擎。用于 Google Chrome(Google 的开源浏览器)及 Node.js 等。

V8 实现了 ECMAScript 与 WebAssembly,能够运行在 Windows 7+、macOS 10.12+ 及使用 x64、IA-32、ARM、MIPS 处理器的 Linux 系统。V8 能独立运行,也能嵌入任何 C++ 应用当中。

V8 编译并执行 JavaScript 源代码,处理对象的内存分配,以及回收不再使用的对象。高效的垃圾收集器是 V8 高性能的关键之一。

图 19.1　ace_engine V8 代码

JavaScript 通常用于编写浏览器中的客户端脚本,例如用于操作文档对象模型(DOM)对象,但是,DOM 通常不是由 JavaScript 引擎提供,而是由浏览器提供的。V8 也是如此,Google Chrome 提供了 DOM,但是 V8 提供了 ECMA 标准中规定的所有数据类型、运算符、对象和函数。

V8 允许 C++ 应用程序将自己的对象和函数公开给 JavaScript 代码。由你来决定要向 JavaScript 公开的对象和函数。

## 19.3　ArkUI JS Engine 框架

ArkUI JS 框架是之前 ACE 框架的新名称,ArkUI JS 整体架构主要由前端框架层、桥接层、引擎层和平台抽象层四大部分组成,如图 19.2 所示。

图 19.2 中各个层的介绍如下。

**1. 前端框架层**

该层包括三套 UI 框架,即类 Web UI 框架、声明式 UI 框架、无须脚本卡片 UI 框架。

1) 类 Web 范式编程(js_frontend)

js_frontend UI 框架采用类 HTML 和 CSS Web 编程语言作为页面布局和页面样式的开发语言,页面业务逻辑则支持 ECMAScript 规范的 JavaScript 语言。JS UI 框架提供了类 Web 编程范式,可以让开发者避免编写 UI 状态切换的代码,从而使视图配置信息更加直观。

2) 卡片 UI 框架(card_frontend)

card_frontend UI 框架主要用在鸿蒙桌面 Widget 服务卡片渲染方面。它不依赖于脚本引擎,由特定格式的 JSON 文件驱动渲染。

卡片中的 JSON 文件的格式主体分为 template、styles、actions、data 等部分,模板中可以写花括号的数据绑定{{product.title}},可以写简单的 JS 表达式,非完全静态。card_

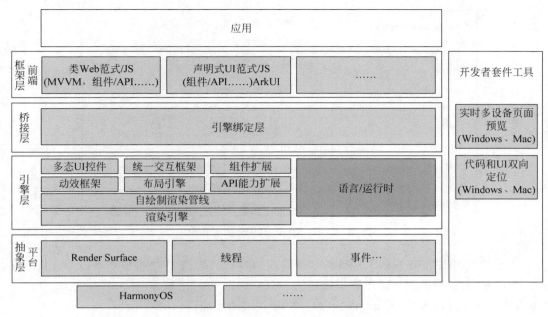

图 19.2 ace_engine Framework 架构图

frontend 类上有 UpdateData() 接口,可以更新模板的数据,具备一定的动态化能力。

### 3）声明式 JS UI 框架（declarative_frontend）

declarative_frontend UI 框架采用类 SwiftUI 的 UI 开发框架,采用 TypeScript 编写代码,该框架在性能方面相比 js_frontend 有较大的提升,与 Flutter 类似,用原子化的布局函数组合 UI,Flutter 里采用 Widget,鸿蒙的声明式 JS UI 则采用 JSView。再结合 jsproxyClass.js 文件里的代码来分析,提供了 ECMAScript 规范中的装饰器/注解辅助编程。

## 2. 桥接层

该层主要作为一个中间层,实现前端开发范式到底层引擎（包括 UI 后端、语言 & 运行时）的对接。

桥接层在 ace_ace_engine 模块的 bridge 目录中,如图 19.3 所示。

### 1）引擎层

该层主要包含两部分：UI 后端引擎和语言执行引擎。

由 C++构建的 UI 后端引擎,包括 UI 组件、布局视图、动画事件、自绘制渲染管线和渲染引擎等。

在渲染方面,尽可能把这部分组件设计得小而灵活。这样的设计,为不同前端框架提供了灵活的 UI 能力,这部分通过 C++组件组合而成。通过底层组件的按需组合,布局计算和渲染并行化,并结合上层开发范式实现了视图变化最小化的局部更新机制,从而实现了高效的 UI 渲染。

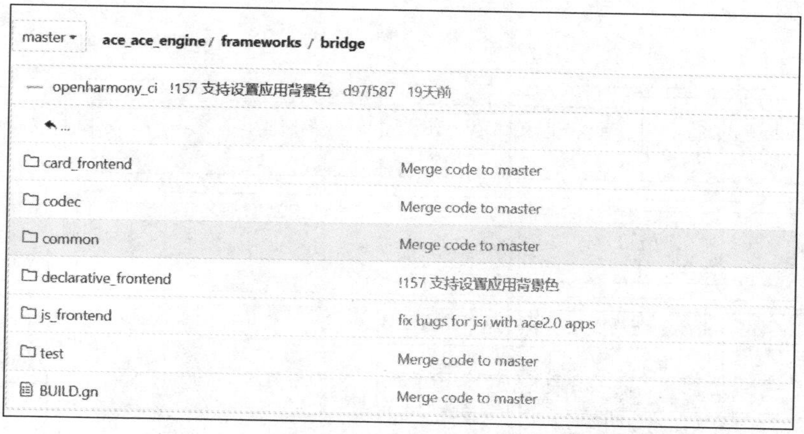

| | |
|---|---|
| master ▼  ace_ace_engine / frameworks / bridge | |
| — openharmony_ci  !157 支持设置应用背景色  d97f587  19天前 | |
| ← ... | |
| 📁 card_frontend | Merge code to master |
| 📁 codec | Merge code to master |
| 📁 common | Merge code to master |
| 📁 declarative_frontend | !157 支持设置应用背景色 |
| 📁 js_frontend | fix bugs for jsi with ace2.0 apps |
| 📁 test | Merge code to master |
| 📄 BUILD.gn | Merge code to master |

图 19.3　ace_engine UI 桥接层目录

除此之外，引擎层还提供了组件的渲染管线、动画、主题、事件处理等基础能力。目前复用了 Flutter 引擎提供的基础图形渲染能力、字体管理、文字排版等能力，底层使用 Skia 或其他图形库实现，并通过 OpenGL 实现 GPU 硬件渲染加速。

在多设备 UI 适配方面，通过多种原子化布局能力（自动折行、隐藏、等比缩放等）和多态 UI 控件（描述统一，表现形式多样），以及统一交互框架（不同的交互方式归一到统一的事件处理）来满足不同设备的形态差异化需求。

另外，引擎层也包含了能力扩展基础设施，实现自定义组件及系统 API 的能力扩展，语言 & 运行时执行引擎。可根据需要切换到不同的运行时执行引擎，满足不同设备的能力差异化需求。

2）平台抽象层

该层主要通过平台抽象，将平台依赖聚焦到底层画布，通用线程及事件机制等少数必要的接口为跨平台打造了相应的基础设施，并能够实现一致化 UI 渲染体验。

3）开发者套件

配套的开发者工具（HUAWEI DevEco Studio）结合 ArkUI UI 的跨平台渲染基础设施，以及自适应渲染，可做到和设备比较一致的渲染体验及多设备上的 UI 实时预览。

Declarative Frontend 声明式 UI 复用了 Flutter 引擎提供的基础图形渲染能力、字体管理、文字排版等能力，底层使用 Skia 或其他图形库实现，并通过 OpenGL 实现 GPU 硬件渲染加速，如图 19.4 所示。

Skia 是一款用 C++ 开发的、性能彪悍的 2D 图像绘制引擎，其前身是一个向量绘图软件。2005 年被 Google 公司收购后，因为其出色的绘制表现被广泛应用在 Chrome 和 Android 等核心产品上。Skia 在图形转换、文字渲染、位图渲染方面都表现卓越，并提供了开发者友好的 API。

Skia 网站的网址为 https://skia.org/。Skia 由谷歌出资管理，任何人都可基于 BSD 免费软件许可证使用。Skia 开发团队致力于开发其核心部分,并广泛采纳各方对于 Skia 的开源贡献。

Skia 是 Android 官方的图像渲染引擎,因此 Flutter Android SDK 无须内嵌 Skia 引擎就可以获得 Skia 支持;iOS 平台,由于 Skia 是跨平台的,因此它作为 Flutter iOS 渲染引擎被嵌入 Flutter 的 iOS SDK 中,替代了 iOS 闭源的 Core Graphics/Core Animation/Core Text,这也正是 Flutter iOS SDK 打包的 App 包体积比 Android 要大一些的原因。

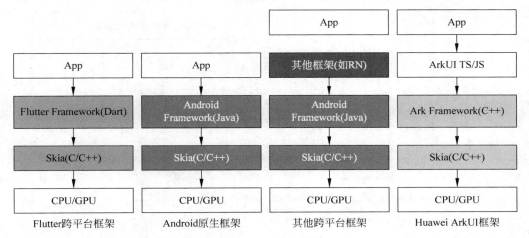

图 19.4 Skia 渲染引擎

## 19.4 新方舟编译器(ArkCompiler 3.0)

ArkCompiler 是华为自研的统一编程平台,包含编译器、工具链、运行时等关键部件,支持高级语言在多种芯片平台的编译与运行,并支撑应用和服务运行在手机、个人计算机、平板、电视、汽车和智能穿戴等多种设备上的需求,如图 19.5 所示。

ArkCompiler 3.0 包含以下几个关键特性:

(1) 前端编译器支持将多种高级语言(包括 JS、TS 和 Java)编译成统一的字节码文件,屏蔽语言的差异,提升运行效率和程序启动性能。

(2) 提供多种端侧执行模式(解释器、JIT 编译器和 AOT 编译器),形成结合设备和应用特征的多层次组合运行策略,满足不同设备硬件规格。

解释器:启动快,但执行性能一般,内存占用小。

JIT 编译器:启动需要预热,执行性能高,内存占用较高。

AOT 编译器:启动快,执行性能高,但内存占用高。

在低端 IoT 设备上,ArkCompiler 3.0 支持纯解释器的执行模式,以满足小设备的内存限制条件。在高端设备上,ArkCompiler 3.0 支持解释器配合 AOT 和 JIT 编译器的执行模式,对大部分应用代码使用 AOT 编译器编译,使程序一开始就可以运行在高质量的优化代码上,获得更好的执行性能。在其他设备上,则根据设备的硬件条件限制来选择策略,设定高频使用需要 AOT 编译的代码范围,其他代码则依靠解释器配合 JIT 编译器运行,使应用执行性能能够得到最大化。

(3) ArkCompiler 3.0 特别对 TS/JS 做了针对性优化规划,其目标是提升执行性能 1 倍。

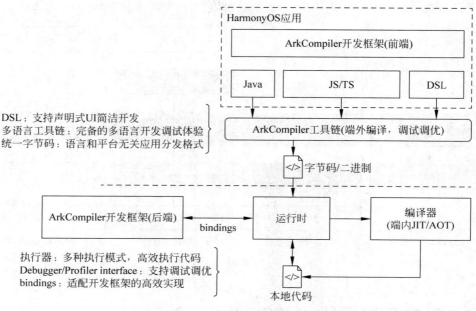

图 19.5　新方舟编译器 3.0

# 第 20 章 类 Web 范式组件设计与开发

如何给 ace_ace_engine 中的 js_frontend 代码仓编写一个新的类 Web 范式组件？本章一步一步带领读者开发一个展示类组件并上传到 Gitee 上。

## 20.1 JavaScript 端组件设计

下面介绍如何设计一个可单击的展示类组件，展示一个圆，支持设置半径、边缘宽度和边缘颜色，可以通过单击事件获得当前圆的半径和边缘宽度。

前端组件设计需要考虑的内容包括：支持哪些设备使用、是否有子组件、组件的输入和输出属性、组件样式、组件事件。

### 20.1.1 前端组件效果

组件的模板定义，代码如下：

```
<div style = "flex-direction: column;align-items: center;">
<text>"MyCircle 的半径为{{radiusOfMyCircle}}"</text>
<text>"MyCircle 的边缘宽度为{{edgeWidthOfMyCircle}}"</text>
<mycircle circleradius = "40vp" style = "circleedge: 2vp red;" @circleclick = "onCircleClick"
></mycircle>
</div>
```

组件的逻辑，代码如下：

```
export default{
    data:{
        radiusOfMyCircle: -1,
        edgeWidthOfMyCircle: -1,
    },
    onCircleClick(event) {
        this.radiusOfMyCircle = event.radius
        this.edgeWidthOfMyCircle = event.edgewidth
    }
}
```

组件的实现效果如图 20.1 所示。

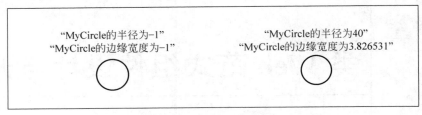

图 20.1　组件实现效果

## 20.1.2　组件的详细设计

前端 JS 组件的设计包含以下几方面：子组件、属性、样式、事件。

### 1. 子组件

这里为了简单演示，该组件不需要设置子组件。

### 2. 属性

该组件的属性只有一个，即 circleradius，如表 20.1 所示。

表 20.1　属性设计

| 属性名称 | 属性类型 | 默认值 | 必填 | 描述 |
| --- | --- | --- | --- | --- |
| circleradius | length | 20vp | 否 | 默认半径 |

### 3. 样式

该组件的样式只有一个，即 circleedge，如表 20.2 所示。

表 20.2　样式设计

| 样式名称 | 样式类型 | 默认值 | 必填 | 描述 |
| --- | --- | --- | --- | --- |
| circleedge | length color | 2vp red | 否 | 默认边缘颜色和宽度 |

### 4. 事件

该组件值包括一个事件，如表 20.3 所示。

表 20.3　事件设计

| 事件名称 | 事件类型 | 描述 |
| --- | --- | --- |
| circleedge | {radius: circle radius, edgewidth: circle edge width} | 单击 MyCircle 组件时触发该回调，返回当前 circle 的半径和边缘宽度，单位为 px |

## 20.2　JS 的界面解析

有了上面 JS 组件的定义后，需要通过桥接层实现界面的真正效果。

## 20.2.1 在 dom_type 中增加新组件的属性定义

这里需要在 dom_type.h 文件和 dom_type.cpp 文件中进行属性定义。

**1. 在 dom_type.h 文件中增加 MyCircle 的属性定义**

文件路径为 frameworks\bridge\common\dom\dom_type.h，代码如下：

```
//node tag defines
/* ................................. */
/* node tag defines of other components */
/* ................................. */
ACE_EXPORT extern const char DOM_NODE_TAG_MYCIRCLE[];

/* ........................ */
/* defines of other components */
/* ........................ */

//mycircle defines
ACE_EXPORT extern const char DOM_MYCIRCLE_CIRCLE_EDGE[];
ACE_EXPORT extern const char DOM_MYCIRCLE_CIRCLE_RADIUS[];
ACE_EXPORT extern const char DOM_MYCIRCLE_CIRCLE_CLICK[];
```

**2. 在 dom_type.cpp 文件中增加 MyCircle 的属性值**

文件路径为 frameworks\bridge\common\dom\dom_type.cpp，代码如下：

```
//node tag defines
/* ................................. */
/* node tag defines of other components */
/* ................................. */
const char DOM_NODE_TAG_MYCIRCLE[] = "mycircle";

/* ........................ */
/* defines of other components */
/* ........................ */

//mycircle defines
const char DOM_MYCIRCLE_CIRCLE_EDGE[] = "circleedge";
const char DOM_MYCIRCLE_CIRCLE_RADIUS[] = "circleradius";
const char DOM_MYCIRCLE_CIRCLE_CLICK[] = "circleclick";
```

## 20.2.2 新增 DOMMyCircle 类

**1. 新增 dom_mycircle.h 文件**

文件路径为 frameworks\bridge\common\dom\dom_mycircle.h，代码如下：

```
class DOMMyCircle final : public DOMNode {
    DECLARE_ACE_TYPE(DOMMyCircle, DOMNode);
```

```cpp
public:
    DOMMyCircle(NodeId nodeId, const std::string& nodeName);
    ~DOMMyCircle() override = default;

    RefPtr<Component> GetSpecializedComponent() override
    {
        return myCircleChild_;
    }

protected:
    bool SetSpecializedAttr(const std::pair<std::string, std::string>& attr) override;
    bool SetSpecializedStyle(const std::pair<std::string, std::string>& style) override;
    bool AddSpecializedEvent(int32_t pageId, const std::string& event) override;

private:
    RefPtr<MyCircleComponent> myCircleChild_;
};
```

DOMMyCircle 继承自 DOMNode，主要功能是解析界面并生成相应的 Component 节点。

### 2. 新增 dom_mycircle.cpp 文件

文件路径为 frameworks\bridge\common\dom\dom_mycircle.cpp

（1）组件属性的解析：SetSpecializedAttr，代码如下：

```cpp
bool DOMMyCircle::SetSpecializedAttr(const std::pair<std::string, std::string>& attr)
{
    if (attr.first == DOM_MYCIRCLE_CIRCLE_RADIUS) { //"circleradius"
        myCircleChild_->SetCircleRadius(StringToDimension(attr.second));
        return true;
    }
    return false;
}
```

这种方法由框架流程调用，我们只需要在这种方法里面实现对应属性的解析，并且设置到 MyCircleComponent 中。

在上面的代码中，如果入参 attr 的格式形如<"circleradius", "40vp">，则我们只需在 attr.first 为"circleradius"时，将 attr.second 转换为 Dimension 格式，并且设置到 MyCircleComponent 中。设置完成后，返回值为 true。

（2）组件样式的解析：SetSpecializedStyle，代码如下：

```cpp
bool DOMMyCircle::SetSpecializedStyle(const std::pair<std::string, std::string>& style)
{
    if (style.first == DOM_MYCIRCLE_CIRCLE_EDGE) { //"circleedge"
        std::vector<std::string> edgeStyles;
```

```cpp
        //The value of [circleedge] is like "2vp red" or "2vp". To parse style value like this,
        //we need 3 steps
        //Step1: Split the string value by ' 'to get vectors like ["2vp", "red"]
        StringUtils::StringSpliter(style.second, ' ', edgeStyles);
        Dimension edgeWidth(1, DimensionUnit::VP);
        Color edgeColor(Color::RED);

        //Step2: Parse edge color and edge width accordingly
        switch(edgeStyles.size()) {
            case 0: //the value is empty
                LOGW("Value for circle edge is empty, using default setting.");
                break;
            case 1: //case when only edge width is set
                //It should be guaranteed by the tool chain when generating js-bundle that
                //the only value is a number type for edge width rather than a color type for
                //edge color
                edgeWidth = StringUtils::StringToDimension(edgeStyles[0]);
                break;
            case 2: //case when edge width and edge color are both set
                edgeWidth = StringUtils::StringToDimension(edgeStyles[0]);
                edgeColor = Color::FromString(edgeStyles[1]);
                break;
            default:
                LOGW("There are more than 2 values for circle edge, please check. The value is
%{private}s",
                    style.second.c_str());
                break;
        }

        //Step3: Set edge color and edge width to [mycircleStyle]
        myCircleChild_->SetEdgeWidth(edgeWidth);
        myCircleChild_->SetEdgeColor(edgeColor);
        return true;
    }
    return false;
}
```

这种方法由框架流程调用，我们只需要在这种方法里面实现对应样式的解析，并且保存到 MyCircleComponent 中。

在上面的代码中，如果入参 style 的格式形如<"circleedge"，"2vp red">，则我们只需在 style.first 为 "circleedge" 时，将 style.second 进行解析，并且设置到 MyCircleComponent 中。设置完成后，返回值为 true。

（3）组件事件的解析：SetSpecializedEvent，代码如下：

```cpp
bool DOMMyCircle::AddSpecializedEvent(int32_t pageId, const std::string& event)
{
```

```
    if (event == DOM_MYCIRCLE_CIRCLE_CLICK) { //"circleclick"
        myCircleChild_ - > SetCircleClickEvent (EventMarker (GetNodeIdForEvent ( ), event,
pageId));
        return true;
    }
    return false;
}
```

这种方法由框架流程调用,我们只需要在这种方法里面实现对应事件的解析,并且保存到 MyCircleComponent 中。

在上面的代码中,只要判断入参 event 的值为"circleclick",我们就需要使用 eventId 和 pageId 构造一个 EventMarker,并设置到 MyCircleComponent 中。设置完成后,返回值为 true。

### 3. 在 dom_document.cpp 文件里增加 mycircle 组件

文件路径为 frameworks\bridge\common\dom\dom_document.cpp,代码如下:

```
RefPtr < DOMNode > DOMDocument::CreateNodeWithId(const std::string& tag, NodeId nodeId, int32_
t itemIndex)
{
    //code block
    static const LinearMapNode < RefPtr < DOMNode >( * )(NodeId, const std::string&, int32_t)>
domNodeCreators[] = {
        //DomNodeCreators of other components
        { DOM_NODE_TAG_MENU, &DOMNodeCreator < DOMMenu > },
        //"mycircle" must be inserted between "menu" and "navigation-bar"
        { DOM_NODE_TAG_MYCIRCLE, &DOMNodeCreator < DOMMyCircle > },
        { DOM_NODE_TAG_NAVIGATION_BAR, &DOMNodeCreator < DomNavigationBar > },
        //DomNodeCreators of other components
    };
    //code block
    return domNode;
}
```

这里尤其需要注意一点,domNodeCreators[]是一个线表,添加{ DOM_NODE_TAG_MYCIRCLE,&DOMNodeCreator < DOMMyCircle > }的地方必须符合字母序,代码如下:

```
DOM_NODE_TAG_MENU[ ] = "menu",
DOM_NODE_TAG_NAVIGATION_BAR[ ] = "navigation-bar",
DOM_NODE_TAG_MYCIRCLE[ ] = "mycircle"
```

所以 DOM_NODE_TAG_MYCIRCLE 的记录必须添加在"menu"之后,并且在"navigation-bar"之前。

## 20.3 后端的布局和绘制

组件在后端的布局和绘制，需要相应地新增以下几个类：MyCircleComponent、MyCircleElement、RenderMyCircle、FlutterRenderMyCircle。

在后端引擎中，Component 树、Element 树和 Render 树为后端引擎维持和更新 UI 最为核心的三棵树。

### 20.3.1 新增 MyCircleComponent 类

**1. 新增 mycircle_component.h 文件**

文件路径为 frameworks\core\components\mycircle\mycircle_component.h，代码如下：

```
class ACE_EXPORT MyCircleComponent : public RenderComponent {
    DECLARE_ACE_TYPE(MyCircleComponent, RenderComponent);

public:
    MyCircleComponent() = default;
    ~MyCircleComponent() override = default;

    RefPtr< RenderNode > CreateRenderNode() override;
    RefPtr< Element > CreateElement() override;

    void SetCircleRadius(const Dimension& circleRadius);
    void SetEdgeWidth(const Dimension& edgeWidth);
    void SetEdgeColor(const Color& edgeColor);
    void SetCircleClickEvent(const EventMarker& circleClickEvent);

    const Dimension& GetCircleRadius() const;
    const Dimension& GetEdgeWidth() const;
    const Color& GetEdgeColor() const;
    const EventMarker& GetCircleClickEvent() const;

private:
    Dimension circleRadius_ = 20.0_vp;
    Dimension edgeWidth_ = 2.0_vp;
    Color edgeColor_ = Color::RED;
    EventMarker circleClickEvent_;
};
```

**2. 新增 mycircle_component.cpp 文件**

文件路径为 frameworks\core\components\mycircle\mycircle_component.cpp。

（1）提供 Set 接口，代码如下：

```cpp
void MyCircleComponent::SetCircleRadius(const Dimension& circleRadius)
{
    circleRadius_ = circleRadius;
}

void MyCircleComponent::SetEdgeWidth(const Dimension& edgeWidth)
{
    edgeWidth_ = edgeWidth;
}

void MyCircleComponent::SetEdgeColor(const Color& edgeColor)
{
    edgeColor_ = edgeColor;
}

void MyCircleComponent::SetCircleClickEvent(const EventMarker& circleClickEvent)
{
    circleClickEvent_ = circleClickEvent;
}
```

(2) 提供 Get 接口,代码如下:

```cpp
const Dimension& MyCircleComponent::GetCircleRadius() const
{
    return circleRadius_;
}

const Dimension& MyCircleComponent::GetEdgeWidth() const
{
    return edgeWidth_;
}

const Color& MyCircleComponent::GetEdgeColor() const
{
    return edgeColor_;
}

const EventMarker& MyCircleComponent::GetCircleClickEvent() const
{
    return circleClickEvent_;
}
```

(3) 实现 CreateRenderNode()和 CreateElement()函数,代码如下:

```cpp
RefPtr<RenderNode> MyCircleComponent::CreateRenderNode()
{
    return RenderMyCircle::Create();
}
```

```cpp
RefPtr<Element> MyCircleComponent::CreateElement()
{
    return AceType::MakeRefPtr<MyCircleElement>();
}
```

## 20.3.2 新增 MyCircleElement 类

新增 mycircle_element.h 文件。

文件路径为 frameworks\core\components\mycircle\mycircle_element.h,代码如下:

```cpp
class MyCircleElement : public RenderElement {
    DECLARE_ACE_TYPE(MyCircleElement, RenderElement);

public:
    MyCircleElement() = default;
    ~MyCircleElement() override = default;
};
```

该组件在 element 层不涉及更多操作,只需定义 MyCircleElement 类。

## 20.3.3 新增 RenderMyCircle 类

### 1. 新增 render_mycircle.h 文件

文件路径为 frameworks\core\components\mycircle\render_mycircle.h,代码如下:

```cpp
using CallbackForJS = std::function<void(const std::string&)>;

class RenderMyCircle : public RenderNode {
    DECLARE_ACE_TYPE(RenderMyCircle, RenderNode);

public:
    static RefPtr<RenderNode> Create();

    void Update(const RefPtr<Component>& component) override;
    void PerformLayout() override;
    void HandleMyCircleClickEvent(const ClickInfo& info);

protected:
    RenderMyCircle();
    void OnTouchTestHit(
        const Offset& coordinateOffset, const TouchRestrict& touchRestrict, TouchTestResult& result) override;

    Dimension circleRadius_;
    Dimension edgeWidth_ = Dimension(1);
    Color edgeColor_ = Color::RED;
```

```cpp
    CallbackForJS callbackForJS_; //callback for js frontend
    RefPtr<ClickRecognizer> clickRecognizer_;
};
```

### 2. 新增 render_mycircle.cpp 文件

文件路径为 frameworks\core\components\mycircle\render_mycircle.cpp。

（1）处理单击事件，代码如下：

```cpp
RenderMyCircle::RenderMyCircle()
{
    clickRecognizer_ = AceType::MakeRefPtr<ClickRecognizer>();
    clickRecognizer_->SetOnClick([wp = WeakClaim(this)](const ClickInfo& info) {
        auto myCircle = wp.Upgrade();
        if (!myCircle) {
            LOGE("WeakPtr of RenderMyCircle fails to be upgraded, stop handling click event.");
            return;
        }
        myCircle->HandleMyCircleClickEvent(info);
    });
}

void RenderMyCircle::OnTouchTestHit(
    const Offset& coordinateOffset, const TouchRestrict& touchRestrict, TouchTestResult& result)
{
    clickRecognizer_->SetCoordinateOffset(coordinateOffset);
    result.emplace_back(clickRecognizer_);
}

void RenderMyCircle::HandleMyCircleClickEvent(const ClickInfo& info)
{
    if (callbackForJS_) {
        auto result = std::string("\"circleclick\",{\"radius\":")
                        .append(std::to_string(NormalizeToPx(circleRadius_)))
                        .append(",\"edgewidth\":")
                        .append(std::to_string(NormalizeToPx(edgeWidth_)))
                        .append("}");
        callbackForJS_(result);
    }
}
```

上面代码的解释如下：

① 创建一个 ClickRecognizer。

② 重写 OnTouchTestHit 函数，注册 RenderMyCircle 的 ClickRecognizer，这样在接收到单击事件时即可触发在创建 ClickRecognizer 时添加的事件回调。

③ 实现在接收到单击事件之后的处理逻辑 HandleMyCircleClickEvent。

(2) 重写 Update()函数,代码如下:

```cpp
void RenderMyCircle::Update(const RefPtr<Component>& component)
{
    const auto& myCircleComponent = AceType::DynamicCast<MyCircleComponent>(component);
    if (!myCircleComponent) {
        LOGE("MyCircleComponent is null!");
        return;
    }
    circleRadius_ = myCircleComponent->GetCircleRadius();
    edgeWidth_ = myCircleComponent->GetEdgeWidth();
    edgeColor_ = myCircleComponent->GetEdgeColor();
    callbackForJS_ =
        AceAsyncEvent<void(const std::string&)>::Create(myCircleComponent->GetCircleClickEvent(), context_);

    //call [MarkNeedLayout] to do [PerformLayout] with new params
    MarkNeedLayout();
}
```

Update()函数负责从 MyCircleComponent 获取所有绘制、布局和事件相关的属性更新。

(3) 重写 PerformLayout()函数,代码如下:

```cpp
void RenderMyCircle::PerformLayout()
{
    double realSize = NormalizeToPx(edgeWidth_) + 2 * NormalizeToPx(circleRadius_);
    Size layoutSizeAfterConstrain = GetLayoutParam().Constrain(Size(realSize, realSize));
    SetLayoutSize(layoutSizeAfterConstrain);
}
```

PerformLayout()函数负责计算布局信息,并且调用 SetLayoutSize()函数设置自己所需要的布局大小。

### 20.3.4　新增 FlutterRenderMyCircle 类

#### 1. 新增 flutter_render_mycircle.h 文件

文件路径为 frameworks\core\components\mycircle\flutter_render_mycircle.h,代码如下:

```cpp
class FlutterRenderMyCircle final : public RenderMyCircle {
    DECLARE_ACE_TYPE(FlutterRenderMyCircle, RenderMyCircle);

public:
```

```
    FlutterRenderMyCircle() = default;
    ~FlutterRenderMyCircle() override = default;

    void Paint(RenderContext& context, const Offset& offset) override;
};
```

**2. 新增 flutter_render_mycircle.cpp 文件**

文件路径为 frameworks\core\components\mycircle\flutter_render_mycircle.cpp。

(1) 实现 RenderMyCircle::Create() 函数, 代码如下:

```
RefPtr< RenderNode > RenderMyCircle::Create()
{
    return AceType::MakeRefPtr< FlutterRenderMyCircle >();
}
```

RenderMyCircle::Create() 函数在基类 RenderMyCircle 中定义, 因为我们当前使用的是 Flutter 引擎, 所以在 flutter_render_mycircle.cpp 文件里面实现, 返回在 Flutter 引擎上自渲染的 FlutterRenderMyCircle 类。

(2) 重写 Paint() 函数, 代码如下:

```
void FlutterRenderMyCircle::Paint(RenderContext& context, const Offset& offset)
{
    auto canvas = ScopedCanvas::Create(context);
    if (!canvas) {
        LOGE("Paint canvas is null");
        return;
    }
    SkPaint skPaint;
    skPaint.setAntiAlias(true);
    skPaint.setStyle(SkPaint::Style::kStroke_Style);
    skPaint.setColor(edgeColor_.GetValue());
    skPaint.setStrokeWidth(NormalizeToPx(edgeWidth_));

    auto paintRadius = GetLayoutSize().Width() / 2.0;
    canvas - > canvas() - > drawCircle(offset.GetX() + paintRadius, offset.GetY() + paintRadius,
        NormalizeToPx(circleRadius_), skPaint);
}
```

Paint() 函数负责调用 Canvas 相应接口进行绘制, 这一步可以认为是新增组件的最后一步, 直接决定在屏幕上绘制什么样的 UI 界面。

## 图 书 推 荐

| 书 名 | 作 者 |
|---|---|
| 鸿蒙应用程序开发 | 董昱 |
| 鸿蒙操作系统开发入门经典 | 徐礼文 |
| 鸿蒙操作系统应用开发实践 | 陈美汝、郑森文、武延军、吴敬征 |
| 华为方舟编译器之美——基于开源代码的架构分析与实现 | 史宁宁 |
| 鲲鹏架构入门与实战 | 张磊 |
| 华为HCIA路由与交换技术实战 | 江礼教 |
| Flutter组件精讲与实战 | 赵龙 |
| Flutter组件详解与实战 | [加]王浩然(Bradley Wang) |
| Flutter实战指南 | 李楠 |
| Dart语言实战——基于Flutter框架的程序开发(第2版) | 亢少军 |
| Dart语言实战——基于Angular框架的Web开发 | 刘仕文 |
| IntelliJ IDEA软件开发与应用 | 乔国辉 |
| Vue+Spring Boot前后端分离开发实战 | 贾志杰 |
| Vue.js企业开发实战 | 千锋教育高教产品研发部 |
| Python人工智能——原理、实践及应用 | 杨博雄主编,于营、肖衡、潘玉霞、高华玲、梁志勇副主编 |
| Python深度学习 | 王志立 |
| Python异步编程实战——基于AIO的全栈开发技术 | 陈少佳 |
| Python数据分析从0到1 | 邓立文、俞心宇、牛瑶 |
| 物联网——嵌入式开发实战 | 连志安 |
| 智慧建造——物联网在建筑设计与管理中的实践 | [美]周晨光(Timothy Chou)著;段晨东、柯吉译 |
| TensorFlow计算机视觉原理与实战 | 欧阳鹏程、任浩然 |
| 分布式机器学习实战 | 陈敬雷 |
| 计算机视觉——基于OpenCV与TensorFlow的深度学习方法 | 余海林、翟中华 |
| 深度学习——理论、方法与PyTorch实践 | 翟中华、孟翔宇 |
| 深度学习原理与PyTorch实战 | 张伟振 |
| ARKit原生开发入门精粹——RealityKit+Swift+SwiftUI | 汪祥春 |
| HoloLens 2开发入门精要——基于Unity和MRTK | 汪祥春 |
| Altium Designer 20 PCB设计实战(视频微课版) | 白军杰 |
| Cadence高速PCB设计——基于手机高阶板的案例分析与实现 | 李卫国、张彬、林超文 |
| Octave程序设计 | 于红博 |
| SolidWorks 2020快速入门与深入实战 | 邵为龙 |
| SolidWorks 2021快速入门与深入实战 | 邵为龙 |
| UG NX 1926快速入门与深入实战 | 邵为龙 |
| 西门子S7-200 SMART PLC编程及应用(视频微课版) | 徐宁、赵丽君 |
| 三菱FX3U PLC编程及应用(视频微课版) | 吴文灵 |
| 全栈UI自动化测试实战 | 胡胜强、单镜石、李睿 |
| pytest框架与自动化测试应用 | 房荔枝、梁丽丽 |
| 软件测试与面试通识 | 于晶、张丹 |
| 深入理解微电子电路设计——电子元器件原理及应用(原书第5版) | [美]理查德·C.耶格(Richard C. Jaeger)、[美]特拉维斯·N.布莱洛克(Travis N. Blalock)著;宋廷强译 |
| 深入理解微电子电路设计——数字电子技术及应用(原书第5版) | [美]理查德·C.耶格(Richard C. Jaeger)、[美]特拉维斯·N.布莱洛克(Travis N. Blalock)著;宋廷强译 |
| 深入理解微电子电路设计——模拟电子技术及应用(原书第5版) | [美]理查德·C.耶格(Richard C. Jaeger)、[美]特拉维斯·N.布莱洛克(Travis N. Blalock)著;宋廷强译 |